NEW FORMAT

T0348156

AJ Sadler

Mathematics Specialist

Student Book

Units 1 & 2

NELSON
A Cengage Company

Australia • Brazil • Japan • Korea • Mexico • Singapore • Spain • United Kingdom • United States

Mathematics Specialist Units 1 and 2
1st Edition
A.J. Sadler

Publishing editor: Robert Yen
Project editor: Alan Stewart
Cover design: Chris Starr (MakeWork)
Text designers: Sarah Anderson, Nicole Melbourne,
Danielle Maccarone
Permissions researcher: Catherine Kerstjens
Answer checker: George Dimitriadis
Production controller: Erin Dowling
Typeset by: Cenveo Publisher Services
Icons made by freepic from www.flaticon.com

Any URLs contained in this publication were checked for
currency during the production process. Note, however, that the
publisher cannot vouch for the ongoing currency of URLs.

For product information and technology assistance,
in Australia call **1300 790 853**;
in New Zealand call **0800 449 725**

For permission to use material from this text or product, please email
aust.permissions@cengage.com

National Library of Australia Cataloguing-in-Publication Data
Sadler, A.J., author.
Mathematics specialist : units 1 and 2 / A.J. Sadler.

1st revised edition
9780170390477 (paperback)
For secondary school age.

Mathematics--Study and teaching (Secondary)--Western Australia.
Mathematics--Textbooks.

510.712

Cengage Learning Australia
Level 7, 80 Dorcas Street
South Melbourne, Victoria Australia 3205

Cengage Learning New Zealand
Unit 4B Rosedale Office Park
331 Rosedale Road, Albany, North Shore 0632, NZ

For learning solutions, visit **cengage.com.au**

Printed in China by 1010 Printing International Ltd
6 7 24

PREFACE

This text targets units one and two of the West Australian course *Mathematics Specialist*. Chapters one to eight cover the content of Unit One and chapters nine to thirteen cover Unit Two.

The West Australian course, *Mathematics Specialist*, is based on the Australian Curriculum Senior Secondary course *Specialist Mathematics*. At the time of writing there are only small differences between the content for units one and two of the West Australian course and units one and two of the Australian Curriculum course, so this text would also be suitable for anyone following units one and two of the Australian Curriculum course, *Specialist Mathematics*. The main differences are that the West Australian units include vertical translations of trigonometric graphs, solving linear equations in two variables by matrix methods, and when considering converse and contrapositive statements the inverse is also covered.

The book contains text, examples and exercises containing many carefully graded questions. A student who studies the appropriate text and relevant examples should make good progress with the exercise that follows.

The book commences with a section entitled **Preliminary work**. This section briefly outlines work of particular relevance to this unit that students should either already have some familiarity with from the mathematics studied in earlier years, or for which the brief outline included in the section may be sufficient to bring the understanding of the concept up to the necessary level.

As students progress through the book they will encounter questions involving this preliminary work in the **Miscellaneous Exercises** that feature at the end of each chapter. These miscellaneous exercises also include questions involving work from preceding chapters to encourage the continual revision needed throughout the unit.

Some chapters commence with a '**Situation**' or two for students to consider, either individually or as a group. In this way students are encouraged to think and discuss a situation, which they are able to tackle using their existing knowledge, but which acts as a fore-runner and stimulus for the ideas that follow. Students should be encouraged to discuss their solutions and answers to these situations and perhaps to present their method of solution to others. For this reason answers to these situations are generally not included in the book.

At times in this series of texts I have found it appropriate to perhaps go a little outside the confines of the syllabus for the unit involved. In this regard readers will find that in unit two I consider
$$\sin P \pm \sin Q \text{ and } \cos P \pm \cos Q,$$
and when considering matrix equations of the form $AX = B$, my considerations go beyond X and B being column matrices. I also include shear transformations and mention proof by exhaustion.

Alan Sadler

ISBN 9780170390477

CONTENTS

UNIT ONE

PRELIMINARY WORK   viii
Number...viii
The absolute value............................viii
Trigonometry.. ix
Use of algebra...................................... ix
Similar triangles.....................................x
Congruent triangles............................ xii
Sets and Venn diagrams xiii

1

TRUE OR FALSE?   2
True or false?.. 4
Converse, inverse and
contrapositive 8
Miscellaneous exercise one 11

2

COUNTING   14
Permutations (arrangements)........... 15
Factorial notation........................... 16
Permutations of objects from a
group of objects, all different 17
Permutations of objects, not
all different............................... 21
Addition principle........................ 22
Inclusion–exclusion principle 23

Arrangements of objects with some
restriction imposed...................... 27
   I. Multiplicative reasoning 27
   II. Additive and multiplicative
      reasoning............................ 32
Combinations 37
^nC_r and Pascal's triangle 46
Miscellaneous exercise two 47

3

VECTORS – BASIC IDEAS   50
Vector quantities 52
Adding vectors............................ 56
Mathematical representation
of a vector quantity 61
Equal vectors 61
The negative of a vector 61
Multiplication of a vector by a
scalar...................................... 62
Parallel vectors........................... 62
Addition of vectors....................... 62
Subtraction of one vector
from another.............................. 63
The zero vector 63
$h\mathbf{a} = k\mathbf{b}$ 64
Miscellaneous exercise three 70

4

VECTORS IN COMPONENT FORM   72
Further examples 82
Position vectors 89
Miscellaneous exercise four 92

5

GEOMETRIC PROOFS   94
Proof... 95
Definition, axioms and theorems..... 95
Circle properties.......................... 98
Angles in circles 98
Tangents and secants 104
Miscellaneous exercise five 109

6

RELATIVE DISPLACEMENT AND RELATIVE VELOCITY   112
Relative displacement................. 115
Relative velocity 118
Miscellaneous exercise six........... 124

7

PROOFS USING VECTORS   128
Miscellaneous exercise seven 133

8

SCALAR PRODUCT   136

Scalar product 137

Algebraic properties of the
scalar product........................... 138

The scalar product from
the components......................... 143

Proofs using the scalar product 147

Miscellaneous exercise eight 150

UNIT TWO

PRELIMINARY WORK   155

Radian measure 155

Unit circle definitions of $y = \sin x$,
$y = \cos x$ and $y = \tan x$ 155

Transformations of $y = \sin x$
(and of $y = \cos x$ and $y = \tan x$) ... 159

Angle sum and angle
difference identities 160

Proof of angle sum and
angle difference identities 161

Solving trigonometric
equations................................ 163

9

**TRIGONOMETRICAL
IDENTITIES AND
EQUATIONS   164**

The double angle identities.......... 170

$a \cos \theta + b \sin \theta$........................ 173

Sec θ, cosec θ and cot θ............ 176

Product to sum and sum to
product................................... 180

General solutions of
trigonometric equations............... 183

Obtaining the rule from the graph 187

Modelling periodic motion 188

Miscellaneous exercise nine 192

10

MATRICES   194

Adding and subtracting matrices   197

Multiplying a matrix by a number   197

Equal matrices 198

Multiplying matrices 201

Zero matrices........................... 209

Multiplicative identity matrices...... 211

The multiplicative inverse of a
square matrix........................... 212

Using the inverse matrix to
solve systems of equations........... 218

Extension activity: Finding the
determinant and inverse of a
3×3 matrix........................... 221

Miscellaneous exercise ten 222

11

**TRANSFORMATION
MATRICES   226**

Transformations and matrices 227

Determining the matrix for a
particular transformation.............. 230

The determinant of a
transformation matrix 230

The inverse of a transformation
matrix..................................... 230

Combining transformations 232

Further examples 233

A general rotation about the
origin 237

A general reflection in a line that
passes through the origin 237

Miscellaneous exercise eleven 239

12

PROOF   242

Proof by exhaustion.................... 246

Proof by induction...................... 248

Extension activity: Investigating
some conjectures...................... 253

Miscellaneous exercise twelve 254

13

COMPLEX NUMBERS   258

Complex numbers...................... 261

Complex number arithmetic 264

The conjugate of a complex
number................................... 265

Equal complex numbers 265

Linear factors of quadratic
polynomials.............................. 266

Argand diagrams 269

Miscellaneous exercise thirteen 271

ANSWERS UNIT ONE   277

ANSWERS UNIT TWO   285

INDEX   296

IMPORTANT NOTE

This text has been written based on my interpretation of the appropriate Mathematics Specialist syllabus documents as they stand at the time of writing.
It is likely that as time progresses some points of interpretation will become clarified and perhaps even some changes could be made to the original syllabus. I urge teachers of the Mathematics Specialist course, and students following the course, to check with the appropriate curriculum authority to make themselves aware of the latest version of the syllabus current at the time they are studying the course.

Acknowledgements

As with all of my previous books I am again indebted to my wife, Rosemary, for her assistance, encouragement and help at every stage.

To my three beautiful daughters, Rosalyn, Jennifer and Donelle, thank you for the continued understanding you show when I am 'still doing sums' and for the love and belief you show in me.

Alan Sadler

ISBN 9780170390477

Mathematics Specialist

Unit One

1

UNIT ONE PRELIMINARY WORK

This book assumes that you are already familiar with a number of mathematical ideas from your mathematical studies in earlier years.

This section outlines the ideas which are of particular relevance to Unit One of the Mathematics Specialist course and for which familiarity will be assumed, or for which the brief explanation given here may be sufficient to bring your understanding of the concept up to the necessary level.

Read this 'Preliminary work' section and if anything is not familiar to you, and you don't understand the brief mention or explanation given here, you may need to do some further reading to bring your understanding of the concepts up to an appropriate level for this unit. (If you do understand the work but feel somewhat 'rusty' with regards to applying the ideas, some of the chapters afford further opportunities for revision, as do some of the questions in the miscellaneous exercises at the end of chapters.)

- Chapters in this book will continue some of the topics from this preliminary work by building on the assumed familiarity with the work.

- The **miscellaneous exercises** that feature at the end of each chapter may include questions requiring an understanding of the topics briefly explained here.

- Familiarity with some of the content of unit one of the *Mathematics Methods* course will be useful for this unit of *Mathematics Specialist*. It is anticipated that students will be studying that unit at the same time as this one and that the useful familiarity with content will be reached in time for its use in this unit.

Number

The set of numbers that you are currently familiar with is called the set of **real numbers**. We use the symbol $\mathbb{R}$ for this set. $\mathbb{R}$ contains many subsets of numbers such as the whole numbers, the integers, the rational numbers, the irrational numbers etc.

The absolute value

A concept that you may have met before is that of the absolute value of a number. This is the distance on the number line that the number is from the origin.

The absolute value of -3, written $\lvert -3 \rvert$, is 3 because -3 is three units from the origin.	The absolute value of 4, written $\lvert 4 \rvert$, is 4 because 4 is four units from the origin.

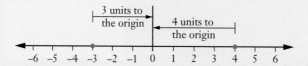

The absolute value makes no distinction between numbers that are to the left of the origin and those that are to the right of the origin.

Thus

$$\lvert 3 \rvert = 3 \qquad \lvert 5 \rvert = 5 \qquad \lvert 3.1 \rvert = 3.1 \qquad \lvert 8.2 \rvert = 8.2$$
$$\lvert -3 \rvert = 3 \qquad \lvert -5 \rvert = 5 \qquad \lvert -3.1 \rvert = 3.1 \qquad \lvert -8.2 \rvert = 8.2$$

ISBN 9780170390477

Trigonometry

It is assumed that you are already familiar with the use of the theorem of Pythagoras and the ideas of sine, cosine and tangent to determine the unknown sides and angles in right triangles. Your answers should be given to the accuracy requested or, if none is specifically requested, to an accuracy that is appropriate for the accuracy of the given data and the nature of the situation.

An understanding of the concept of a bearing being measured clockwise from North, e.g. 080°, and using compass points, e.g. N 80° E, is also assumed.

With sine $= \dfrac{\text{opp}}{\text{hyp}}$, cosine $= \dfrac{\text{adj}}{\text{hyp}}$ and tangent $= \dfrac{\text{opp}}{\text{adj}}$, it follows that

$$\tan A = \frac{\sin A}{\cos A}$$

This is a result you may not be familiar with but that we will use in this unit.

You may be familiar with:

- the formula $A = \dfrac{1}{2} ab \sin C$ to determine the area of a triangle

- the sine rule $\dfrac{a}{\sin A} = \dfrac{b}{\sin B} = \dfrac{c}{\sin C}$

- the cosine rule $c^2 = a^2 + b^2 - 2ab \cos C$

- the use of the unit circle to give meaning to the sine and cosine of angles bigger than 90°

- expressing the trigonometric ratios of some angles as exact values,

or you will become familiar with these concepts as your concurrent study of unit one of the *Mathematics Methods* course progresses.

It is also assumed that you are familiar with the idea that *similar triangles* have corresponding sides in the same ratio – the underlying idea on which the trigonometrical ratios of right triangles are based.

Use of algebra

It is assumed that you are already familiar with manipulating algebraic expressions, in particular:

- Expanding and simplifying:

For example	$4(x + 3) - 3(x + 2)$	expands to	$4x + 12 - 3x - 6$
		which simplifies to	$x + 6$
	$(x - 7)(x + 1)$	expands to	$x^2 + 1x - 7x - 7$
		which simplifies to	$x^2 - 6x - 7$
	$(2x - 7)^2$, i.e. $(2x - 7)(2x - 7)$	expands to	$4x^2 - 28x + 49$

- Factorising:

For example,	$21x + 7$	factorises to	$7(3x + 1)$
	$x^2 - 6x - 7$	factorises to	$(x - 7)(x + 1)$
	$x^2 - y^2$	factorises to	$(x - y)(x + y).$

the last one being referred to as the *difference of two squares*.

- Solving equations: In particular linear, simultaneous and quadratic equations.

Similar triangles

Whilst it is assumed that you are familiar with the concept of two triangles being similar (as mentioned at the end of the trigonometry section of this *Preliminary work* section) it is further assumed that you can apply these ideas in the formulation of a proof. Read through the following by way of a reminder of these ideas.

If two triangles are similar, corresponding sides are in the same ratio. (One triangle is like a 'photographic enlargement' of the other.)

For example, if the two triangles ABC and DEF, shown below, are similar then

$$AB:DE = AC:DF = BC:EF$$

i.e.
$$\frac{AB}{DE} = \frac{AC}{DF} = \frac{BC}{EF}$$

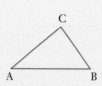
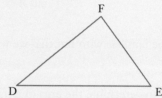

Note   If two triangles are similar, the corresponding angles are equal.

   i.e., in the similar triangles shown above

$\angle CAB = \angle FDE,$

$\angle ABC = \angle DEF,$

$\angle ACB = \angle DFE.$

We can use the symbol '~' to indicate that two shapes are similar.

Thus △ABC ~ △DEF

The order of the letters *is* significant and indicates corresponding sides and angles. Thus, for triangles ABC and DEF above, it would **not** be correct to say △ABC ~ △FED.

Knowing that two triangles are similar can sometimes allow us to determine the length of some of the sides. For example if we were told that for the triangles sketched below,

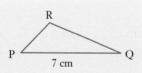
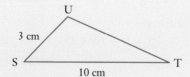

   △PQR ~ △STU

Then $\dfrac{PQ}{ST} = \dfrac{PR}{SU}$

i.e. $\dfrac{7}{10} = \dfrac{x}{3}$ where x cm is the length of PR.

Solving gives $x = 2.1$.   Thus PR is of length 2.1 cm.

SPECIALIST MATHEMATICS Units 1 & 2

ISBN 9780170390477

To determine whether two triangles are similar we can:

See if the three angles of one triangle are equal to the three angles of the other triangle.

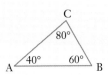 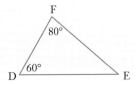

△ABC and △EDF are similar.   Reason: Corresponding angles equal.

Or: See if the lengths of corresponding sides are in the same ratio.

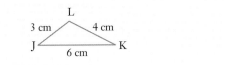 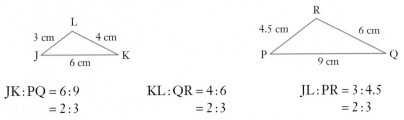

$$JK:PQ = 6:9 \qquad KL:QR = 4:6 \qquad JL:PR = 3:4.5$$
$$= 2:3 \qquad\qquad = 2:3 \qquad\qquad = 2:3$$

△JKL and △PQR are similar.   Reason: Corresponding sides in same ratio.

Or: See if the lengths of two pairs of corresponding sides are in the same ratio and the angles between the sides are equal.

 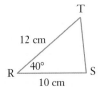

$$XZ:RS = 5:10 \qquad\qquad XY:RT = 6:12$$
$$= 1:2 \qquad\qquad\qquad = 1:2$$

The angle between XY and XZ = the angle between RT and RS.

△XZY and △RST are similar.   Reason: Two pairs of corresponding sides in same ratio and the *included* angles equal.

Note: If the two triangles are right angled, the corresponding sides that are in the
same ratio need not *include* the right angle.

ISBN 9780170390477

Example

Given that in the diagram shown

FG = 3 m,   GH = 6 m,   JG = 2 m.

prove that IH is of length 6 m.

Solution

Given: Diagram as shown.

To prove: IH = 6 m.

Proof: In triangles FGJ and FHI:

$$\angle FGJ = \angle FHI \qquad \text{(Each angle = 90°.)}$$

$$\angle JFG = \angle IFH \qquad \text{(Same angle.)}$$

$$\therefore \qquad \angle FJG = \angle FIH \qquad \text{(Third angle in each triangle.)}$$

Thus $\triangle FGJ \sim \triangle FHI$ (Corresponding angles equal.)

$$\therefore \qquad \frac{JG}{IH} = \frac{FG}{FH} \qquad \text{i.e., if IH is } x \text{ m in length,} \qquad \frac{2}{x} = \frac{3}{9}$$

$$x = 6$$

Thus IH is of length 6 m, as required.

Notice how known truths are referred to in order to justify statements. Such justifications do not need to be 'essays' but are instead a brief statement to clearly indicate which truth justifies the statement.

Congruent triangles

If two triangles are identical, i.e. *congruent*, the three angles and three sides of one will match the three angles and three sides of the other. For example, the triangles ABC and DEF shown below are congruent.

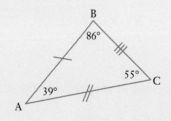
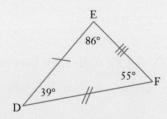

We write: $\triangle ABC \cong \triangle DEF$

However, we do not need to know that all six pieces of information,

A A A S S S,

'match' before being able to say that two triangles are congruent.

In fact, to prove two triangles congruent, we need only to show one of the sets of facts shown on the next page.

ISBN 9780170390477

To prove two triangles congruent we need only show one of the following:

SSS   The lengths of the three sides of one triangle are the same as the lengths of the three sides of the other triangle.

SAS   The lengths of two sides of one triangle are the same as the lengths of two sides of the other triangle and the included angles are equal.

AA corresponding S Two angles of one triangle equal two angles of the other and one side in one triangle is the same length as the corresponding side in the other.

RHS   Both triangles are right angled, the two hypotenuses are of the same length and one other side from each triangle are of equal length.

Example

If we define an isosceles triangle to be a triangle that has two of its sides of the same length, prove that two of its angles must also be equal.

Solution

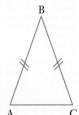

Given:	$\triangle ABC$ with $AB = CB$.
To prove:	$\angle BAC = \angle BCA$
Construction:	Draw the line from B to D, the midpoint of AC. (See second diagram on the right.)
Proof:	In triangles ADB and CDB

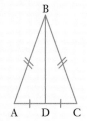

      $AB = CB$ (Given.)

      $AD = CD$ (D is the midpoint of AC.)

      $BD = BD$ (Common to both triangles.)

$\therefore$ $\triangle ADB \cong \triangle CDB$ (SSS)

Hence $\angle BAD = \angle BCD$ (Corresponding angles)

and so $\angle BAC = \angle BCA$ as required.

Sets and Venn diagrams

Some familiarity with sets and **Venn diagrams** is assumed. The **Venn diagram** on the right shows the **universal set**, U, which contains all of the **elements** currently under consideration, and the sets A and B contained within it.

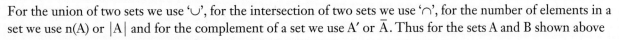

We use 'curly brackets' to list a set. Thus for the sets shown

  $U = \{1, 2, 3, 4, 5, 6, 7, 8, 9, 10\}$, $A = \{2, 3, 5, 7\}$ and $B = \{1, 3, 5, 7, 9\}$.

Using the symbol $\in$ for 'is a member of': $3 \in A, 9 \in B, 9 \notin A$.

For the union of two sets we use '$\cup$', for the intersection of two sets we use '$\cap$', for the number of elements in a set we use n(A) or $|A|$ and for the complement of a set we use A' or $\bar{A}$. Thus for the sets A and B shown above

$$A \cap B = \{3, 5, 7\}, \qquad\qquad A \cup B = \{1, 2, 3, 5, 7, 9\},$$
$$n(A) = 4 \qquad\qquad \text{and} \qquad\qquad A' = \{1, 4, 6, 8, 9, 10\}.$$

Sometimes we use the Venn diagram to show the number of elements in the particular regions, rather than the elements themselves.

1.

True or false?

- True or false?
- Converse, inverse and contrapositive
- Miscellaneous exercise one

Situation: Stairs

Let us suppose that the descent of a set of stairs can be made by taking the stairs

one at a time   or   two at a time.

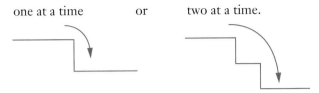

If the stairs involve just **one** step there is just **one** way of descending – one step at a time.

If the stairs involve **two** steps there are **two** ways of descending.

If **three** steps are involved there are **three** ways of descending.

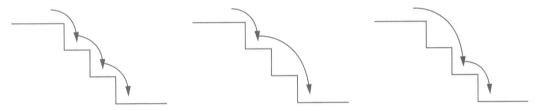

Seeing the above pattern John conjectures that if we use 'one step at a time' or 'two steps at a time' then there will be

> Four ways of descending stairs with four steps.
> Five ways of descending stairs with five steps.
> Six ways of descending stairs with six steps. Etc.

Is John's conjecture correct? If you think it is not correct explain why you think this and try to come up with a correct conjecture yourself for this situation.

Note

A *conjecture* is an opinion. It is what someone is suggesting they think to be the case. It is likely that John has formed this opinion based on observation and thought – in this case by observing and thinking about the pattern of numbers from the 1-step, 2-step and 3-step situations.

Shutterstock.com/tranman111

True or false?

Consider each of the following statements and for each one decide if the statement is true or false.

Note: In this activity only accept something as being true if there are no circumstances in which it can be false. Thus if $x^2 = 16$ the conclusion that $x = 4$ will be adjudged to be false because there is a circumstance when x does not have to be 4, it could be –4. Hence the statement *If $x^2 = 16$ then $x = 4$* should be adjudged to be a false conclusion because x is not necessarily 4.

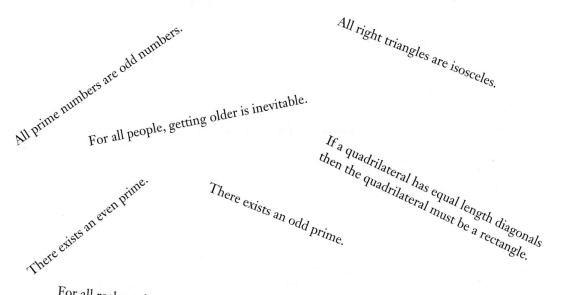

All prime numbers are odd numbers.

All right triangles are isosceles.

For all people, getting older is inevitable.

If a quadrilateral has equal length diagonals then the quadrilateral must be a rectangle.

There exists an even prime.

There exists an odd prime.

For all real numbers, the square of the number is non-negative, i.e. for all $x \in \mathbb{R}$, $x^2 \geq 0$.

If $x = 2$ then $x^2 = 4$

For all people in Australia, the equator is to the North.

If a particular animal is an eagle then the animal is a bird.

If $x^2 = 4$ then $x = 2$.

If a particular animal is a bird then the animal is an eagle.

ISBN 9780170390477

Every living dog has a heart.
My cat has a heart;
therefore, my cat is a dog.

A triangle with angles of 20°, 80° and 80°
must be an isosceles triangle.

An isosceles triangle must have
angles of 20°, 80° and 80°.

If a quadrilateral is cyclic, then a circle
can be drawn through its four vertices.

If $x^2 \neq 4$ then $x \neq 2$.

If a circle can be drawn through the four vertices of a quadrilateral,
then the quadrilateral is cyclic.

If a triangle has two sides of the same length then it has two angles equal in size.

If a triangle has two angles of the same size
then it has two sides of the same length.

If a polygon has exactly four sides
then the polygon is a quadrilateral.

If a postman has seven letters
to deliver to six letterboxes,
at least one letterbox must
get more than one letter.

If a polygon is not a quadrilateral then
it does not have exactly four sides.

If it is not 'not raining' then it must be raining.

You cannot have a right triangle with
one side of length $3x$ cm,
another side of length $(4x + 5)$ cm
and the longest side of length $(5x + 4)$ cm.

1. True or false? ●●●●●●●●●●●●●●

Let us now recall the situation at the beginning of this chapter which involved the various ways we could negotiate a set of stairs consisting of various numbers of steps which we could take 'one step at a time' or 'two steps at a time'.

You might recall that, having spotted a pattern, John made the conjecture that there were:

Four ways of descending stairs with four steps.
Five ways of descending stairs with five steps.
Six ways of descending stairs with six steps. Etc.

Did you agree with the conjecture?

Counter example

Counter-examples

If you checked the four step situation (or indeed any of the other situations after that of three steps) you should have found that the facts contradicted John's conjecture. Because John's conjecture, as stated above, makes a claim that applies to all situations of four steps or more we need find only one **counter example** to show the general conjecture to be false.

To test the validity of a generalisation we could systematically check specific cases to see if the statement holds true for them. The more examples we find for which the statement holds the more confident we might become with regard to the validity of the statement. However we need only find one **counter example** to show the generalisation to be false.

Did you use counter examples to convince yourself that the following statements on the previous two pages were not true?

All prime numbers are odd numbers.
All right triangles are isosceles.
If a quadrilateral has equal length diagonals, then the quadrilateral must be a rectangle.

Quantifiers

If we are making a statement about **all** members of a set then the statement must apply to **all** members of that set. Words like 'all' or 'many' are **quantifiers** – they indicate quantity. If we claim that something is true for **all** members of the set then just one counter example proves our claim wrong. The following statements on the previous two pages involved the word all but in each case they were correct as the statement did indeed apply to all members of the set:

For all people, getting older is inevitable.
For all real numbers, the square of the number is non-negative, i.e. for all $x \in \mathbb{R}$, $x^2 \geq 0$.
For all people in Australia the equator is to the North.

(The symbol $\forall$ can be used to represent 'for all'. Thus: $\forall$ real numbers x, $x^2 \geq 0$.)

Another quantifier would be the phrase 'at least one'. The following statements from the previous two pages used the idea that there was at least one of something by using the phrase '**there exists**'.

There exists an even prime.
There exists an odd prime.

For such statements we only have to show the existence of one such item and the statement is true.

(The symbol $\exists$ can be used to represent 'there exists'. Thus: $\exists$ an even prime.)

Some statements are true **by definition**. If we define a cyclic quadrilateral to be one for which the four vertices lie on a circle it follows that a circle can be drawn through the four vertices of a cyclic quadrilateral.

Some of the statements that you were asked to consider as to whether they were true or false suggested that 'if ' something was true 'then' this **implied** that something else was true.

For example: *If x = 2 then x^2 = 4.*
If a particular animal is an eagle then the animal is a bird.

Both of which are true.

If event P implies event Q, then we write '**If P then Q**' or, using the symbol '⇒' we can write P ⇒ Q.

Thus $x = 2 \Rightarrow x^2 = 4$

and   an animal is an eagle ⇒ the animal is a bird

Converse

The **converse** of 'If P then Q' is 'If Q then P', i.e. Q ⇒ P. However, just because P ⇒ Q it does not automatically follow that Q ⇒ P.

For example: *If x = 2 then x^2 = 4*   is true

but the **converse**:   *If x^2 = 4 then x = 2*   is false (because *x* could be –2).

Similarly, from the statements given earlier, it may be true that every living dog has a heart but I cannot conclude that just because my cat has a heart it too must be a dog. Neither can I conclude from the fact that an eagle is a bird that a bird is necessarily an eagle. Similarly a triangle with angles of 20°, 80° and 80° must indeed be an isosceles triangle but one counterexample would be sufficient to prove that the converse, i.e. that an isosceles triangle must have angles of 20°, 80° and 80°, is not the case.

The converse of a true statement need not be true.

However if it is the case that P ⇒ Q and Q ⇒ P the symbol ⇔ can be used, i.e. P ⇔ Q. Statements P and Q are then said to be **equivalent**.

For example:

If a triangle has two sides of the same length then it has two angles equal in size.
I.e.: two sides of triangle the same length ⇒ two angles of triangle equal.

If a triangle has two angles of the same size then it has two sides of the same length.
I.e.: two angles of triangle equal ⇒ two sides of triangle the same length.

Hence: Two sides of triangle the same length ⇔ two angles of triangle equal.

The statement, P ⇔ Q, can also be written as 'P **if and only if** Q' (or as 'P iff Q'):

A triangle has two sides of the same length if and only if it has two angles equal in size.

Contrapositive

The **contrapositive** of 'If P then Q' is 'If not Q then not P'.

Consider the true statement   *If x = 2 then x^2 = 4.*

The contrapositive statement is   *If x^2 ≠ 4 then x ≠ 2.*   Also a true statement.

The contrapositive of a true statement is also true.

For example:

If a polygon has exactly four sides then the polygon is a quadrilateral.

is a true statement, as is the contrapositive:

If a polygon is not a quadrilateral then it does not have exactly four sides.

Some of the statements that you were asked to judge as being true or false were probably quite obviously true such as the statement about the postman with seven letters to post in six letter boxes. However the logic behind it, in this case called the **pigeon-hole principle**, can be useful.

The pigeon-hole principle states:

If there are *n* pigeon holes, *n* ≥ 1, and *n* + 1 pigeons to go in them, then at least one pigeon hole must get two or more pigeons.

Negation

Some of the statements involved the **negation** of a statement. If P is the statement

It is raining.

Then the negation of P is the statement

It is not raining.

The statement '*it is not not raining*' given on an earlier page involves a double negative. If it is not not raining it must be raining.

Some statements are not so immediately obvious as being true or false and require careful thought. How, for example, did you decide upon the truth or otherwise of the statement:

You cannot have a right triangle with one side of length 3x cm, another side of length (4x + 5) cm and the longest side of length (5x + 4) cm.

One approach is to assume the opposite, i.e. assume that we can indeed have a right triangle with the given side lengths and then prove that this assumption leads to something that cannot be true. This is called **proof by contradiction** – we assume what we are asked to prove is not the case, and then show that this assumption leads to a contradiction.

Converse, inverse and contrapositive

The statement 'if P then Q' also has an **inverse** statement which is 'if not P then not Q'.

For example, the inverse of the statement: if $x = 2$ then $x^2 = 4$

is if $x \neq 2$ then $x^2 \neq 4$.

this inverse being false as the counterexample $x = -2$ shows.

Thus for the statement: if P then Q,

    the converse statement is   if Q then P,

    the inverse statement is   if not P then not Q,

and   the contrapositive statement is   if not Q then not P.

The contrapositive statement involves both the effect of the converse, in its switch of P and Q, and of the inverse, with its negation of both P and Q.

If the original statement, if P then Q, is true then the contrapositive is also true but the converse and the inverse may not be.

(Readers may notice that the inverse is the contrapositive of the converse.)

Exercise 1A

Using the idea behind the pigeon-hole principle, what can be concluded for each of the situations described in numbers **1** to **7**?

1 A teacher of *Mathematics Specialist* asked eight students each to choose one question to do from numbers 1 to 7 of this exercise.

2 The Singh triplets are all in year three at the same primary school and this school only has two year three classes.

3 Let us suppose that a particular genetic marker possessed by every human has two billion different variations and there are over six billion people in the world.

4 Peter has 12 socks in a drawer, six red socks and six blue socks but other than their colour the socks are indistinguishable. Peter reaches in and pulls out three socks.

5 On average the number of hairs on a young adult human's head is about 100 000 and we would not expect to find any human with as many as one million hairs on their head. However there are over twenty million people in Australia.

6 If we count our ancestors as our parents, our parents' parents, our parents' parents' parents etc., then if we go back one generation we have 2 ancestors in that generation, go back two generations and we have 4 ancestors in that generation, go back three generations and we have 8 ancestors in that generation. If we go back 40 generations the number of ancestors we have is theoretically greater than the number of people that have ever lived.

7 Fifteen people at a gathering start to mix and mingle. Each person shakes hands with however many of the other 14 people that they wish to. (Shaking hands with oneself does not count!)

   a What is the largest number of different people a person in this group could shake hands with?

   b What is the smallest number of different people a person in this group could shake hands with?

   c Could one person have shaken hands with the largest number possible (i.e. the answer for part **a**) AND someone else in the group have shaken hands with the smallest number possible (i.e. the answer for part **b**)?

The **converse** of 'if P then Q' is 'if Q then P'.

Each of the following statements should be assumed to be true. Write the converse of each statement, state whether the converse is true or false, and write P ⇔ Q or P ⇎ Q.

8 If a polygon has exactly three sides then the polygon is a triangle.

9 If Jenny is talking then her mouth is open.

10 If the animal is a platypus then it is a mammal.

1. True or false? ●●●●●●●●●●●●●●●

11 If the car is out of fuel it will not start.

12 If points lie on the same straight line then they are collinear points.

The **contrapositive** of 'if P then Q' is 'if not Q then not P'.

Each of the following statements should be assumed to be true. Write the contrapositive of each statement and then check that each contrapositive is also true.

13 If today is Thursday then tomorrow is Friday.

14 If a number is even then it is a multiple of two.

15 If a triangle is scalene then it has three different length sides.

16 If my sprinklers are on then my lawn is wet.

17 If it is not a school day then Armand does not get up before 8 am.

The **inverse** of 'if P then Q' is 'if not P then not Q'.

For each of questions **18** to **22**:
a state whether the initial statement is true or false,
b write the inverse of the given statement,
c state whether the inverse statement is true or false.

18 If a polygon is a triangle then its angles add up to 180°.

19 If a positive integer has exactly two factors then it is a prime number.

20 If the car battery is flat then the car will not start.

21 If there are letters in my mail box then the postperson has been to our road.

22 If a number is even then it is a multiple of 4.

23 For the statement

If a polygon is five sided then the polygon is a pentagon

write the converse statement, the inverse statement and the contrapositive statement and in each case state whether it is true or false.

24 For the statement

If a quadrilateral is a square then the four angles of the quadrilateral are all right angles.

write the converse statement, the inverse statement and the contrapositive statement and in each case state whether it is true or false.

ISBN 9780170390477

Prove each of the following using the method of proof by contradiction.

25 A triangle with side lengths of 8 cm, 9 cm and 10 cm is not right angled.

26 There are no integers p and q such that $6p + 10q = 151$.

27 If a and b are any two positive real numbers:

$$\frac{a}{b} + \frac{b}{a} \geq 2$$

28 A triangle with one side of length $5x$, another of length $12x + 13$ and the longest side of length $13x + 12$, cannot be a right angled triangle.

Miscellaneous exercise one

This miscellaneous exercise may include questions involving the work of this chapter and the ideas mentioned in the Preliminary work section at the beginning of the book.

1 (Revision of right triangle trigonometry.)

A ladder stands with its base on horizontal ground and its top against a vertical wall. When the base of the ladder is a metres from the wall the ladder makes an angle of 75° with the ground.

What angle will the ladder make with the ground if the base of the ladder is $\frac{5a}{4}$ metres from the wall?

Questions **2** to **8** (Revision of proofs using similar and congruent triangles).

Questions **2** to **8** each ask you to prove some geometrical fact. You should set out your proofs as in the examples given in the *Preliminary work* sections involving similar triangles and congruent triangles. I.e. draw a clear diagram, state what you are given, what you have to prove and any constructions made. Then set out your proof clearly and with statements justified.

In your proofs the following may be stated as fact, without proof:

• Angles that together form a straight line have a sum of 180°.
I.e., in the diagram on the right, A + B = 180°.
(And, conversely, if angles have a sum of 180° then they together form a straight line.)

• The angles of a triangle add up to 180°.
I.e., in the diagram on the right, A + B + C = 180°.
(And conversely, if the angles of a polygon sum to 180° then the polygon is a triangle.)

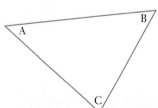

2 Prove that the angles of a quadrilateral add up to 360°.

3 In the diagram shown on the right,

E is a point on AD and B is a point on AC.

$$AB = 8 \text{ m}$$
$$BC = 8 \text{ m}$$
$$CD = 10 \text{ m}$$

and right angles are as indicated.

Prove that $\quad EB = 5$ m.

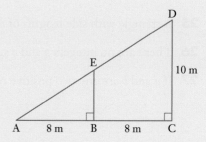

4 In the diagram shown on the right,

B is a point on AC.

$$AB = 3 \text{ m}$$
$$BC = 5 \text{ m}$$
$$CD = 4 \text{ m}.$$
$$\angle EBD = 180° - 2 \times \angle EBA$$

and right angles are as indicated.

Prove that $\quad AE = 2.4$ m.

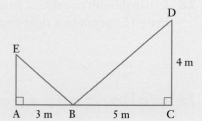

5 In the diagram shown on the right,

B is a point on AC and E is a point on AD.

$$\angle AEB = \angle ACD$$
$$AB = 4 \text{ cm}$$
$$AD = 7 \text{ cm}$$
$$BE = 6 \text{ cm}.$$

Prove that $\quad CD = 10.5$ cm.

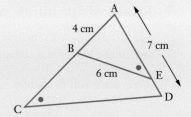

Each of questions **6**, **7** and **8** involve an isosceles triangle. For each question, assume only that the triangle has two sides of equal length. Any other facts already proved for isosceles triangles should not be assumed.

6 If we define an isosceles triangle to be a triangle that has two of its sides of the same length, prove that the line drawn from the vertex that is common to the equal sides, to the midpoint of the third side, is perpendicular to that third side.

7 △XYZ is an isosceles triangle with XZ = YZ.

Prove that if a line is drawn from Z to meet XY at right angles then this meeting point will be the midpoint of XY.

8 △PQR is an isosceles triangle with PQ = PR.

Prove that if a line is drawn from P, bisecting ∠QPR and meeting QR at the point M then M is the midpoint of QR.

ISBN 9780170390477

Counting

- Permutations (arrangements)

- Factorial notation

- Permutations of objects from a group of objects, all different

- Permutations of objects, not all different

- Addition principle

- Inclusion–exclusion principle

- Arrangements of objects with some restriction imposed

  I. Multiplicative reasoning

  II. Additive and multiplicative reasoning

- Combinations

- ^nC_r and Pascal's triangle

- Miscellaneous exercise two

Permutations (arrangements)

Asked to determine the number of possible **arrangements** (or **permutations**) there are of the letters of the word POST, with each letter being used once, we could list them:

POST	POTS	PSOT	PSTO	PTOS	PTSO
OPST	OPTS	OSPT	OSTP	OTPS	OTSP
SPOT	SPTO	SOPT	SOTP	STPO	STOP
TPOS	TPSO	TOPS	TOSP	TSPO	TSOP

to arrive at the answer of 24.

Alternatively we could choose a tree diagram form of display, as shown on the right, to again arrive at the answer of 24

However this listing is tedious, even with just the four letters of the word POST, so instead we develop techniques for counting the number of possible arrangements without having to list them all. Hence the title of this chapter, *Counting*.

One useful principle we can use in this situation is the **multiplication principle**:

> If there are *a* ways an activity can be performed, and for each of these there are *b* ways that a second activity can be performed after the first, and for each of these there are *c* ways that a third activity can be performed after the second, and so on, then there are $a \times b \times c \times \ldots$ ways of performing the successive activities.

In the situation given above, there were 4 choices of first letter, followed by 3 choices of second letter etc.

Number of arrangements $= 4 \times 3 \times 2 \times 1$
$$= 24$$

No. of ways for each letter			
1st	2nd	3rd	4th
4	3	2	1

Alamy Stock Photo/BERANGER/BSIP

Factorial notation

Use of the multiplication principle frequently involves us in evaluating expressions like

$$2 \times 1$$
$$3 \times 2 \times 1$$
$$4 \times 3 \times 2 \times 1$$
$$5 \times 4 \times 3 \times 2 \times 1$$
$$6 \times 5 \times 4 \times 3 \times 2 \times 1$$

We write $n!$, pronounced 'n **factorial**', to represent

$$n \times (n-1) \times (n-2) \times \ldots \times 3 \times 2 \times 1 \qquad \text{where } n \text{ is a positive integer.}$$

For example

$$3! = 3 \times 2 \times 1$$
$$= 6$$

$$5! = 5 \times 4 \times 3 \times 2 \times 1$$
$$= 120$$

$$10! = 10 \times 9 \times 8 \times 7 \times 6 \times 5 \times 4 \times 3 \times 2 \times 1$$
$$= 3\,628\,800$$

3!	
	6
5!	
	120
10!	
	3628800

EXAMPLE 1

Evaluate　　**a**　6!　　　　　**b**　5! ÷ 3!　　　　　**c**　100! ÷ 98!

Solution

a　　$6! = 6 \times 5 \times 4 \times 3 \times 2 \times 1$
$$= 720$$

b　　$5! \div 3! = \dfrac{5 \times 4 \times 3!}{3!}$
$$= 5 \times 4$$
$$= 20$$

c　　$100! \div 98! = \dfrac{100 \times 99 \times 98!}{98!}$
$$= 100 \times 99$$
$$= 9900$$

Using this factorial notation we can say:

There are $n!$ ways of arranging n different objects in a row. I.e. $n!$ permutations.

EXAMPLE 2

How many different five letter 'words' can be formed using the letters of the word MATHS if

a each letter is used just once?

b each letter can be used any number of times?

Solution

a The first letter can be chosen in 5 ways, the second can then be chosen in 4 ways, the third in 3 ways etc.

Total number of words $= 5 \times 4 \times 3 \times 2 \times 1$
$= 120$

No. of ways for each letter				
5	4	3	2	1

b The first letter can be chosen in 5 ways, the second can then be chosen in 5 ways, the third in 5 ways etc.

Total number of words $= 5 \times 5 \times 5 \times 5 \times 5$
$= 3125$

No. of ways for each letter				
5	5	5	5	5

Permutations of objects from a group of objects, all different

Permutation calculations

Permutations

In our consideration of the arrangements of the letters of the word POST, on an earlier page, we were using *4 letters* to make *4 letter* words.

Similarly in example 2 above, we were arranging *5 letters* to make *5 letter* words.

Both of these situations involved using all of the available letters in each word.

How then do we determine the number of 5 letter words that can be created if we have more than 5 available? Again the multiplication principle comes to our aid.

Let us consider the case of making 2 letter words when the 5 letters of the word CAKES are available, with no repeat use of letters allowed.

We have **5** choices for first letter: C, A, K, E, S.

Having chosen the first letter we then have 4 choices of second letter.

Total number of choices $= 5 \times 4$
$= 20$

The complete listing is shown on the right.

Thus given n different objects, and wanting to arrange r of them:

CA	AC	KC	EC	SC
CK	AK	KA	EA	SA
CE	AE	KE	EK	SK
CS	AS	KS	ES	SE

The 1st could be chosen in n ways, the 2nd in $(n-1)$ ways, the 3rd in $(n-2)$ ways ... until we get to the rth object which could be chosen in $(n-r+1)$ ways.

Number of permutations $= n \times (n-1) \times (n-2) \times ... \times (n-r+1)$
$= \dfrac{n!}{(n-r)!}$

Hence:

> The number of permutations of r objects taken from n different objects is $\dfrac{n!}{(n-r)!}$.
> We write this as $^{n}P_{r}$.

EXAMPLE 3

How many different three-letter 'words' can be formed using the letters of the word MAKER if each letter is used just once?

Solution

The first letter can be chosen in 5 ways, the second can then be chosen in 4 ways and the third in 3 ways.

No. of ways for each letter		
5	4	3

Total number of words $= 5 \times 4 \times 3$
$$= 60$$

Or, using the ^nP_r notation:

Number of 3 letter arrangements chosen from 5 different letters is $^5P_3$.

$$^5P_3 = \frac{5!}{(5-3)!}$$
$$= \frac{5!}{2!}$$
$$= 5 \times 4 \times 3$$
$$= 60$$

```
5P3
                              60
```

EXAMPLE 4

One student is to be selected from the six students: Armand, Chris, Jennifer, Kate, Tony, Varun, to be student of the week for week 1. A different student from the group is then selected to be student of the week for week 2 and another is selected for week 3. In how many ways can this be done?

Solution

The student for week 1 can be chosen from any one of the six. The student for week two can then be chosen from the remaining 5 and the student for week 3 can be chosen from the remaining 4.

No. of ways for each week		
6	5	4

Total number of ways $= 6 \times 5 \times 4$
$$= 120$$

Note: ^nP_r gives the number of permutations of r objects taken from n different objects.

However we know that there are $n!$ permutations if we use all n objects.

Thus $^nP_n = n!$

I.e. $\dfrac{n!}{(n-n)!} = n!$

Thus we need to define $(n - n)!$ i.e. $0!$, to equal 1. Check that your calculator agrees with this value for zero factorial.

Exercise 2A

1 Evaluate each of the following, without the assistance of a calculator.

a $3!$

b $3! + 2!$

c $(3 + 2)!$

d $\dfrac{11!}{10!}$

e $\dfrac{11!}{9!}$

f $\dfrac{6!}{4!2!}$

g $^5P_2$

h $^7P_3$

i $^8P_2$

2 Three roads are available for the journey from town A to town B and two are available from town B to town C.

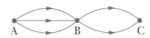

How many different routes are possible from A to C, via B?

3 How many different ham rolls are possible if we have four different types of roll, just one type of ham, lettuce or no lettuce, mustard or no mustard?

4 A father places six marbles, each of a different colour, in a row on the floor. His two-year-old daughter tells him he has put them in the wrong order! How many different orders are possible, other than the one he has already used?

5 Each day a student can make the journey to school in one of three ways:

by bus, by bike or on foot.

How many different ways can the student arrange his travel to school in the course of one week (five school days)?

For example, one arrangement could be:

bus, bike, foot, foot, bike.

6 How many different seven-letter 'words' can be formed using the letters of the word FASHION if:

a each letter is used just once in each word?

b each letter can be used any number of times?

7 How many different five-letter 'words' can be formed using the letters of the word FASHION if:

a each letter is used just once in each word?

b each letter can be used any number of times?

8 The names of the fifteen finalists in a competition are each written on a piece of paper and the fifteen pieces are placed in a hat. The first name drawn out then wins the first prize, the second out wins second prize and the third out wins third prize. How many different ways can the three prizes be awarded if, after a name is drawn from the hat, it is:

a returned to the hat?

b not returned to the hat?

9 A science teacher develops a computer testmaker program for her class. The program contains a bank of science questions from which a number of different tests can be constructed. The questions are classified in terms of difficulty using a ten-point scale, I to X.

To run the program the user specifies x, the number of questions in the test ($x \leq 10$) and the program creates a test with question 1 of difficulty level I, question 2 of difficulty level II, question 3 of difficulty level III etc., up to question x of difficulty level x. The number of questions available at each level of difficulty is shown in this table:

Level	I	II	III	IV	V	VI	VII	VIII	IX	X
No. of questions	12	10	10	6	8	4	9	6	5	5

How many different tests can be created if the test is to contain:

a five questions?   **b** eight questions?   **c** ten questions?

10 A 'straight exacta' bet on a horse race requires the person making the bet to state, in order, the horse they think will come first and the horse they think will come second. How many different 'straight exactas' could be made on a race involving 12 horses?

11 How many different four-digit security numbers can be made using the digits 0, 1, 2, 3, 4, 5, 6, 7, 8, 9 if the numbers may start with 0 but must not contain any repeated digits. i.e. numbers like 4436 (two 4s), 1281 (two 1s), 3533 (three 3s) are not permitted.

> Security numbers are frequently referred to as PINs. Why?

12 Let us suppose that a 'first six' bet on a horse race requires the person making the bet to state, in the correct order, the horses that will finish in the first six places. How many different 'first six' bets could be made on a race involving 12 horses?

13 A test consists of ten multiple-choice questions with each question having four possible choices. Assuming all ten questions are attempted, in how many ways can the multiple-choice test be answered?

14 A ballot paper lists fifteen candidates and requires the voter to write 1st next to one of these fifteen, 2nd against another and 3rd against another. In how many ways can this be done?

15 A television soccer program invites viewers to enter the *Goal of the month* competition by listing a 1st, 2nd and 3rd place from a choice of eight goals. How many different entries are possible?

How many different entries are possible if instead all eight had to be listed in order of preference?

16 A test comprises ten questions each requiring a response of either yes or no. Assuming all questions are responded to, in how many ways can the test be answered?

Permutations of objects, not all different

Permutations with repetitions

Consider now arranging the following into a row:

$$a_1, \quad a_2, \quad a_3, \quad b.$$

This involves arranging 4 different objects so there will be 4!, or 24, arrangements:

$a_1 a_2 a_3 b$	$a_1 a_2 b a_3$	$a_1 b a_2 a_3$	$b a_1 a_2 a_3$
$a_1 a_3 a_2 b$	$a_1 a_3 b a_2$	$a_1 b a_3 a_2$	$b a_1 a_3 a_2$
$a_2 a_1 a_3 b$	$a_2 a_1 b a_3$	$a_2 b a_1 a_3$	$b a_2 a_1 a_3$
$a_2 a_3 a_1 b$	$a_2 a_3 b a_1$	$a_2 b a_3 a_1$	$b a_2 a_3 a_1$
$a_3 a_1 a_2 b$	$a_3 a_1 b a_2$	$a_3 b a_1 a_2$	$b a_3 a_1 a_2$
$a_3 a_2 a_1 b$	$a_3 a_2 b a_1$	$a_3 b a_2 a_1$	$b a_3 a_2 a_1$

However, if the three letter 'a's were indistinguishable the above list would shrink to show just 4 arrangements

a a a b	a a b a	a b a a	b a a a

Each of the 3! (= 6) arrangements of a_1, a_2 and a_3 forming a column in the first listing now giving rise to just one arrangement in the second list.

Hence there are $\dfrac{4!}{3!}$ arrangements of the letters a, a, a, b.

> If n objects contain p of one kind, q of another, r of another etc., then there are
>
> $$\frac{n!}{p!q!r!\ldots}$$ arrangements of the n objects.

EXAMPLE 5

How many ways can the nine letters of the word ISOSCELES be arranged in a row?
How many of these start with the L?

Solution

The word ISOSCELES involves 9 letters including 3 Ss and 2 Es.

Number of arrangements $= \dfrac{9!}{3!2!}$

$= 30\,240$

There are 30 240 arrangements of the nine letters of the word ISOSCELES.

If we start with the L then we are arranging the other 8 letters, which include 3 Ss and 2 Es, and then placing an L at the front of each arrangement.

Number of arrangements starting with the L $= \dfrac{8!}{3!2!}$

$= 3360$

There are 3360 arrangements that start with the L.

Addition principle

Consider the following question:

How many code numbers can be made from the digits 1, 2, 3 and 4 if the code numbers can either be two digit numbers or four digit numbers but no code may use the same digit more than once?

The 36 possible code numbers are shown in the two tree diagrams on the right.

However we know that to work out the number of possible 2 digit codes and the number of possible 4 digit codes we do not need to display them all.

Instead we can use *multiplicative reasoning*:

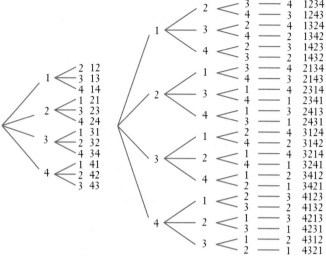

Number of 2 digit codes = 4×3

$\qquad\qquad\qquad = 12$

Number of 4 digit codes = $4 \times 3 \times 2 \times 1$

$\qquad\qquad\qquad = 24$

Hence the total number of possible 2 or 4 digit codes will be 12 + 24, i.e. 36.

Notice that in the last step we **added** the number of two digit codes to the number of four digit codes. We used *additive reasoning*.

This reasoning is formalised in the *addition principle*:

> **The addition principle**
>
> If there are *a* ways that event A can occur and *b* ways that event B can occur, then, provided A and B are mutually exclusive (i.e. A and B cannot occur together), *a* + *b* is the number of ways either A or B can occur.

EXAMPLE 6

A three-character code can either consist of three digits from 1, 2, 3, 4 and 5 or it can consist of three letters from A, B, C, D and E. How many codes are possible if digits can be used more than once in a code but letters cannot?

Solution

Number of possible three digit codes $= 5 \times 5 \times 5$

$\qquad\qquad\qquad\qquad\qquad = 125$

No. of ways for each digit		
5	5	5

Number of possible three letter codes $= 5 \times 4 \times 3$

$\qquad\qquad\qquad\qquad\qquad = 60$

No. of ways for each letter		
5	4	3

These events are mutually exclusive because a three character code cannot be both three digits and three letters. Thus 185 (= 125 + 60) codes can be formed.

ISBN 9780170390477

Care must be taken regarding the requirement that the two events cannot occur together, i.e. the *mutually exclusive* requirement. In the previous situation that prompted the tree diagrams, and in example 6, the two types of code were mutually exclusive. A code could not be both 2 digits and 3 digits, and in example 6 a 3 digit code could not be both 3 digits and 3 letters. The two events were mutually exclusive. If this requirement for *mutually exclusivity* is not met we will find we are counting some arrangements twice.

For example, suppose instead the two events involved making a two digit code number from the digits 1, 2, 3, 4 and making a two digit code number from the digits 3, 4, 5, 6, no repeat digits allowed. These events are not mutually exclusive because the code numbers 34 and 43 appear in both, as we can see from the tree diagrams below.

In this case   Number of codes ≠ 12 + 12

Instead: Number of codes = 22

          i.e. 12 + 12 − 2 (the 2 being subtracted to compensate for the two repeated codes)

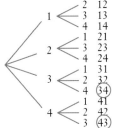
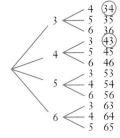

Inclusion–exclusion principle

The 2 digit code and 3 digit code situation shown as tree diagrams on the previous page involved mutual exclusivity thus allowing us to add together the numbers 12 and 24. The Venn diagram on the right shows this mutually exclusive situation.

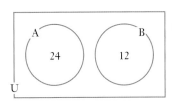

$$n(A \cup B) = n(A) + n(B)$$
$$= 24 + 12$$
$$= 36$$

The Venn diagram on the right shows the non-mutually exclusive situation shown in the tree diagrams considered at the top of this page. Now our addition uses the more general rule for the union of two sets:

$$n(A \cup B) = n(A) + n(B) - n(A \cap B)$$
$$\text{Number of codes} = 12 + 12 - 2$$
$$= 22$$

Extending this idea to three sets gives the following expression for $n(A \cup B \cup C)$:

$$n(A) + n(B) + n(C) - n(A \cap B) - n(A \cap C) - n(B \cap C) + n(A \cap B \cap C)$$

Notice that in this rule we first

          include the numbers in each set,

and then   take away (**exclude**) the numbers in the 'two set intersections',

and then   **include** the number in the 'three set intersection'.

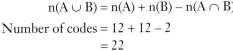
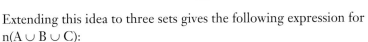

This alternating 'inclusion–exclusion' idea can be extended to 4 or more sets and is known as the **inclusion–exclusion principle**.

The inclusion-exclusion principle

EXAMPLE 7

a How many multiples of 2 are there between 1 and 89?

b How many multiples of 3 are there between 1 and 89?

c How many numbers between 1 and 89 are multiples of 2 or 3?

Note: In mathematics we take the word 'or' to mean 'one or the other or both'.
 In other words we take 'or' to mean 'at least one of'.
 Thus for two events A and B we take 'A or B' to mean $A \cup B$.

Solution

a The multiples of 2 between 1 and 89 are: 2, 4, 6, 8, ... 88.
 44 numbers in total.

b The multiples of 3 between 1 and 89 are: 3, 6, 9, 12, ... 87.
 29 numbers altogether.

c The numbers between 1 and 89 that are multiples of both 2 and 3 are:

$$6, 12, 18, 24, ... 84$$

14 numbers altogether.

Thus the required number $\begin{aligned}&= 44 + 29 - 14\\&= 59\end{aligned}$

There are 59 numbers between 1 and 89 that are multiples of 2 or 3.

Exercise 2B
The first six questions involve: Arrangements of objects, not all different.

1 How many ways can the 6 letters of the word REPEAT be arranged in a row?

2 How many ways can the 7 letters of the word CLASSES be arranged in a row?

3 How many ways can the 7 letters of the word TROTTER be arranged in a row?

4 How many 12 letter permutations are there of the word PERMUTATIONS?

5 How many ways can the 11 letters of the word MISSISSIPPI be arranged in a row?
 How many of these start with the M?

6 How many ways can the 10 letters of the word WOLLONGONG be arranged in a row?
 How many of these start with the W?
 How many do not start with the W?

The remaining questions involve an understanding of: n(A ∪ B).

7 How many two or three digit numbers can be formed using the digits 1, 2, 3, 4 and 5 if no digit may be used more than once in a number?

8 How many two or three letter codes can be formed using letters from the alphabet and allowing any letter to be used more than once?

9 How many two or three letter codes can be formed using letters from the alphabet if each code must not feature any letter more than once?

10 The digits 1, 2 and 3 are to be used to make two digit numbers and three digit numbers. How many such numbers are possible if
  a repeated use of digits in a number is permitted?
  b repeated use of digits in a number is not permitted?

11 The digits 1, 2, 3, 4 and 5 are to be used to make two digit numbers and three digit numbers. How many such numbers are possible if
  a repeated use of digits in a number is permitted?
  b repeated use of digits in a number is not permitted?

12 A four character code can either consist of four digits from 1, 2, 3, 4 and 5, or it can consist of four letters from A, B, C, D, E, F and G. How many codes are possible if digits can be used more than once in a code but letters cannot?

13 A three letter code comprising different letters is to be made either by arranging three letters chosen from the word MATHS or by obtaining the first letter in the arrangement from this source and then obtaining the next two letters from the word FUN. How many such codes are possible?

14 Sanshi is at the race track and wants to place one more bet, either on race 8 or race 9, but not on both. He is not sure whether to randomly predict 1st, 2nd and 3rd for race 8, in which there are 8 horses racing or randomly predict a 1st and 2nd for race 9 with 12 horses racing.

How many different bets does this choice involve altogether?

15 A lock-making company makes a number of different locks each with their own key design. The keys are made by cutting one of eight 'cut styles' in each of the five positions in the long key base model or in each of the three positions in the short key base model.

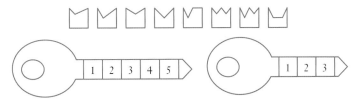

How many different keys are possible in each of the following situations?
  a Each key can feature the same 'cut style' in more than one of its positions.
  b Each key must not feature any one 'cut style' more than once.

16 Repeat question **15** but now include the possibility of each position in a key remaining uncut.

17 A two digit code number is to be made either by using two different digits from the digits 1, 2, 3, 4, 5 and 6 or by using two different digits from the digits 6, 7 and 8.

How many different two digit codes are possible?

18 A national credit company decides to give each of its employees a security card for computer access. When the card is placed into the reader the employee will be asked to type in their code 'word'. These code words will consist of four different letters either all taken from the word CREDIT or all taken from the word COMPANY. How many different code 'words' are possible?

19 When choosing her new car Shahani has narrowed her choice down to two possible models: *The Nifty Townabout* or *The Sedate Tourer*.

The *Nifty* is available in 4 colours, 2 engine sizes, with or without air conditioning, with or without automatic transmission and with or without power steering.

The *Sedate* is available in 5 colours, 3 engine sizes and all models feature air conditioning, automatic transmission and power steering.

How many 'different' cars is Shahani actually considering?

20 Code numbers consisting of two different digits are to be made.

The numbers will consist of two digits chosen from the digits 1, 2, 3, 4 and 5 or two digits chosen from the digits 4, 5, 6 and 7.

How many different two digit codes are possible?

21 Code numbers consisting of three different digits are to be made.

The numbers will consist of three digits chosen from the digits 1, 2, 3, 4 and 5 or three digits chosen from the digits 3, 4, 5 and 6.

How many different three digit codes are possible?

22 a How many multiples of five are there between 1 and 999?

   b How many multiples of seven are there between 1 and 999?

   c How many numbers between 1 and 999 are multiples of five or seven?

   (Remember that we take 'or' to mean 'at least one of'.)

23 Use the inclusion–exclusion principle for three sets, i.e.

$$|A \cup B \cup C| = |A| + |B| + |C| - |A \cap B| - |A \cap C| - |B \cap C| + |A \cap B \cap C|$$

to determine $|A \cup B \cup C|$ given the following information:

$$n(A) = 27 \qquad\qquad n(B) = 17 \qquad\qquad n(C) = 29$$
$$n(A \cap B) = 12 \qquad\qquad n(A \cap C) = 18 \qquad\qquad n(B \cap C) = 10$$
$$n(A \cap B \cap C) = 7$$

Check your answer by drawing a Venn diagram of the situation.

24 How many numbers between 1 and 101 are multiples of 2, 3 or 5?

25 How many numbers between 1 and 1001 are multiples of 3, 10 or 25?

26 Use the alternating inclusion, exclusion nature of the inclusion–exclusion principle to write the rule for $|A \cup B \cup C \cup D|$ for the four sets A, B, C and D.

Arrangements of objects with some restriction imposed

Some of the questions encountered so far have involved arrangements with some restriction imposed, for example, arranging the letters of the word ISOSCELES with the restriction that the first letter must be the L. This section considers such ideas further.

I. Multiplicative reasoning

If asked to determine the number of three letter codes that can be formed using the letters A, B, C, D and E, given that a code cannot use a letter more than once, we would proceed as follows:

There are five choices for the first letter, four for the second and three for the third.

No. of ways for each letter		
5	4	3

Number of three letter codes
$$= 5 \times 4 \times 3$$
$$= 60 \quad \text{i.e. } {}^5P_3$$

Suppose now that some restriction is involved, for example, that the middle letter must be a vowel. How many different three letter arrangements are there now?

In such cases we can still use multiplicative reasoning to determine the total number of arrangements but it is advisable to **consider the restriction first**, as follows.

There are two choices for the middle letter, A or E.

No. of ways for each letter		
?	2	?

There then remain four choices for the first, B, C and D and whichever of A and E was not used for the middle letter, and then three for the last.

Total number of arrangements
$$= 4 \times 2 \times 3$$
$$= 24$$

4	2	3

EXAMPLE 8

How many different *five* letter 'words' can be formed using the letters of the word FASHION if the middle letter in each arrangement must be the S and no word may feature a letter used more than once?

Solution

There is only one choice for the middle letter, an S.

No. of ways for each letter				
?	?	1	?	?

There then remain six choices for the first, F, A, H, I, O, N, five for the second, four for the fourth and three for the last.

Total number of arrangements
$$= 6 \times 5 \times 1 \times 4 \times 3$$
$$= 360$$

6	5	1	4	3

ISBN 9780170390477

EXAMPLE 9

a How many five digit *even* numbers can be made using the digits 3, 4, 5, 6 and 7 if no digit may feature more than once in a number?

b How many of the numbers from **a** are greater than 70 000?

Solution

a Start by choosing the last digit, which must be either the 4 or the 6, to ensure an even number:

No. of ways for each digit				
?	?	?	?	2

Left hand digit is then chosen from remaining 4, next digit from remaining 3 and so on:

4	3	2	1	2

The number of possible five digit even numbers
$$= 4 \times 3 \times 2 \times 1 \times 2$$
$$= 48$$

b The last digit must be either the 4 or the 6 and the first digit must be the 7.

No. of ways for each digit				
1	?	?	?	2

The next digit is then chosen from the remaining 3 and so on:

1	3	2	1	2

Number that are even and greater than 70 000
$$= 1 \times 3 \times 2 \times 1 \times 2$$
$$= 12.$$

EXAMPLE 10

A security code consists of four digits chosen from 0, 1, 2, …, 9 followed by a capital letter chosen from A, B, C, …, Z. For example

3	7	0	9	E

How many such codes are possible in each of the following cases?

a No digit is to feature more than once in a code.

b There is no restriction on the number of times a digit may feature in a code.

c The code must not start with zero and no digit must be used more than once in a code.

d The code must not start with zero, must end with a vowel, and no digit may feature more than once in a code.

Solution

a If we start by choosing the first digit and then work across, the number of ways each entry can be chosen will be:

10	9	8	7	26

The number of possible codes is $10 \times 9 \times 8 \times 7 \times 26 = 131\,040$

b Number of ways each entry can be chosen will be:

10	10	10	10	26

The number of possible codes is $10 \times 10 \times 10 \times 10 \times 26 = 260\,000$

ISBN 9780170390477

c The first digit can be chosen in 9 ways (zero is *not* allowed), the second can then be chosen in 9 ways (now zero *is* allowed), the third in 8 etc. The number of ways each entry can be chosen will be:

| 9 | 9 | 8 | 7 | 26 |

The number of possible codes is $9 \times 9 \times 8 \times 7 \times 26 = 117\,936$

d Number of ways each entry can be chosen will be:

| 9 | 9 | 8 | 7 | 5 |

The number of possible codes is $9 \times 9 \times 8 \times 7 \times 5 = 22\,680$

EXAMPLE 11

Seven files, A, B, C, D, E, F and G are to be arranged on a shelf.

a In how many ways can this be done?

b In how many of these arrangements is file A next to file B?

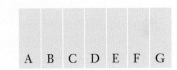
A B C D E F G

Solution

a The first file can be chosen in 7 ways, the next in 6, the next in 5 and so on.

Total number of arrangements = $7 \times 6 \times 5 \times 4 \times 3 \times 2 \times 1$ i.e. 7!, or 5040.

b If we imagine files A and B tied together we then have six things to arrange. However, A and B could be tied together in 2 ways (AB or BA).

Total number of arrangements = $(6 \times 5 \times 4 \times 3 \times 2 \times 1) \times 2$ i.e. $6! \times 2$, or 1440.

Exercise 2C

1 How many different *five* letter 'words' can be formed using the letters of the word GREAT if the second letter in each arrangement must be the G and

   a no word may feature a letter used more than once?

   b letters may be used more than once in a word?

2 a How many six digit *odd* numbers can be made using the digits 1, 2, 3, 4, 5 and 6 if no digit may feature more than once in a number?

   b How many of the numbers from **a** are greater than 600 000?

3 Six files, A, B, C, D, E and F are to be arranged on a shelf.

   a In how many ways can this be done?

   b In how many of the arrangements is file D next to file F?

   c In how many of the arrangements are files ABC together in that order?

   d In how many of the arrangements are files ABC together but in no particular order?

A B C D E F

4 How many ways can the five letters

$$X, A, E, Q, R$$

be written in a line if no letter can be used more than once?

How many of these arrangements of the five letters

a start with a consonant?

b start with a vowel?

5 In how many different orders can the five names Alex, Dennis, Jack, Jill and Kris be written? In how many of these is

a Jack listed first?

b Jill listed second?

c Jack listed first and Jill listed second?

6 a How many seven digit numbers can be made using the digits 1, 2, 3, 4, 5, 6 and 7 if no digit may be used more than once in a number?

b How many seven digit *even* numbers can be made using the digits 1, 2, 3, 4, 5, 6 and 7 if no digit may be used more than once in a number?

c How many of the even numbers from **b** are bigger than 7 000 000?

7 A restaurant offers a special 3-course lunch deal. Customers choose one starter, one main course and one dessert from:

Starter	Main course	Dessert
Prawn cocktail	Chicken Kiev	Apple pie
Paté	Steak	Ice cream
Soup	Lasagne	
	Cottage pie	

a How many different three course meals are possible?

b How many of these involve lasagne?

c How many involve lasagne and ice cream?

8 A security code consists of three digits chosen from 0, 1, 2, …, 9 followed by two letters chosen from A, B, C, …, Z. For example:

2	4	1	Z	E

How many such codes are possible in each of the following cases?

a No digit is to feature more than once in a code but the letters can be repeated.

b There is no restriction on the number of times each digit may feature in a code but no repeat letters are allowed.

c The code must not start with zero and no digit nor letter may feature in a code more than once.

d The code must not start with zero, letters and digits can be used more than once but each code must end with a vowel.

ISBN 9780170390477

9 A coin is tossed, then a normal die is rolled, then another coin is tossed and another normal die is rolled. One possible sequence of results is H, 5, T, 1.

a How many possible sequences are there?

In how many of the sequences do

b the two dice show the same number?

c the two coins show the same result?

10 A security code consists of three letters chosen from the 26 in the alphabet followed by two digits chosen from 0, 1, 2, ..., 9.

For example:

| Z | E | K | 4 | 3 |

How many such codes are possible in each of the following cases?

a There are no restrictions on the choice of letters and digits.

b No letter and no digit may feature more than once in the code.

c The initial letter must not be a vowel and no letters or digits are to be used more than once.

d The first and third letters must be the same as each other and different to the second, and the two digits must be the same as each other.

e There is no restriction on the two digits but the letters must be consecutive letters of the alphabet and must feature in alphabetical order ('reverse alphabetical' not permitted).

The following example is acceptable.

| P | Q | R | 5 | 2 |

f The final digit must be one more than the digit before it and the letters must be consecutive letters of the alphabet featuring in alphabetical order ('reverse alphabetical' not permitted).

The following example is acceptable.

| L | M | N | 5 | 6 |

11 Ten books are to be arranged on a shelf. Three of the ten books are by one author and the other books are all by different authors.

How many arrangements are possible if the three by the same author:

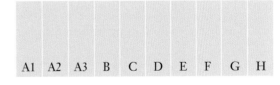

| A1 | A2 | A3 | B | C | D | E | F | G | H |

a need not be kept together?

b must be kept together and in a particular order?

c must be kept together but in no particular order?

d must be kept together at the left end of the shelf and in a particular order?

II. Additive and multiplicative reasoning

As has been mentioned before, the use of *additive reasoning* needs care. If we are adding the number of ways an event A can occur to the number of ways an event B can occur this will only give the number of ways A or B can occur if A and B cannot occur together, i.e. provided A and B are **mutually exclusive**. If this is not the case the addition will mean that some events will be counted more than once. In such cases we can either break the task into separate mutually exclusive situations, and then it will be safe to add, or we can adjust for the 'double counting' by using $n(A \cup B) = n(A) + n(B) - n(A \cap B)$.

(Remember in mathematics we interpret 'A or B' to mean 'at least one of'.)

For example, suppose we are forming four letter codes from the letters

$$A, B, C, D, E, F,$$

with no code using a letter more than once.

	No. of ways for each letter			

Number of codes starting with an A $\quad = 1 \times 5 \times 4 \times 3$
$\qquad\qquad\qquad\qquad\qquad\qquad\qquad = 60$

1	5	4	3

Number of codes starting with an F $\quad = 1 \times 5 \times 4 \times 3$
$\qquad\qquad\qquad\qquad\qquad\qquad\qquad = 60$

1	5	4	3

Number starting with an A or an F $\qquad = 60 + 60$
$\qquad\qquad\qquad\qquad\qquad\qquad\qquad = 120$

The use of additive reasoning is valid because a code cannot start with both an A and with an F. The two events *are* mutually exclusive.

Contrast this with the 'start with an A or end with an F' situation.

	No. of ways for each letter			

Number of codes starting with an A $\quad = 1 \times 5 \times 4 \times 3$
$\qquad\qquad\qquad\qquad\qquad\qquad\qquad = 60$

1	5	4	3

Number of codes ending with an F $\quad = 5 \times 4 \times 3 \times 1$
$\qquad\qquad\qquad\qquad\qquad\qquad\qquad = 60$

5	4	3	1

But the number of codes starting with A or ending with F $\neq 60 + 60$ because this would count codes like ABCF, ACDF, ADCF, etc., twice. The events are *not* mutually exclusive as a code *can* start with an A and at the same time end with an F.

Instead we either use $n(A \cup B) = n(A) + n(B) - n(A \cap B)$:

Number of codes starting with an A and end with an F $\qquad = 1 \times 4 \times 3 \times 1$
$\qquad\qquad\qquad\qquad\qquad\qquad\qquad\qquad\qquad\qquad\qquad\qquad = 12$

Thus number that start with an A or end with an F $\qquad = 60 + 60 - 12$
$\qquad\qquad\qquad\qquad\qquad\qquad\qquad\qquad\qquad\qquad\qquad\qquad = 108$

Or we consider the mutually exclusive situations:

Start with an A and end with an F: $\qquad\qquad\quad 1 \times 4 \times 3 \times 1 = 12$
Start with an A and not end with an F: $\qquad\quad 1 \times 4 \times 3 \times 4 = 48$
Do not start with an A but end with an F: $\qquad 4 \times 4 \times 3 \times 1 = 48$

Thus number that start with an A or end with an F: $\qquad = 12 + 48 + 48$
$\qquad\qquad\qquad\qquad\qquad\qquad\qquad\qquad\qquad\qquad\qquad\qquad\qquad = 108$, as before.

EXAMPLE 12

How many three digit codes can be made using the digits 0, 1, 2, 3, 4 and 5 if there are no restrictions on using a digit more than once?

How many of these codes **a** start with a 3? **b** start with a 4?

c do not start with a 4? **d** start with a 3 or a 4?

e start with a 3 and end with a 4? **f** start with a 3 or end with a 4?

Solution

Number of ways each digit of code can be chosen:

6	6	6

The number of possible codes is $6 \times 6 \times 6 = 216$

a If the code must start with a 3:

1	6	6

The number of possible codes is $1 \times 6 \times 6 = 36$

b If the code must start with a 4:

1	6	6

The number of possible codes is $1 \times 6 \times 6 = 36$

c If the code must not start with a 4:

5	6	6

The number of possible codes is $5 \times 6 \times 6 = 180$

Alternatively: 36 of the 216 codes start with a 4 therefore 180 (= 216 – 36) do not start with a 4. (This alternative method uses the idea of the *complementary* event.)

d If the code must start with a 3 or a 4:

2	6	6

The number of possible codes is $2 \times 6 \times 6 = 72$

Alternatively, using additive reasoning because start with 3 and start with a 4 are mutually exclusive:

Number of codes that start with a 3 or a 4 is $36 + 36 = 72$

e If the code must start with a 3 and end with a 4:

1	6	1

The number of possible codes is $1 \times 6 \times 1 = 6$

f If the code must start with a 3:

1	6	6

The number of possible codes is $1 \times 6 \times 6 = 36$

If the code must end with a 4:

6	6	1

The number of possible codes is $6 \times 6 \times 1 = 36$

'Start with 3' and 'end with a 4' are not mutually exclusive. They can occur together, as in the code 314. The answer will *not* be 36 + 36.

Using n(A ∪ B) = n(A) + n(B) – n(A ∩ B):

Number of codes starting with 3 or end with 4 is 36 + 36 – 6 = 66

Or, considering 'start with a 3 or end with a 4' as three mutually exclusive events:

- Start with a 3 and end with a 4:

1	6	1

- Start with a 3 and not end with a 4:

1	6	5

- End with a 4 but not start with a 3:

5	6	1

Number of codes starting with 3 or end with 4 is $6 + 30 + 30 = 66$, as before.

EXAMPLE 13

Five members of a basketball team all have to stand in a line for a photograph.

The players are: Alex, Keith, Mark, Rani and Steve.

How many arrangements are there in which

a   Rani is in the middle?

b   Alex is at the left end?

c   Mark is at the right end?

d   At least one of **b** and **c** occur?

e   Keith and Steve are next to each other?

f   Keith and Steve are not next to each other?

Solution

a   Middle place filled in one way:

		1		

Then fill left most space from remaining 4 players, next from remaining 3 etc.
Number of possible arrangements is $4 \times 3 \times 1 \times 2 \times 1 = 24$.

4	3	1	2	1

b   Alex at left end then fill spaces from left:
Number of possible arrangements is $1 \times 4 \times 3 \times 2 \times 1 = 24$.

1	4	3	2	1

c   Mark at right end then fill spaces from left:
Number of possible arrangements is $4 \times 3 \times 2 \times 1 \times 1 = 24$.

4	3	2	1	1

d   Alex at left end and Mark not at right end:

1	3	2	1	3

Mark at right end and Alex not at left end:

3	3	2	1	1

Alex at left end and Mark at right end:
Number of possible arrangements is $18 + 18 + 6 = 42$.

1	3	2	1	1

Or: Add answers from **b** and **c** then subtract the 'and situation' that would otherwise be counted twice: $24 + 24 - 6 = 42$, as before.

e   'Tie' Keith and Steve together (2 ways). We now have 4 'things' to arrange.
Number of possible arrangements is $2 \times 4 \times 3 \times 2 \times 1 = 48$.

f   There are $5 \times 4 \times 3 \times 2 \times 1$ (= 120) arrangements altogether and 48 have Keith and Steve together. Thus 72 (= 120 – 48) must have Keith and Steve not together.
There are 72 arrangements in which Keith and Steve are not together.

Note the use of the complementary event in part **f**.

EXAMPLE 14

a How many six digit even numbers can be made using the digits 1, 2, 3, 4, 5 and 6 if no digit may be used more than once in a number?

b How many of the numbers from **a** are bigger than 400 000?

Solution

a Start by choosing the last digit, which must be
2, 4 or 6, to ensure the number is even:

| ? | ? | ? | ? | ? | 3 |

Left hand digit is then chosen from remaining 5,
next digit from remaining 4 and so on:

| 5 | 4 | 3 | 2 | 1 | 3 |

The number of possible 6 digit even numbers $= 5 \times 4 \times 3 \times 2 \times 1 \times 3$
$$= 360$$

There are 360 six digit even numbers that can be made using the digits 1, 2, 3, 4, 5 and 6.

b The problem now is that if we choose the last digit from the 2, 4 or 6 the number of choices for the first digit is 3, if the last digit was the 2, and 2 if the last digit was the 4 or the 6.

To cope with this we consider mutually exclusive events.

Ending with 2 and bigger than 400 000:

| 3 | 4 | 3 | 2 | 1 | 1 |

Ending with 4 and bigger than 400 000:

| 2 | 4 | 3 | 2 | 1 | 1 |

Ending with 6 and bigger than 400 000:

| 2 | 4 | 3 | 2 | 1 | 1 |

The number of possible 6 digit even numbers $> 400\,000 = 72 + 48 + 48$
$$= 168$$

There are 168 six digit even numbers bigger than 400 000 that can be made using the digits 1, 2, 3, 4, 5 and 6.

Exercise 2D

1 The digits 1, 2, 3, 4 and 5 are to be used to make three digit numbers and four digit numbers. How many such numbers are possible if

  a repeated use of digits in a number is permitted?

  b repeated use of digits in a number is not permitted?

  c repeated use of digits in a number is not permitted and only odd numbers are to be formed?

2 The letters U, V, W, X, Y and Z are to be used to make three letter and five letter code words. Determine how many such code words are possible in each of the following situations.

  a Repeated use of letters in a code is permitted.

  b Repeated use of letters in a code is not permitted.

  c Repeated use of letters is not permitted and no code word is allowed to start with the Z.

3 a How many seven digit even numbers can be made using the digits 1, 2, 3, 4, 5, 6 and 7 if no digit may be used more than once in a number?

b How many of the numbers from **a** are greater that 6 000 000?

4 In how many ways can the five digits 1, 2, 3, 4 and 5 be arranged to form a five digit code number, with each code using all five digits once each?

How many of these arrangements

a start with 3?

b end with 5?

c start with 3 and end with 5?

d start with 3 or end with 5?

5 a How many eight letter arrangements are there of the letters of the word FORECAST if in each arrangement each letter must be used just once?

How many of these arrangements

b have the E and the O next to each other?

c have the E and the O separated?

d start with A and have E and O next to each other?

e start with A or end with S?

6 Three digit numbers are to be made using the digits 1, 2, 3, 4, 5, 6 and 7 with no digit used more than once in any particular three digit number.

a How many three digit numbers are possible?

How many of these

b start with 4?

c end with 5?

d start with 4 and end with 5?

e start with 4 or end with 5?

f are odd numbers?

g are greater than 700?

h are greater than 500?

i are even and greater than 500?

7 Four members of a sports team have to arrange themselves in a line for a photograph.

The members are: Terri, Jen, Diane, May.

How many arrangements are there in which

a Terri is at the left end?

b Diane is at the right end?

c Terri is at the left end and Diane is at the right end?

d Terri is at the left end or Diane is at the right end?

e Jen and Diane occupy the middle two positions?

f Jen and Diane are not next to each other?

8 a How many seven digit odd numbers can be made using the digits 0, 1, 2, 3, 4, 5, 6, 7 and 8 if no digit may be used more than once in a number and zero cannot be the first digit?

b How many of these numbers from **a** are less than 4 000 000?

9 The staff at a particular college consist of 1 principal (P), 2 deputies (D), 10 teachers (T) and 2 support staff (S). The two styles of arranging the staff that are being considered for the staff photograph are shown below.

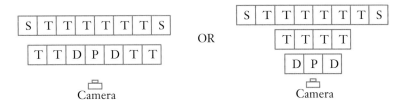

How many different arrangements of the 15 staff are there?

10 To make a two letter code the first letter is chosen from the set of vowels and the second letter is chosen from the 26 letters of the alphabet.

a How many different codes are possible?

How many of the codes

b start with an E?

c end with a D?

d start with an E and end with a D?

e start with an E or end with a D?

f start and end with the same letter?

g consist of two different letters?

Combinations

There are $4 \times 3 \times 2$, i.e. $^4P_3 = 24$, arrangements of three letters taken from the letters A, B, C and D:

ABC	ABD	ACD	BCD
ACB	ADB	ADC	BDC
BAC	BAD	CAD	CBD
BCA	BDA	CDA	CDB
CAB	DAB	DAC	DBC
CBA	DBA	DCA	DCB

Combination calculations

Combinations

Suppose now that A, B, C and D represented four people, Alex, Bonnie, Chris and Dan and that we are wishing to select three of these four to attend a meeting for which the order of the three is unimportant. There are now just four **selections**, or **combinations** of three items chosen from four different items:

ABC	ABD	ACD	BCD

> A **combination** is a **selection** – the order does not matter.
>
> A **permutation** is an **arrangement** – the order does matter.

We use the notation ^nC_r for the number of **combinations** of r different objects taken from a set of n different objects.

In the example of choosing a group of 3 letters from 4 letters our initial listing of 24 arrangements reduced to just 4 selections because, when order became unimportant, each column of 6 (= 3!) arrangements reduced to just one selection.

$$^4C_3 = \frac{^4P_3}{3!}$$

$$= \frac{4!}{(4-3)!\,3!}$$

To apply this thinking to combinations of r objects chosen from n different objects

$$^nC_r = \frac{^nP_r}{r!}$$

$$= \frac{n!}{(n-r)!\,r!}$$

There are ^nC_r combinations of r objects chosen from n different objects

where $^nC_r = \dfrac{n!}{(n-r)!\,r!}$.

EXAMPLE 15

How many combinations are there of two objects chosen from five different objects that we will call A, E, I, O and U?

Solution

Number of combinations $= {}^5C_2$

$$= \frac{5!}{(5-2)!\,2!}$$

$$= \frac{5!}{3!\,2!}$$

$$= 10$$

```
nCr(5,2)
                    10
```

The ten combinations are: AE,   AI,   AO,   AU,

EI,   EO,   EU,

IO,   IU,

OU.

Note
- $^{n}C_{r}$ is also written $\begin{pmatrix} n \\ r \end{pmatrix}$. For example $\begin{pmatrix} 7 \\ 2 \end{pmatrix} = {}^{7}C_{2}$.

- $^{n}C_{r}$ can be thought of as 'from *n choose r*'.

- If asked the question 'How many ways can five people be arranged in a line for a photograph given that the five can themselves be selected from a larger group of eight' we can solve this as before:

$$\text{Number of arrangements} = 8 \times 7 \times 6 \times 5 \times 4$$
$$= 6720$$

or we could consider this as 'from 8 people choose 5 and then arrange them'.

$$\text{Number of arrangements} = {}^{8}C_{5} \times 5!$$
$$= 6720$$

EXAMPLE 16

How many combinations are there of five people to attend a particular conference if the five are to be selected from twelve people?

Solution

From   12
choose   5

$$\text{Number of combinations} = {}^{12}C_{5}$$

$$= \frac{12!}{(12-5)!\,5!}$$

$$= \frac{12!}{7!\,5!}$$

$$= 792$$

EXAMPLE 17

How many different ways can a group of 6 people, 3 male and 3 female, be selected from 8 males and 9 females?

Solution

	Male			Female	

From   8   and   from   9 $\text{Number of ways} = {}^{8}C_{3} \times {}^{9}C_{3}$
choose 3   choose 3 $= 4704$

A group of three males and three females can be selected from eight males and nine females in 4704 ways.

Note that in $^{8}C_{3} \times {}^{9}C_{3}$ the upper numbers add to 17 and the lower to 6. This makes sense because we are choosing 6 people from 17 but we are being careful to select the correct number of people from each subgroup.

EXAMPLE 18

A normal pack of playing cards consists of 52 cards arranged in four suits.

Hearts	A♥	2♥	3♥	4♥	5♥	6♥	7♥	8♥	9♥	10♥	J♥	Q♥	K♥
Diamonds	A♦	2♦	3♦	4♦	5♦	6♦	7♦	8♦	9♦	10♦	J♦	Q♦	K♦
Spades	A♠	2♠	3♠	4♠	5♠	6♠	7♠	8♠	9♠	10♠	J♠	Q♠	K♠
Clubs	A♣	2♣	3♣	4♣	5♣	6♣	7♣	8♣	9♣	10♣	J♣	Q♣	K♣

These cards are well shuffled and 7 cards are randomly dealt to form a 'hand'.

a How many different 7 card hands are there?

How many of these hands contain:

b the ace of hearts (A♥)?

c the ace and two of hearts (A♥ and 2♥)?

d exactly 3 of the 4 kings?

e at least one ace?

Solution

It is the cards that make up the hand that is important, not the order in which they are dealt. Thus this situation involves combinations.

a

From	52
Choose	7

Number of hands = $^{52}C_7$
= 133 784 560

b

	A♥	Others
From	1	51
Choose	1	6

Number of hands = $^1C_1 \times {}^{51}C_6$
= 18 009 460

c

	A♥	2♥	Others
From	1	1	50
Choose	1	1	5

Number of hands = $^1C_1 \times {}^1C_1 \times {}^{50}C_5$
= 2 118 760

d

	K	Others
From	4	48
Choose	3	4

Number of hands = $^4C_3 \times {}^{48}C_4$
= 778 320

e Number of hands with *no aces*:

	A	Others
From	4	48
Choose	0	7

Number of hands = $^4C_0 \times {}^{48}C_7$
= 73 629 072

Number with at least one ace = 133 784 560 − 73 629 072
= 60 155 488

Note carefully the use of the complementary event in part **e**.

Alternatively we could consider hands with 1 ace + hands with 2 aces + etc. i.e.
$^4C_1 \times {}^{48}C_6 + {}^4C_2 \times {}^{48}C_5 + {}^4C_3 \times {}^{48}C_4 + {}^4C_4 \times {}^{48}C_3 = 60\,155\,488$, as before.

EXAMPLE 19

A subcommittee of five people is to be chosen from the following twelve people:

Alex	Ben	Chris	Dave	Eric	Frank
Gemma	Hetti	Icolyn	Jenny	Kym	Louise

How many different subcommittees are possible in each of the following cases?

a There are no restrictions as to the make up of the subcommittee.

b Jenny and Eric must both be on the subcommittee.

c Ben and Gemma must either both be on the subcommittee or neither be on the subcommittee.

d Dave and Icolyn must not both be on the subcommittee. Dave can be, or Icolyn can be, but not both.

Solution

a From   12   No. of subcommittees $= {}^{12}C_5$
Choose   5 $= 792$

b

	J & E	Others
From	2	10
Choose	2	3

No. of subcommittees $= {}^{2}C_2 \times {}^{10}C_3$
$= 120$

c <u>B and G both on</u>   OR   <u>B and G both not on</u>

	B & G	Others			B & G	Others
From	2	10		From	2	10
Choose	2	3		Choose	0	5

No. of subcommittees $= {}^{2}C_2 \times {}^{10}C_3 + {}^{2}C_0 \times {}^{10}C_5$
$= 372$

d

	<u>D not I</u>			<u>I not D</u>			<u>Neither</u>		
	D	I	Others	D	I	Others	D	I	Others
From	1	1	10	1	1	10	1	1	10
Choose	1	0	4	0	1	4	0	0	5

No. of subcommittees $= {}^{1}C_1 \times {}^{1}C_0 \times {}^{10}C_4 + {}^{1}C_0 \times {}^{1}C_1 \times {}^{10}C_4 + {}^{1}C_0 \times {}^{1}C_0 \times {}^{10}C_5$
$= 672$

Alternatively, for part **d**, we could find the number of committees containing both Dave and Icolyn and take this from the total number of subcommittees:

No. of subcommittees $= {}^{12}C_5 - {}^{2}C_2 \times {}^{10}C_3$
$= 792 - 120$
$= 672$ as before.

EXAMPLE 20

How many different 5 letter arrangements can be made each consisting of 5 different letters of the alphabet, with exactly one of the 5 being a vowel?

Solution

Method 1: Suppose the vowel is first.

Number of arrangements with vowel first $= 5 \times 21 \times 20 \times 19 \times 18$
$= 718\,200$

But the vowel could be in any of the five positions
Number of arrangements with one vowel $= 5 \times 718\,200$
$= 3\,591\,000$

Method 2: Choose one vowel. Number of ways $= {}^5C_1$
Choose four consonants. Number of ways $= {}^{21}C_4$
Arrange the five items. Number of ways $= 5!$
Number of arrangements with one vowel $= {}^5C_1 \times {}^{21}C_4 \times 5!$
$= 3\,591\,000$ as before.

• Note especially method 2 above in which the number of arrangements have been determined using a 'choose and then arrange' approach.

EXAMPLE 21

Including A itself and the empty set how many subsets can be made using the elements of set A where A = {a, e, i, o, u}?

Solution

There exists
1 subset of A with no elements. ${}^5C_1$ subsets with 1 element.
${}^5C_2$ subsets with 2 elements. ${}^5C_3$ subsets with 3 elements.
${}^5C_4$ subsets with 4 elements. ${}^5C_5$ subsets with 5 elements.

Total number of subsets $= 1 + {}^5C_1 + {}^5C_2 + {}^5C_3 + {}^5C_4 + {}^5C_5$
$= 1 + 5 + 10 + 10 + 5 + 1$
$= 32$

Alternatively, when forming a subset of {a, e, i, o, u} we can consider each element on an 'include it or not include it' basis. We can include the letter a or not, we can then include the letter e or not, we can then include the letter i or not, etc. Thus there are 2 ways of dealing with each letter.

Thus number of subsets $= 2 \times 2 \times 2 \times 2 \times 2$
$= 2^5$
$= 32$

If we include the empty set and the set itself then a set with n different elements has 2^n possible subsets.

EXAMPLE 22

How many three letter permutations are there of the letters of the word PARALLEL?

Solution

Parallel involves 8 letters including 2 As and 3 Ls.

P	A	R	L	E
	A		L	
			L	

Consider mutually exclusive situations:

Arrangements with

3 letters, all different: $\qquad\qquad\qquad$ $5 \times 4 \times 3 = 60$

2 As and one other, AA?, A?A, ?AA $\qquad$ $3 \times 4 = 12$

2 Ls and one other, LL?, L?L, ?LL $\qquad$ $3 \times 4 = 12$

3 Ls $\qquad\qquad\qquad\qquad\qquad\qquad$ $1 = 1$

$\qquad\qquad\qquad$ $60 + 12 + 12 + 1 = 85$

There are 85 permutations altogether.

Or, using a 'choose and then arrange' approach:

Arrangements with

3 letters, all different: $\qquad\qquad\qquad$ $^5C_3 \times 3! = 60$

2 As and one other, AA?, A?A, ?AA $\qquad$ $^4C_1 \times \dfrac{3!}{2!} = 12$

2 Ls and one other, LL?, L?L, ?LL $\qquad$ $^4C_1 \times \dfrac{3!}{2!} = 12$

3 Ls $\qquad\qquad\qquad\qquad\qquad\qquad$ $= 1$

$\qquad\qquad\qquad$ $60 + 12 + 12 + 1 = 85$

There are 85 permutations altogether, as before.

Exercise 2E

1 Is a combination lock correctly named?

2 How many combinations are there of 3 different letters chosen from the set {a, b, c, d, e}?

3 How many combinations are there of four people to attend a particular conference if the four are to be selected from twenty people?

4 How many different ways can a group of six people, 3 male and 3 female, be selected from 7 males and 10 females?

5 How many different soccer teams each consisting of 1 goalkeeper, 3 defenders, 4 midfielders and 3 strikers are possible if selection is made from 2 goalkeepers, 5 defenders, 7 midfielders and 3 strikers?

6 A college offers its year 11 mathematics students a choice of twelve units:

List I	List II	List III	List IV
Calculus I	Vectors	Equations	Correlation
Statistics I	Trigonometry	Networks	Time series
	Matrices	Optimisation	Counting
		Sets	

Students must choose six units in all, the two from list I, one from list II, one from list III and two from list IV. How many different allowable six unit combinations are there?

7 Four people are selected from a committee of 12 to represent the committee at a particular function. How many different groups of four are possible?

How many groups of four are possible if either the chairperson or the vicechairperson, but not both, must be in the group?

8 Including A itself and the empty set how many subsets can be made using the elements of set A where A = {a, b, c, d, e, f, g}.

9 If we define the *proper* subsets of a set as those subsets other than the empty set and the set itself, how many proper subsets are there for the set {1, 2, 3, 4, 5, 6, 7, 8, 9}?

10 A company wishes to send a team of seven people overseas to investigate possible markets for its products. The team is to comprise 1 manager, 1 engineer, 3 marketing people and 2 legal experts. These are chosen from 3 managers, 12 engineers, a marketing team of 15 and a legal team of 5.

How many different such teams of seven are possible?

Joe is one of the 12 engineers and Sue is one of the 5 in the legal team.

How many of the possible teams include

a Joe? **b** Sue? **c** at least one of these 2 people?

11 A subcommittee of four people is to be chosen from the following ten people:

Alice Betty Connie Daisy Ennis Fred Gerry Henry Indi Javed

How many different subcommittees are there in *each* of the following cases?

a There are no restrictions as to the make up of the subcommittee.

b Javed and Ennis must both be on the subcommittee.

c Connie and Fred must either both be on the subcommittee or neither of them be on the subcommittee.

d Betty and Henry must not both be on the subcommittee. Betty can be, or Henry can be, but not both.

12 A group of seven people is to be formed from a larger group comprising 8 women and 6 men. How many different groups of seven people are there if the group must consist of

a 4 women and 3 men? **b** all women?

c all men? **d** more than five women?

e more men than women?

13 A normal pack of playing cards consist of 52 cards arranged in four suits.

Hearts	A♥	2♥	3♥	4♥	5♥	6♥	7♥	8♥	9♥	10♥	J♥	Q♥	K♥
Diamonds	A♦	2♦	3♦	4♦	5♦	6♦	7♦	8♦	9♦	10♦	J♦	Q♦	K♦
Spades	A♠	2♠	3♠	4♠	5♠	6♠	7♠	8♠	9♠	10♠	J♠	Q♠	K♠
Clubs	A♣	2♣	3♣	4♣	5♣	6♣	7♣	8♣	9♣	10♣	J♣	Q♣	K♣

Suppose these cards are well shuffled and eight cards are randomly dealt to form a 'hand', the order the cards are dealt being irrelevant.

a How many different 8 card hands are there?

How many of these hands contain:

b the jack of hearts (J♥)?

c 5 red cards and the rest black?

d exactly two queens?

e at least two queens?

Shutterstock.com/Michael D Brown

14 Four places exist on a course for fire officers. The four are to be chosen from 13 officers: 6 from A division, 4 from B division and 3 from C division.

How many different groups of four are there?

How many different groups of four are there if each group must contain at least one officer from each division?

15 Points A, B, C, D, E, F, G, H and I are arranged in two rows as shown on the right. How many triangles can be formed having vertices chosen from these points?

A   B   C   D   E
•   •   •   •   •

•   •   •   •
F   G   H   I

How many of these triangles have point A as one of their vertices?

16 How many three letter permutations are there of the letters of the word EQUILATERAL?

17 How many four letter permutations are there of the letters of the word CANDLEPOWER?

iStock.com/Samarskaya

ISBN 9780170390477

$^{n}C_{r}$ and Pascal's triangle

The arrangement of numbers shown below is known as **Pascal's triangle**:

Notice that each line starts and finishes with a 1 and other entries are obtained by adding the two numbers 'above left' and 'above right'.

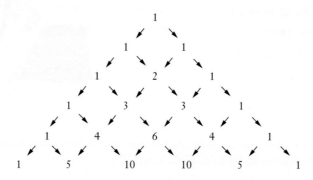

Not immediately obvious is the fact that the entries in Pascal's triangle can be expressed in $^{n}C_{r}$ notation:

$$1$$
$$^{1}C_{0} \qquad ^{1}C_{1}$$
$$^{2}C_{0} \qquad ^{2}C_{1} \qquad ^{2}C_{2}$$
$$^{3}C_{0} \qquad ^{3}C_{1} \qquad ^{3}C_{2} \qquad ^{3}C_{3}$$
$$^{4}C_{0} \qquad ^{4}C_{1} \qquad ^{4}C_{2} \qquad ^{4}C_{3} \qquad ^{4}C_{4}$$
$$^{5}C_{0} \qquad ^{5}C_{1} \qquad ^{5}C_{2} \qquad ^{5}C_{3} \qquad ^{5}C_{4} \qquad ^{5}C_{5}$$

Exercise 2F

1 Use the fact that $^{n}C_{r} = \dfrac{n!}{(n-r)!\,r!}$ to show that:

a $\quad ^{n}C_{r} = {}^{n}C_{n-r}$

b $\quad \dfrac{n}{r} \times {}^{n-1}C_{r-1} = {}^{n}C_{r}$

2 Show that the way in which the entries in Pascal's triangle are formed by adding 'above left' and 'above right' is consistent with the following statement:

$$^{n}C_{r-1} + {}^{n}C_{r} = {}^{n+1}C_{r}$$

Use the fact that $^{n}C_{r} = \dfrac{n!}{(n-r)!\,r!}$ to show that the above statement is true.

ISBN 9780170390477

Miscellaneous exercise two

This miscellaneous exercise may include questions involving the work of this chapter, the work of any previous chapters, and the ideas mentioned in the Preliminary work section at the beginning of the book.

For each of questions **1** and **2** use trigonometry to determine x, correct to one decimal place, clearly showing the use of trigonometry in your working. (Diagrams not to scale.)

1

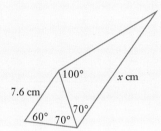

2

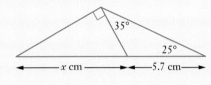

3 Given the true statement   *If $x = 8$ then $x^2 = 64$*

write both the converse statement and the contrapositive statement and for each one state whether it is true or false.

4 As an introduction to counting techniques, Ms Jackson, a Mathematics teacher, gives each of the 25 students in her class a piece of paper and asks each to write down a permutation of the letters of the word FISH that uses all four letters once and once only.

Every one of the students performs this class correctly and Ms Jackson then collects in the 25 pieces of paper.

What can we conclude about the responses, and why?

5 If you did **question 14** in **exercise 2B** you encountered Sanshi, at the race track, wanting to place just one more bet, either on race 8 or race 9. In that question he was not sure whether to randomly predict 1st, 2nd and 3rd for race 8, in which there were 8 horses racing or randomly predict a 1st and 2nd for race 9 with 12 horses racing.

Suppose instead that he makes his one bet that of attempting to randomly select the correct order for all five horses, i.e. 1st in race 8, 2nd in race 8, 3rd in race 8, 1st in race 9, 2nd in race 9. How many different bets of this type are there?

6 How many arrangements of five different letters can be made from the letters a, b, c, d, e, f, g, h and i?

How many of these arrangements contain 2 vowels and 3 consonants?

(Hint: *Choose* 2 vowels, then *choose* 3 consonants and then *arrange* the 5 letters.)

7 A child is told she can bring five toys with her on holiday. The child decides to choose the five from

                6 jigsaws,   8 dolls,   4 balls,   2 trucks.

How many different sets of five toys are there?

How many of these sets have at least one from each of the four categories of toy listed above?

ISBN 9780170390477

Questions **8** to **11** each ask you to prove some geometrical fact. You should set out your proofs as in the examples given in the *Preliminary work* section involving similar triangles and congruent triangles. I.e. draw a clear diagram, state what you are given, what you have to prove and any conclusions made. Then set out your proof clearly and with statements justified.

In your proofs the following may be stated as fact, without proof:

- Angles that together form a straight line have a sum of 180°.
  I.e., in the diagram on the right, A + B = 180°. (And conversely, if the angles have a sum of 180° then they together form a straight line.)

  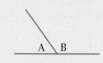

- When a transversal cuts parallel lines, corresponding angles are equal.
  I.e., in the diagram on the right, A = E, B = F, C = G, D = H.

  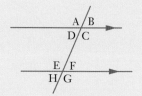

- When two straight lines intersect the vertically opposite angles are equal. I.e., in the diagram on the right, A = B.

- When a transversal cuts parallel lines, alternate angles are equal.
  I.e., in the diagram on the right, Q = R, P = S.
  (And, conversely, if alternate angles are equal we have parallel lines.)

- When a transversal cuts parallel lines, co-interior angles are supplementary.
  I.e., in the diagram on the right,   P + R = 180°,
                                            Q + S = 180°.

- The angles of a triangle add up to 180°.
  I.e., in the diagram on the right,   A + B + C = 180°.

  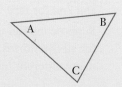

8 Prove that if a quadrilateral has its opposite sides parallel, then its opposite sides must be of equal length.

9 Given that a parallelogram has opposite sides parallel and equal, prove that the diagonals of the parallelogram bisect each other.

10 If a quadrilateral has all four sides of equal length we call it a rhombus. Prove that a rhombus must also have its opposite sides parallel (i.e. that a rhombus must also be a parallelogram).

11 Prove that if two lines, AB and CD, bisect each other at right angles then AD = DB = BC = CA.

3.

Vectors – basic ideas

- Vector quantities
- Adding vectors
- Mathematical representation of a vector quantity
- Equal vectors
- The negative of a vector
- Multiplication of a vector by a scalar
- Parallel vectors
- Addition of vectors
- Subtraction of one vector from another
- The zero vector
- $h\mathbf{a} = k\mathbf{b}$
- Miscellaneous exercise three

Situation One

A mother duck shelters by the bank of a pond with her young family of ducklings. One brave duckling ventures out for a swim. The duckling swims 5 metres on a bearing 050° followed by 7 metres on a bearing 170°.

- How far is the duckling from its mother at the end of this two stage excursion?

- What is the bearing of the duckling from the mother?

- What is the bearing of the mother from the duckling?

Shutterstock.com/Michael Leidel

Situation Two

An orienteering competition is to involve ten stages. The organisers print the instuctions for travelling from each check point to the next. These instructions commence as follows:

From the start travel **400** metres on a bearing **040°** to 1st checkpoint.

From 1st checkpoint travel **300** metres on a bearing **100°** to 2nd checkpoint.

From 2nd checkpoint travel **450** metres on a bearing **180°** to 3rd checkpoint.

The organisers are concerned that if heavy rain falls the night before the event then checkpoint 2 will be unapproachable. In case of this they need alternative instructions for a competition involving nine stages. This alternative competition would not involve the original checkpoint 2 but re-numbers checkpoints 3 to 10 as 2 to 9. Copy and complete the following instructions for the first two stages of this alternative race.

From the start travel _____ metres on a bearing _____ to 1st checkpoint.

From 1st checkpoint travel _____ metres on a bearing _____ to 2nd checkpoint.

Vector quantities

Each of the situations on the previous page involved the addition of two stages of a journey. Hopefully, you did not simply add the two distances together to give the final distance from the starting point. Such an approach would not give the correct answer because, whilst 5 m + 7 m = 12 m,

$$5 \text{ m in direction } 050° \ + \ 7 \text{ m in direction } 170° \ \neq \ 12 \text{ m in a direction } 220°.$$

Our normal rules for the addition of two quantities will not necessarily hold when the two quantities to be added involve *direction*. For example, if a person walks 5 km due North followed by 3 km due South, the total distance walked will be 8 km but the person will not be 8 km from the starting point.

1st leg of journey	2nd leg of journey	Dist travelled	Final location
5 km N	3 km S	8 km	2 km N of starting point

In this case, the directions of each leg of the journey are such that the final location could be determined mentally. If this is not the case we must either:

- sketch the situation and use trigonometry to determine the final location relative to the initial position,

or
- make an accurate scale drawing to determine the final location relative to the initial position.

EXAMPLE 1

A boat sails 15 km on a bearing 170° followed by 9 km due East. Find the distance and bearing of the boat's final position from its initial position.

Solution

Method One: By calculation (i.e. sketch and use trigonometry)

If A is the initial position and C is the final position, then the distance from A to C is given by AC.

Now $AC^2 = 15^2 + 9^2 - 2 \times 15 \times 9 \times \cos 100°$ (cosine rule)

∴ $AC \approx 18.79$

The bearing of C from A is $(170° - \angle BAC)$

By the sine rule $\dfrac{9}{\sin \angle BAC} = \dfrac{AC}{\sin 100°}$

giving $\angle BAC \approx 28.2°$ Obtuse solution to equation not applicable as triangle already has an obtuse angle.

The final position of the boat is approximately 19 km from the initial position and on a bearing 142°.

Important note: Had there not already been an obtuse angle in △BAC, finding $\angle BAC$, rather than $\angle ACB$, is a wise strategy because $\angle BAC$ is not opposite the longer side and so cannot be obtuse.

Method Two: By scale drawing

If A is the initial position and C is the final position then the distance from A to C is given by AC.

From the scale drawing shown on the right:

$$AC \approx 9.4 \text{ cm}$$

and $\angle BAC \approx 28°$.

The final position of the boat is approximately 19 km from the initial position and on a bearing 142°.

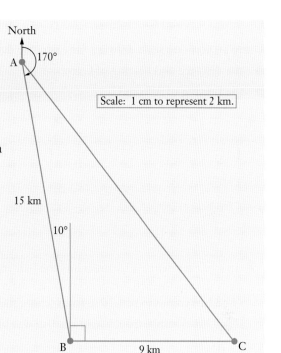

North

A 170°

Scale: 1 cm to represent 2 km.

15 km

10°

B 9 km C

Rather than creating the scale drawing with pencil, ruler and protractor, it is possible to create it using the drawing facility of some calculators. Explore the capability of your calculator in this regard.

As we have seen above, adding quantities which have magnitude and direction needs special care. Such quantities are called **vector** quantities. Some common examples of **vector** quantities are:

Displacement,	e.g. 5 km South.
Velocity,	e.g. 5 m/s North.
Force,	e.g. 6 Newtons upwards.
Acceleration,	e.g. 5 m/s^2 East.

Quantities which have magnitude only are not vectors. Such quantities are called **scalars**. Some common examples of scalar quantities are:

Distance,	e.g. 5 km.
Speed,	e.g. 5 m/s.
The magnitude of a force,	e.g. 6 Newtons.
The magnitude of the acceleration,	e.g. 5 m/s^2.
Energy,	e.g. 50 Joules.

Exercise 3A

Questions **1**, **2** and **3** each show a sketch of a journey from A to C via B.

Use trigonometry to calculate   **a** the distance and bearing of C from A,
   b the bearing of A from C.

1

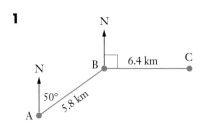

2

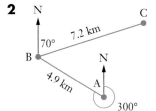

3

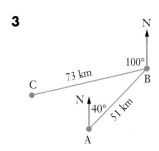

Questions **4**, **5** and **6** each show a sketch of a journey from A to C via B.

Using an accurate scale drawing for each case determine

a the distance and bearing of C from A,

b the bearing of A from C.

4

5

6

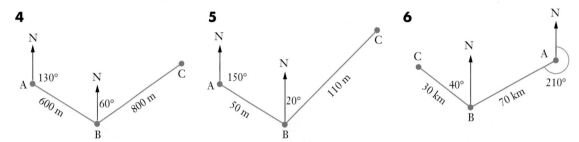

7 In a cricket match a batsman hits the ball and a fielder, 30 m away, fails to stop the ball but deflects it (see diagram) and slows it down. The ball comes to rest 20 m from where the fielder deflected it.

Find the distance from where the ball was hit to where it came to rest.

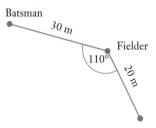

8 A yacht sails 5.2 km on a bearing 190° followed by 6. 4 km on a bearing 110°. Calculate the distance and bearing of the yacht's final position from its initial position.

9 A hiker travels 2.6 km due North followed by 4.3 km on a bearing 132°. Calculate the distance and bearing of the hiker's final position from his initial position.

10 In an orienteering race the first checkpoint is 500 m from the start and on a bearing of 030°. The second checkpoint is 400 m from the first checkpoint and is 600 m from the start. Find, to the nearest degree, the two possible bearings of checkpoint 2 from the start.

11 A particular hole on a golf course is as shown on the right. A positive thinking golfer plans to complete the hole in two shots as shown by the 'planned route'. His first shot is to be 280 m at 020° and the second 200 m at 050°.

Unfortunately he miss-hits his first shot and it goes 250 m due North.

Calculate the distance (nearest metre) and direction (nearest degree) that his second shot needs to be if he is still to complete the hole in two shots.

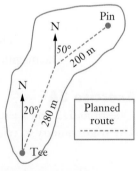

As was explained on the previous page, quantities which have magnitude and direction are called **vector** quantities and special care needs to be taken when such quantities are added. The situations at the beginning of this chapter, Example 1 and all of the questions in Exercise 3A involved the vector quantity **displacement**. The situation on the next page involves the vector quantity **force**.

Situation Three

Each of the following diagrams show one or two tugboats about to tow a floating platform into position. The force that each tug is exerting, in Newtons, N, is as indicated. Find the direction in which the platform will begin to move in each case, giving your answer as a bearing, and state the total force acting in that direction.

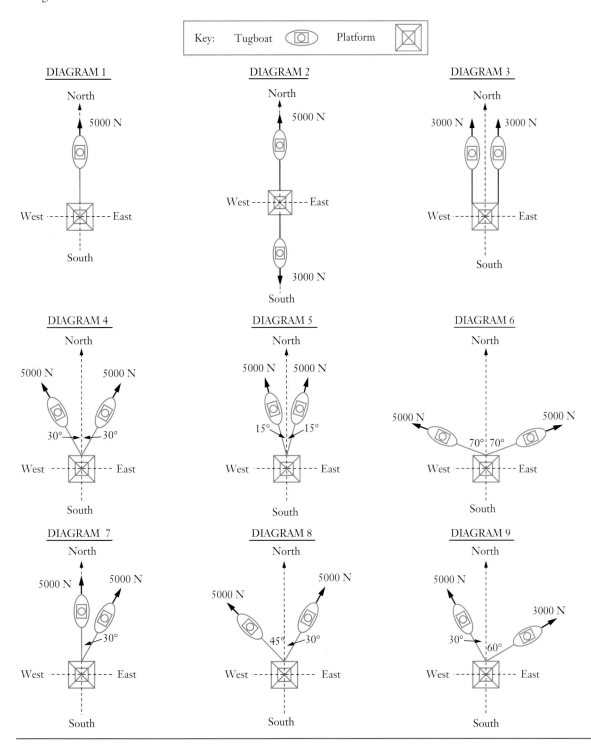

Adding vectors

Situation Three involved the addition of vector quantities, in each case, two forces. Diagrams 1, 2 and 3 should not have caused you any problems but how did you get on with diagrams 4 to 9? Did you develop any methods for adding two vector quantities, (in this case, forces)?

As has been previously mentioned care needs to be taken when adding vectors because we cannot simply sum the magnitudes and sum the directions:

5 N in direction 050° + 7 N in direction 170° ≠ 12 N in a direction 220°.

In example 1, encountered earlier, we added displacement vectors by sketching and using trigonometry or by scale drawing.

To determine the combined or **resultant** effect of 15 km in a direction 170° and 9 km due East we made a sketch like that shown on the right and then used trigonometry to determine the magnitude and direction of AC.

Notice that the two vectors to be added are drawn 'nose to tail', i.e. the 'tail' of the second vector takes over from the 'nose' of the first.

This was clearly the logical thing to do with displacement vectors because the second leg of the journey did indeed start from where the first leg finished. However, we can use this same 'nose to tail' method to determine the resultant of vector quantities other than displacement.

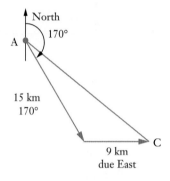

Note also in the above diagram that the resultant vector is given by the magnitude and direction of the line AC, the side that completes the triangle from the tail of the first vector to the tip of the second vector.

EXAMPLE 2

Forces of 6.2 Newtons vertically upwards and 8.9 Newtons acting at 30° to the vertical act on a body, see diagram. Determine the magnitude and direction of the single force that could replace these two forces (i.e. determine the **resultant** of the two forces).

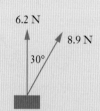

Solution

First sketch the vector triangle remembering the 'nose to tail' idea for vector addition. By the cosine rule:

$$\text{(Magnitude of resultant)}^2 = 6.2^2 + 8.9^2 - 2(6.2)(8.9)\cos 150°$$

∴   Magnitude of resultant ≈ 14.602 Newtons

By the sine rule: $\dfrac{\text{Magnitude of resultant}}{\sin 150°} = \dfrac{8.9}{\sin \theta}$

∴ $\theta \approx 17.7°$

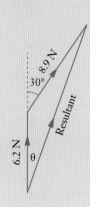

(Obtuse solution not applicable, θ not opposite longest side, and triangle already has obtuse angle.)

The resultant of the two forces is a force of magnitude 14.6 Newtons (correct to one decimal place) acting at 18° to the vertical (to the nearest degree).

Some questions refer to the angle between two forces. This refers to the angle between the forces when they are either both directed away from a point or both directed towards it.

Angle between forces is 40°.

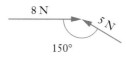

Angle between forces is 150°.

Angle between forces is 60°.

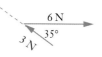

Angle between forces is 145°.

EXAMPLE 3

Two forces have magnitudes of 6 N and 8 N and the angle between them is 50°. Find the magnitude of the resultant and the angle it makes with the smaller of the two forces.

Solution

First sketch the situation and then draw the vector triangle:

By the cosine rule: Magnitude of resultant $= \sqrt{6^2 + 8^2 - 2 \times 6 \times 8 \times \cos 130°}$

$$\approx 12.7$$

By the sine rule: $\dfrac{\text{Magnitude of resultant}}{\sin 130°} = \dfrac{8}{\sin \theta}$

$$\theta \approx 28.8°$$

The resultant of the two forces has magnitude 12.7 N and makes an angle of approximately 29° with the smaller of the two forces.

Exercise 3B

Questions **1** to **4** each show two forces acting on a body. In each case determine the magnitude, correct to one decimal place, and direction, to the nearest degree, of the resultant of the two forces. (Give the direction in terms of the angle made with the vertical – shown as a broken line).

1

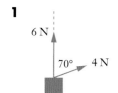

2

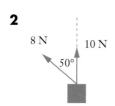

3

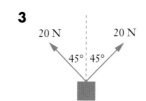

4

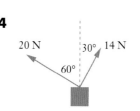

Find the magnitude and direction of the resultant of each of the following pairs of forces.
(Give magnitude in exact form for questions **5** and **6** and correct to one decimal place for **7** and **8**
and directions as three figure bearings to the nearest degree.)

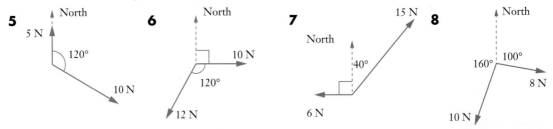

When a body slides down a rough inclined plane it experiences a normal reaction, R, perpendicular to
the surface and a frictional force F, parallel to the surface. The resultant of R and F is called the total
reaction. Determine the magnitude (to the nearest Newton) and direction (to the nearest degree) of the
total reaction in each of questions **9**, **10** and **11** giving the direction as the acute angle the total reaction
makes with the slope.

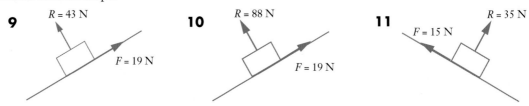

12 Two forces have magnitudes of 8 N and 12 N and the angle between them is 130°. Find the
magnitude of the resultant and the angle it makes with the larger of the two forces.

13 Two forces have magnitudes of 10 N and 15 N and the angle between them is 45°. Find the
magnitude of the resultant and the angle it makes with the smaller of the two forces.

We have so far used this 'nose to tail' approach for adding two displacement vectors and for adding
two forces. The following example, and **Exercise 3C** that follows it, involve this same process used
for velocity vectors.

EXAMPLE 4

A canoeist wishes to paddle her canoe across a river from
point A on one bank to the opposite bank. The canoeist can
maintain a constant 5 km/h in still water. However the river
is flowing at 3 km/h (see diagram).

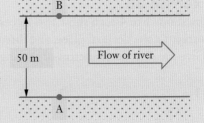

a If the canoeist wishes to journey across the river as quickly
as possible, how long will the journey take and how far
down river will the canoeist travel?

b If instead she wishes to reach point B on the opposite bank, directly opposite A,
in which direction should she paddle and how long will the journey take?

Solution

a If she wishes to journey across the river as quickly as possible, she must put all her efforts into getting across and let the current take her down river.

She travels at 5 km/h across the river and the current takes her down river at 3 km/h. Her resultant velocity is as shown on the right.

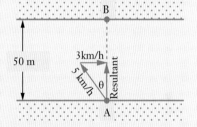

Now speed $= \dfrac{\text{distance}}{\text{time}}$.

Thus to travel the 50 m *across* the river when her speed *across* is 5 km/h will take t_1 seconds where

$$\frac{5 \times 1000}{60 \times 60} = \frac{50}{t_1} \qquad \therefore \quad t_1 = 36.$$

The distance travelled downstream will be given by BC, see diagram.

By similar triangles $\qquad \dfrac{3}{5} = \dfrac{BC}{50} \qquad \therefore \quad BC = 30.$

If she wishes to journey across the river as quickly as possible it will take her 36 seconds and she will travel 30 metres down river.

b To reach point B on the far bank she needs to set a course such that the combined effect of her paddling and the flow of the river takes her directly across the river (see diagram).

By trigonometry: $\qquad\qquad\qquad \theta \approx 37°$

By Pythagoras: $\qquad$ Resultant speed $= 4$ km/h.

To travel 50 metres at 4 km/h will take 45 seconds.

To reach point B on the opposite bank, directly opposite A, she should paddle upstream at 53° to the bank. The combined effect of this paddling and the current will cause her to travel to B in 45 seconds.

Exercise 3C

Find the resultant of each of the following pairs of velocities stating the magnitude of the resultant, correct to one decimal place, and the acute angle it makes with the bank, to the nearest degree. (In each question the 2 m/s is parallel to the bank).

1
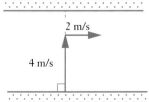

2
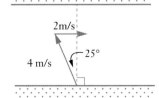

3
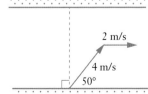

4 A boat starts out at 20 km/h in direction 030°. However the boat is blown off this direction by a 12 km/h wind blowing from 080°. What direction does the boat now travel in and how far does it travel in one hour?

5 A bird flying south at 50 km/h encounters a 24 km/h wind from 330°. In what direction and with what speed will the bird now be travelling?

If the bird wishes to continue flying south, in what direction should it attempt to fly so that the wind causes it to travel south?

6 A hot-air balloon is gaining height at 3 m/s and the wind is blowing horizontally at a steady 1 m/s.

Find   **a**   the height of the balloon one minute after take off,

b   the speed of the balloon,

c   the angle the balloon's travel makes with the horizontal.

7 A boat has a speed of 10 km/h in still water. It is to be driven directly across a river of width 80 m. At what angle to the bank must the boat be directed, to the nearest degree, and how long will the journey take, to the nearest second, if the current flows at:

a   3 km/h?   **b**   4 km/h?   **c**   6 km/h?

8 In still air an aircraft can maintain a speed of 400 km/h. In what direction should the aircraft be pointing if it is to travel due north and a wind of 28 km/h is blowing from the west?

9 In still air an aircraft can maintain a speed of 300 km/h. In what direction should the aircraft be pointing if it is to travel due north and a 28 km/h wind is blowing from 070°?

10 a   In what direction must a plane head if it can maintain 350 km/h in still air, wishes to fly 500 km in a direction 040° and the wind is blowing at 56 km/h from 100°?

b   How long would the journey take (to the nearest minute)?

c   How long would the return journey take, again to the nearest minute, if the wind is still 56 km/h from 100°?

11 The diagram shows a canoeing course in a river. The canoeist has to go from the start at A to B to C to the finish at D. The river is flowing at 2 m/s and a particular canoeist can paddle at a rate that would produce a speed of 6 m/s in still water. Find, to the nearest second, the least time this canoeist takes to complete the course.

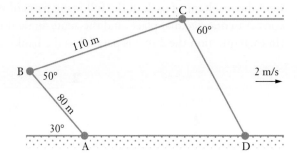

Mathematical representation of a vector quantity

Representing vectors

The vectors considered so far in this chapter have been displacement, force and velocity. When considering the manipulation of vector quantities we need not have a particular vector context, such as force or velocity, in mind. We could simply consider a vector of given magnitude, say 5 units, and with given direction, say due East. We represent such a vector diagrammatically by a line segment in the given direction and whose length represents the magnitude of the vector.

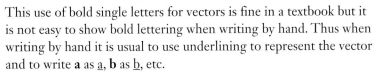

The above vector is from A to B and we write it as $\underline{AB}$ or $\overrightarrow{AB}$. The order of the letters indicates the direction, i.e. **from** A **to** B. The arrow above the letters emphasises this direction and this arrow, or the line underneath, distinguishes the vector $\underline{AB}$ from the scalar AB, the distance from A to B.

From our 'nose to tail' method of addition it follows from the diagram on the right that:

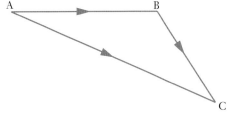

$$\overrightarrow{AB} + \overrightarrow{BC} = \overrightarrow{AC}$$

Vectors are also frequently written using lower case letters and in such cases bold type is used. Thus from the diagram on the right it follows that:

$$\mathbf{p} + \mathbf{q} = \mathbf{r}.$$

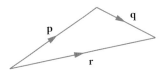

This use of bold single letters for vectors is fine in a textbook but it is not easy to show bold lettering when writing by hand. Thus when writing by hand it is usual to use underlining to represent the vector and to write **a** as $\underline{a}$, **b** as $\underline{b}$, etc.

For the magnitude of vector **a** we write $|\mathbf{a}|$ or $|\underline{a}|$.

Equal vectors

Two vectors are equal if they have the same magnitude *and* the same direction.

a = **b**
Same magnitude,
same direction.

c ≠ **d**
Same magnitude,
different directions.

e ≠ **f**
Same direction,
different magnitudes.

The negative of a vector

The vectors, **a** and −**a**, will have the same magnitudes but will be in opposite directions.

If $\mathbf{a} = \overrightarrow{AB}$ then $-\mathbf{a} = -\overrightarrow{AB}$
$\qquad\qquad\quad = \overrightarrow{BA}$

Multiplication of a vector by a scalar

If **b** = 2**a** then **b** is in the same direction as **a** but twice the magnitude.

If **c** = 3**a** then **c** is in the same direction as **a** but three times the magnitude.

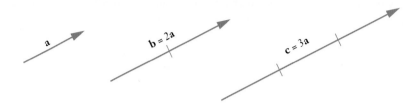

If for some positive scalar k, **b** = k**a**, then **b** is in the same direction as **a** and k times the magnitude.

If k is negative, then **b** will be in the opposite direction to **a** and k times the magnitude.

Parallel vectors

Two vectors are parallel if one is a scalar multiple of the other.

If the scalar multiple is positive, the vectors are said to be *like* parallel vectors, i.e. in the *same* direction.

e.g. **a** and 2**a** are like parallel vectors.

If the scalar multiple is negative the vectors are said to be *unlike* parallel vectors, i.e. in *opposite* directions.

e.g. **a** and −2**a** are unlike parallel vectors.

Addition of vectors

To add two vectors means to find the single, or resultant, vector that could replace the two. We add the vectors using a vector triangle in which the two vectors to be added follow 'nose to tail' and form two of the sides of the triangle. The resultant is then the third side of the triangle with its direction 'around the triangle' being in the opposite sense to the other two.

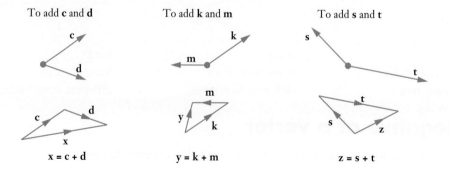

This vector addition is sometimes referred to as the *parallelogram law*:

> If the two vectors to be added are represented in magnitude and direction by $\vec{AB}$ and $\vec{AD}$ in the parallelogram ABCD then the resultant of the two vectors will be represented in magnitude and direction by the diagonal $\vec{AC}$.

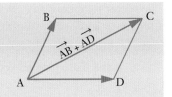

Parallelogram rule for vector addition

Considering the triangle ADC and remembering that AB and DC are opposite sides of a parallelogram, i.e. $\vec{AB} = \vec{DC}$, it can be seen that the parallelogram method of vector addition and the triangle method are equivalent.

With $|\mathbf{a}|$ as the magnitude of $\mathbf{a}$ it also follows that

$$|\mathbf{a} + \mathbf{b}| \leq |\mathbf{a}| + |\mathbf{b}|$$

the triangle inequality.

Subtraction of one vector from another

To give meaning to vector subtraction we consider $\mathbf{a} - \mathbf{b}$ as $\mathbf{a} + (-\mathbf{b})$ and then use our technique for adding vectors.

With the parallelogram approach we see that whilst $\mathbf{a} + \mathbf{b}$ is one diagonal, $\mathbf{a} - \mathbf{b}$ is the other.

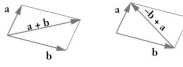

The zero vector

If we add a vector to the negative of itself we obtain the zero vector: $\mathbf{p} + (-\mathbf{p}) = \mathbf{0}$.

The zero vector has zero magnitude and an undefined direction. We can write the zero vector as $\mathbf{0}$ or $\underline{0}$ or simply as 0 because it is usually clear from the context whether 0 is a number or a vector.

ISBN 9780170390477

ha = kb

The vector statement $h\mathbf{a} = k\mathbf{b}$ means that some scalar multiple of $\mathbf{a}$ has the same magnitude and direction as some scalar multiple of $\mathbf{b}$.

If this is the case then either

- $\mathbf{a}$ and $\mathbf{b}$ are parallel vectors (because one is a scalar multiple of the other),

or
- $h = k = 0$.

Thus if $\mathbf{a}$ and $\mathbf{b}$ are *not* parallel and we have a vector expression of the form

$$p\mathbf{a} + q\mathbf{b} = r\mathbf{a} + s\mathbf{b} \quad [1]$$

i.e.
$$(p - r)\mathbf{a} = (s - q)\mathbf{b},$$

it follows that $p - r = 0$ and $s - q = 0$.

i.e.
$$p = r \quad \text{and} \quad s = q.$$

Thus in equation [1], with $\mathbf{a}$ and $\mathbf{b}$ not parallel, we can equate the coefficients of $\mathbf{a}$ to give $p = r$, and we can equate the coefficients of $\mathbf{b}$ to give $q = s$.

Now that we are able to represent a vector mathematically using line segments and letters, we are able to consider abstract vector questions in which the context e.g. velocity, force etc. is not known. Examples 5 and 6 demonstrate this manipulation of abstract vectors.

EXAMPLE 5

If $\mathbf{a}$ is a vector of magnitude 5 units in direction 040° and $\mathbf{b}$ is a vector of magnitude 3 units in direction 100°, find the magnitude and direction of:

a $\mathbf{a} + \mathbf{b}$ **b** $\mathbf{a} - \mathbf{b}$.

Solution

First sketch $\mathbf{a}$ and $\mathbf{b}$:

a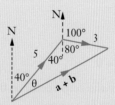

By the cosine rule $|\mathbf{a} + \mathbf{b}|^2 = 3^2 + 5^2 - 2 \times 3 \times 5 \times \cos 120°$

Thus $|\mathbf{a} + \mathbf{b}| = 7$

By the sine rule $\dfrac{|\mathbf{a} + \mathbf{b}|}{\sin 120°} = \dfrac{3}{\sin \theta}$

Thus $\theta \approx 21.8°$

$\mathbf{a} + \mathbf{b}$ has a magnitude of 7 units and direction 062° (to the nearest degree).

b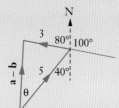

By the cosine rule $|\mathbf{a} - \mathbf{b}|^2 = 3^2 + 5^2 - 2 \times 3 \times 5 \times \cos 60°$

Thus $|\mathbf{a} - \mathbf{b}| \approx 4.36$

By the sine rule $\dfrac{|\mathbf{a} - \mathbf{b}|}{\sin 60°} = \dfrac{3}{\sin \theta}$

Thus $\theta \approx 36.6°$

$\mathbf{a} - \mathbf{b}$ has magnitude 4.4 units (correct to 1 decimal place) and direction 003° (to nearest degree).

Alternatively scale versions of the triangles used to determine **a** + **b** and **a** – **b** could be drawn by hand or using the ability of some calculators to draw geometrical figures. Again you are encouraged to explore the capability of your calculator in this regard.

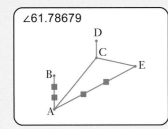

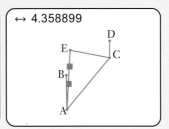

EXAMPLE 6

In parallelogram ABCD, $\overrightarrow{AB}$ = **a** and $\overrightarrow{AD}$ = **b**.

E is a point on DC such that DE : EC = 1 : 2.

Express each of the following vectors in terms of **a** and/or **b**.

a $\overrightarrow{DC}$ **b** $\overrightarrow{CB}$ **c** $\overrightarrow{DE}$ **d** $\overrightarrow{BE}$

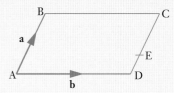

Solution

a $\overrightarrow{DC}$ is the same magnitude and direction as $\overrightarrow{AB}$. Thus $\overrightarrow{DC}$ = **a**.

b $\overrightarrow{CB}$ is the same magnitude but opposite direction to $\overrightarrow{AD}$. Thus $\overrightarrow{CB}$ = –**b**.

c DE : EC = 1 : 2.   Thus $\overrightarrow{DE} = \frac{1}{3}\overrightarrow{DC}$

$$\therefore \quad \overrightarrow{DE} = \frac{1}{3}\mathbf{a}.$$

d $\overrightarrow{BE} = \overrightarrow{BC} + \overrightarrow{CE}$ or $\overrightarrow{BE} = \overrightarrow{BA} + \overrightarrow{AD} + \overrightarrow{DE}$

$\quad\quad = \mathbf{b} + \left(-\frac{2}{3}\mathbf{a}\right)$ $= -\mathbf{a} + \mathbf{b} + \frac{1}{3}\mathbf{a}$

$\quad\quad = \mathbf{b} - \frac{2}{3}\mathbf{a}$ $= -\frac{2}{3}\mathbf{a} + \mathbf{b}$

Exercise 3D

1 For the vectors **a** to **f** shown on the right, state:

   a two like parallel vectors that are unequal,

   b two unlike parallel vectors that are unequal,

   c two vectors that are the same magnitude but not equal,

   d two equal vectors.

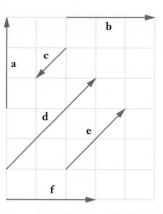

2 For each of the following diagrams select the appropriate vector equation from:

   $\mathbf{a} + \mathbf{b} = \mathbf{c}$, $\mathbf{a} + \mathbf{c} = \mathbf{b}$, $\mathbf{b} + \mathbf{c} = \mathbf{a}$.

a

b
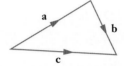

c
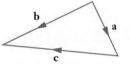

3 With **a** as defined in the diagram below, express each of the vectors **b**, **c**, **d**, **e**, **f**, **g** and **h** in terms of **a**.

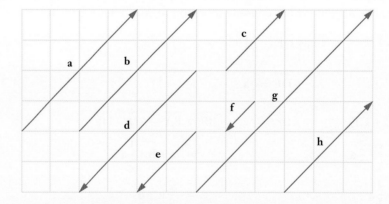

4 With **m** and **n** as defined in the diagram below, express each of the vectors **p**, **q**, **r**, **s**, **t**, **u** and **v** in terms of **m** and/or **n**.

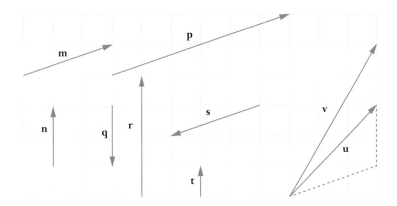

5 With **a** and **b** as defined in the diagram below, express each of the vectors **c**, **d**, **e**, **f** and **g** in terms of **a** and **b**.

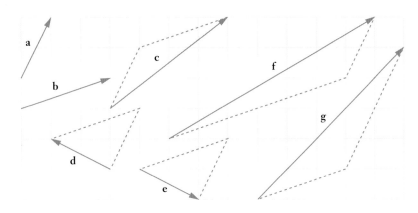

6 With **s** and **t** as defined in the diagram below, express each of the vectors **u**, **v**, **w**, **x**, **y** and **z** in terms of **s** and **t**.

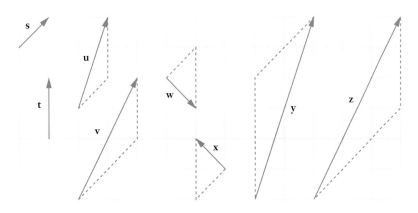

7 With **a** and **b** as defined in the diagram on the right, draw on grid paper each of the following:

a	2**a**	**b**	3**b**
c	−**a**	**d**	−**b**
e	**a** + **b**	**f**	**a** − **b**
g	**b** − **a**	**h**	**a** + 2**b**

8 If **a** is a vector of magnitude 5 units in direction 070° and **b** is a vector of magnitude 4 units in direction 330°, find the magnitude (correct to one decimal place) and direction (to the nearest degree) of the following.

    a   **a** + **b**   **b**   **a** − **b**

9 If **e** is a vector of magnitude 40 units in direction 130° and **f** is a vector of magnitude 30 units in direction 260°, find the magnitude (correct to the nearest unit) and direction (to the nearest degree) of the following.

    a   2**e** + **f**   **b**   **e** − 2**f**

10 The velocity of a body changes from **u**, 5.4 m/s due North, to **v**, 7.8 m/s due West, in 5 seconds.

The acceleration of the body is given by $\mathbf{a} = \dfrac{\mathbf{v} - \mathbf{u}}{\text{time taken}}$.

Find the magnitude and direction of **a**.

11 The velocity of a body changes from an initial velocity **u**, 10.4 m/s in direction 020°, to a final velocity **v**, 12.1 m/s due West, in 4 seconds.

The body's acceleration is given by $\mathbf{a} = \dfrac{\mathbf{v} - \mathbf{u}}{\text{time taken}}$.

Find the magnitude and direction of **a**.

12 Find the values of λ and μ in each of the following cases given that **a** and **b** are non-parallel vectors.

a	λ**a** = μ**b**	**b**	3λ**a** = 5μ**b**
c	(λ − 3)**a** = (μ + 4)**b**	**d**	λ**a** − 2**a** = 5**b** − μ**b**
e	λ**a** − 2**b** = μ**b** + 5**a**	**f**	(λ + μ − 4)**a** = (μ − 3λ)**b**
g	2**a** + 3**b** + μ**b** = 2**b** + λ**a**	**h**	λ**a** + μ**b** + 2λ**b** = 5**a** + 4**b** + μ**a**
i	λ**a** − **b** + μ**b** = 4**a** + μ**a** − 4λ**b**	**j**	2λ**a** + 3μ**a** − μ**b** + 2**b** = λ**b** + 2**a**

13 OABC is a rectangle with $\overrightarrow{OA}$ = **a** and $\overrightarrow{OC}$ = **c**. P and Q are the midpoints of AB and BC respectively. Express each of the following in terms of **a** and/or **c**.

a	$\overrightarrow{CB}$	**b**	$\overrightarrow{BC}$	**c**	$\overrightarrow{AB}$	**d**	$\overrightarrow{BA}$
e	$\overrightarrow{AP}$	**f**	$\overrightarrow{OQ}$	**g**	$\overrightarrow{OP}$	**h**	$\overrightarrow{PQ}$

ISBN 9780170390477

14 OAB is a triangle with C a point on AB such that AC is three-quarters of AB.

If $\overrightarrow{OA} = \mathbf{a}$ and $\overrightarrow{OB} = \mathbf{b}$, express each of the following in terms of $\mathbf{a}$ and/or $\mathbf{b}$.

a $\overrightarrow{AB}$ **b** $\overrightarrow{AC}$ **c** $\overrightarrow{CB}$ **d** $\overrightarrow{OC}$

15 ABCD is a parallelogram with $\overrightarrow{AB} = \mathbf{a}$ and $\overrightarrow{AD} = \mathbf{b}$.

E is a point on BC such that BE : EC = 1 : 2.

F is a point on CD such that CF : CD = 1 : 2.

Express each of the following in terms of $\mathbf{a}$ and/or $\mathbf{b}$.

a $\overrightarrow{AC}$ **b** $\overrightarrow{BE}$ **c** $\overrightarrow{DF}$ **d** $\overrightarrow{AE}$

e $\overrightarrow{AF}$ **f** $\overrightarrow{BF}$ **g** $\overrightarrow{DE}$ **h** $\overrightarrow{EF}$

16 OABC is a trapezium with OC parallel to, and twice as long as, AB. D is the midpoint of BC.

If $\overrightarrow{OA} = \mathbf{a}$ and $\overrightarrow{AB} = \mathbf{b}$ express each of the following in terms of $\mathbf{a}$ and/or $\mathbf{b}$.

a $\overrightarrow{OB}$ **b** $\overrightarrow{OC}$ **c** $\overrightarrow{BC}$

d $\overrightarrow{BD}$ **e** $\overrightarrow{OD}$

17 OAB is a triangle with C the midpoint of OA and D a point on AB such that AD is two thirds of AB. CD continued meets OB continued at E.

If $\overrightarrow{OA} = \mathbf{a}$ and $\overrightarrow{OB} = \mathbf{b}$ express each of the following in terms of $\mathbf{a}$ and/or $\mathbf{b}$.

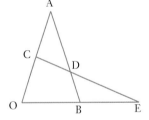

a $\overrightarrow{OC}$ **b** $\overrightarrow{AB}$ **c** $\overrightarrow{AD}$ **d** $\overrightarrow{CD}$

e If $\overrightarrow{CE} = h\overrightarrow{CD}$ and $\overrightarrow{OE} = k\overrightarrow{OB}$ use the fact that $\overrightarrow{OC} + \overrightarrow{CE} = \overrightarrow{OE}$ to determine h and k.

18 In the trapezium OABC, $\overrightarrow{OA} = \mathbf{a}$, $\overrightarrow{OC} = \mathbf{c}$ and $\overrightarrow{AB} = 2\mathbf{c}$. D is a point on CB such that $\overrightarrow{CD} = \dfrac{2}{3}\overrightarrow{CB}$. OD continued meets AB continued at E.

If $\overrightarrow{OE} = h\overrightarrow{OD}$ and $\overrightarrow{AE} = k\overrightarrow{AB}$ determine h and k.

Miscellaneous exercise three

This miscellaneous exercise may include questions involving the work of this chapter, the work of any previous chapters, and the ideas mentioned in the Preliminary work section at the beginning of the book.

1 Each switch in a system of six switches can be either on or off.

How many different settings are there for the system?

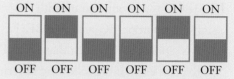

2 Jack walks 2.4 km on a bearing 060° followed by 4.4 km on a bearing 190°. On what bearing and for what distance should he now walk to return directly to his starting point?

3 Members of a wine club are invited to select a combination of eight different bottles of wine from a list of one dozen wines. How many different combinations are possible?

4 What is the least number of people needed to be in a room before we could guarantee that at least three of them would have the same birth month?

5 Let us suppose that for the purposes of this question I teach Specialist Mathematics at, and only at, XYZ high school, a year 7 to 12 high school.

Under the above assumption, the following would be a true statement:

*If you are in my Specialist Mathematics class
then you attend XYZ high school.*

Write the converse of this statement and the contrapositive of this statement and in each case state whether it is true or false.

6 Eight members of a rugby team are to be arranged in a line for a photograph. If the eight can themselves be chosen from the fifteen in the team how many different possible arrangements are there?

7 If **c** is a vector of magnitude 10 units in direction 160° and **d** is a vector of magnitude 12 units due East, find the magnitude (correct to one decimal place) and direction (to the nearest degree) of

   a $\mathbf{c} + \mathbf{d}$ **b** $\mathbf{c} - \mathbf{d}$ **c** $\mathbf{c} + 2\mathbf{d}$

8 Given that **a** and **b** are not parallel, determine h and k in each of the following vector statements.

   a $h\mathbf{a} = k\mathbf{b}$ **b** $h\mathbf{a} + \mathbf{b} = k\mathbf{b}$

   c $(h - 3)\mathbf{a} = (k + 1)\mathbf{b}$ **d** $h\mathbf{a} + 2\mathbf{a} = k\mathbf{b} - 3\mathbf{a}$

   e $3h\mathbf{a} + k\mathbf{a} + h\mathbf{b} - 2k\mathbf{b} = \mathbf{a} + 5\mathbf{b}$ **f** $h(\mathbf{a} + \mathbf{b}) + k(\mathbf{a} - \mathbf{b}) = 3\mathbf{a} + 5\mathbf{b}$

9 A group of 5 people is to be formed from a group comprising 10 women and 6 men. How many different groups of 5 people are there if the group must consist of

   a 3 women and 2 men?   **b** all women?

   c more women than men?   **d** more men than women?

4.

Vectors in component form

- Further examples
- Position vectors
- Miscellaneous exercise four

Situation

A crippled oil tanker, with its engines out of action, is drifting towards some rocks. Tow lines have been attached to the tanker and three tugs are attempting to pull the boat away from the rocks. The forces exerted by the tugs and by the wind and water conditions are as shown below.

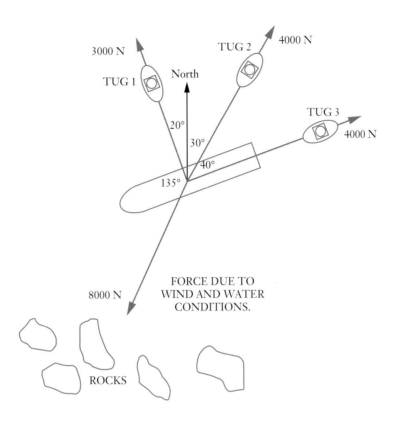

Find the magnitude and direction of the resultant force acting on the tanker.

How did you get on with the situation on the previous page? It involved finding the resultant of a number of vectors. Did you sum the vectors using one of the methods from chapter 3 i.e. scale drawing or by using trigonometry?

To use scale drawing we can extend the 'nose to tail' idea of the triangle of vectors:

$$p = a + b$$

to a polygon of vectors:

$$q = a + b + c + d + e$$

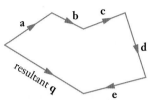

Resolution of forces

EXAMPLE 1

A body experiences five forces as shown in the diagram on the right. Use a scale drawing to determine the magnitude and direction of the resultant of these forces.

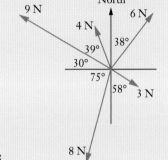

Solution

Choose a suitable scale and draw the forces 'nose to tail'. The resultant will be given by the line segment completing the polygon and in the opposite sense to the other forces:

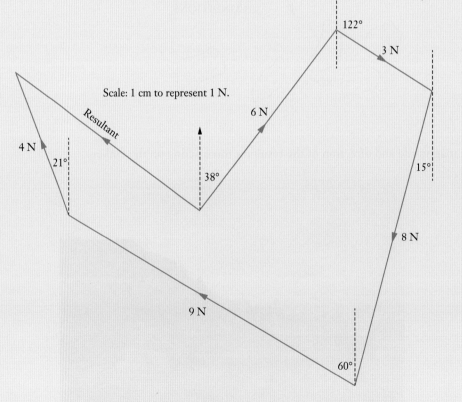

Scale: 1 cm to represent 1 N.

The resultant has a magnitude of ~6.2 N at ~306°.

Alternatively we could create the scale diagram using the drawing capability of some calculators.

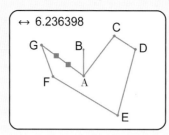

An alternative to using a scale drawing to determine the resultant of the five forces involves expressing each vector in terms of its **components** in two mutually perpendicular directions e.g. horizontal and vertical.

Component and polar forms of vectors

Consider the vector **a** shown on the right.

The vector triangle illustrated shows **a** expressed as the sum of the horizontal vector **x** and the vertical vector **y**.

Thus $\qquad\qquad\qquad\qquad\qquad \mathbf{a} = \mathbf{x} + \mathbf{y}$

By trigonometry: $\qquad\qquad \cos\theta = \dfrac{|\mathbf{x}|}{|\mathbf{a}|} \qquad$ i.e. $\quad |\mathbf{x}| = |\mathbf{a}|\cos\theta$

and $\qquad\qquad\qquad \sin\theta = \dfrac{|\mathbf{y}|}{|\mathbf{a}|} \qquad$ i.e. $\quad |\mathbf{y}| = |\mathbf{a}|\sin\theta$

Thus **a** can be expressed as the sum of $|\mathbf{a}|\cos\theta$ units horizontally and $|\mathbf{a}|\sin\theta$ units vertically.

e.g.

The 10 N force shown in the diagram can be expressed as

$10\cos 30°$ N horizontally $\quad+\quad 10\sin 30°$ N vertically

i.e. $\quad 5\sqrt{3}$ N horizontally $\quad+\quad 5$ N vertically.

However:
- It is rather tedious to have to write 'horizontally' and 'vertically' every time.

- Horizontally could mean to the left or to the right and vertically could mean up or down (at this stage we are only considering vectors in two dimensions so do not have to consider vectors coming out of the page).

To avoid these problems we use **i** and **j** to represent horizontal and vertical **unit vectors** (i.e. vectors of unit length) as shown in the diagram on the right.

The vectors **a**, **b**, **c**, **d**, **e** and **f** shown in the diagram can be expressed in terms of these unit vectors as follows.

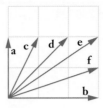

$$\mathbf{a} = 2\mathbf{j} \qquad\qquad \mathbf{b} = 3\mathbf{i} \qquad\qquad \mathbf{c} = \mathbf{i} + 2\mathbf{j}$$

$$\mathbf{d} = 2\mathbf{i} + 2\mathbf{j} \qquad\qquad \mathbf{e} = 3\mathbf{i} + 2\mathbf{j} \qquad\qquad \mathbf{f} = 3\mathbf{i} + \mathbf{j}$$

For the vector $a\mathbf{i} + b\mathbf{j}$, '$a$' is said to be the **horizontal component** and 'b' the **vertical component**.

Once vectors are expressed in this $a\mathbf{i} + b\mathbf{j}$ form it becomes easy to manipulate them.

For example, consider again the vectors **c** and **d**.

As before, $\mathbf{c} = \mathbf{i} + 2\mathbf{j}$ and $\mathbf{d} = 2\mathbf{i} + 2\mathbf{j}$.

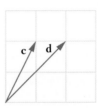

The working and diagrams below show **c** + **d**, **c** − **d** and 2**c**.

$$\begin{aligned}\mathbf{c} + \mathbf{d} &= (\mathbf{i} + 2\mathbf{j}) + (2\mathbf{i} + 2\mathbf{j}) \\ &= 3\mathbf{i} + 4\mathbf{j}\end{aligned} \qquad \begin{aligned}\mathbf{c} - \mathbf{d} &= (\mathbf{i} + 2\mathbf{j}) - (2\mathbf{i} + 2\mathbf{j}) \\ &= -\mathbf{i}\end{aligned} \qquad \begin{aligned}2\mathbf{c} &= 2(\mathbf{i} + 2\mathbf{j}) \\ &= 2\mathbf{i} + 4\mathbf{j}\end{aligned}$$

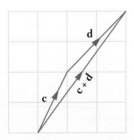
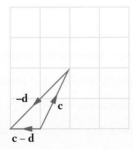

Now that we have seen how vectors can be expressed in the form $a\mathbf{i} + b\mathbf{j}$ how can this be used to sum a number of vectors, as the situation at the beginning of this chapter required us to do?

We were required to sum four vectors as shown in the diagram on the right.

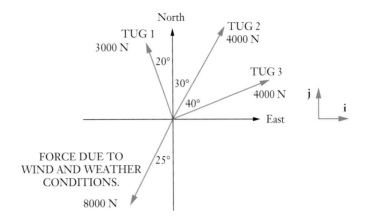

Expressing each in the form $a\mathbf{i} + b\mathbf{j}$:

Force from Tug 1: $\qquad$ $(-3000 \cos 70° \, \mathbf{i} + 3000 \sin 70° \, \mathbf{j})$ N

Force from Tug 2: $\qquad$ $(4000 \cos 60° \, \mathbf{i} + 4000 \sin 60° \, \mathbf{j})$ N

Force from Tug 3: $\qquad$ $(4000 \cos 20° \, \mathbf{i} + 4000 \sin 20° \, \mathbf{j})$ N

Force due to wind and water: $\qquad$ $(-8000 \cos 65° \, \mathbf{i} - 8000 \sin 65° \, \mathbf{j})$ N

Thus the resultant of these forces will be

$$[(-3000 \cos 70° + 4000 \cos 60° + 4000 \cos 20° - 8000 \cos 65°) \, \mathbf{i}$$
$$+ \, (3000 \sin 70° + 4000 \sin 60° + 4000 \sin 20° - 8000 \sin 65°) \, \mathbf{j}] \text{ N}$$
$$\approx (1352\mathbf{i} + 401\mathbf{j}) \text{ N}$$

i.e. a vector of magnitude $\sqrt{1352^2 + 401^2} \approx 1410$ Newtons, in a direction ~073°.

Note:
- To find the magnitude and direction of the resultant from the component form we used Pythagoras and trigonometry.

  In the general case, if $\mathbf{p} = a\mathbf{i} + b\mathbf{j}$ then **magnitude**, or **modulus**, of $\mathbf{p}$ is given by

  $$|\mathbf{p}| = \sqrt{a^2 + b^2}$$

  and θ is found using $\qquad \tan \theta = \dfrac{b}{a}.$

- The vector $p\mathbf{i} + q\mathbf{j}$ is sometimes written as an ordered pair (p, q), or perhaps $<p, q>$,

  and sometimes as a *column matrix* $\begin{pmatrix} p \\ q \end{pmatrix}$.

  Whilst the reader needs to be aware of these alternative ways of writing the vector $p\mathbf{i} + q\mathbf{j}$, at this stage this book will tend to use the $p\mathbf{i} + q\mathbf{j}$ form most frequently.

- Some calculators have built in routines for changing a vector given in component form to its magnitude and direction. Does your calculator have this ability?

Exercise 4A

For numbers **1** to **4** use a scale drawing to determine the magnitude and direction of the resultant of each system of vectors.

1
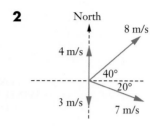

2
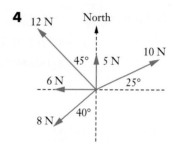

3

4

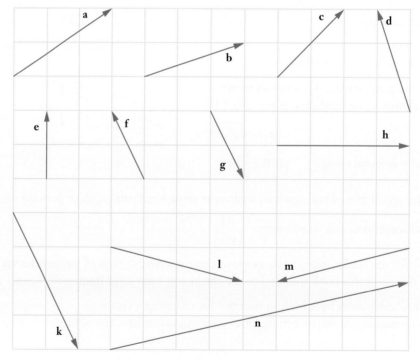

5 Express each of the vectors shown below in the form $a\mathbf{i} + b\mathbf{j}$ where $\mathbf{i}$ is a unit vector to the right and $\mathbf{j}$ is a unit vector up.

6 Calculate the magnitude of each of the vectors shown in question five.

7 Calculate the magnitude of the vector $(-7\mathbf{i} + 24\mathbf{j})$ Newtons.

8 Express each of the following vectors in the form $a\mathbf{i} + b\mathbf{j}$ where $\mathbf{i}$ is a unit vector to the right and $\mathbf{j}$ is a unit vector up. (Give a and b correct to one decimal place.)

a

b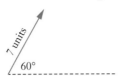

c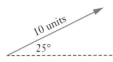

d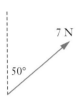

e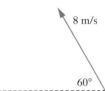

f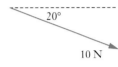

g

h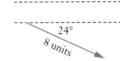

i

j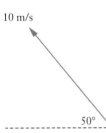

k

l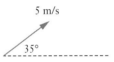

9 For each of the vectors **a** to **f** shown in this question, calculate the magnitude (as an exact value) and determine the angle θ, to the nearest 0.1°.

a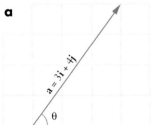

b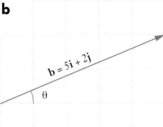

c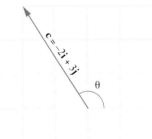

d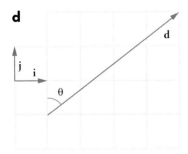

e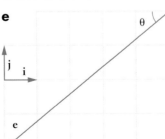

f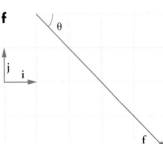

10 An aircraft is flying at 350 km/h on a bearing 160°.

Find   **a**   the northerly component of the velocity,

         b   the easterly component of the velocity.

11 A vector has a westerly component of 5 units and a northerly component of 8 units. Find the magnitude and direction of the vector.

12 If $\mathbf{a} = 2\mathbf{i} + 3\mathbf{j}$ and $\mathbf{b} = \mathbf{i} + 4\mathbf{j}$, find

a $\mathbf{a} + \mathbf{b}$	**b** $\mathbf{a} - \mathbf{b}$	**c** $\mathbf{b} - \mathbf{a}$	**d** $2\mathbf{a}$
e $3\mathbf{b}$	**f** $2\mathbf{a} + 3\mathbf{b}$	**g** $2\mathbf{a} - 3\mathbf{b}$	**h** $-2\mathbf{a} + 3\mathbf{b}$
i $\lvert\mathbf{a}\rvert$	**j** $\lvert\mathbf{b}\rvert$	**k** $\lvert\mathbf{a}\rvert + \lvert\mathbf{b}\rvert$	**l** $\lvert\mathbf{a} + \mathbf{b}\rvert$

13 If $\mathbf{c} = \mathbf{i} - \mathbf{j}$ and $\mathbf{d} = 2\mathbf{i} + \mathbf{j}$, find

a $2\mathbf{c} + \mathbf{d}$	**b** $\mathbf{c} - \mathbf{d}$	**c** $\mathbf{d} - \mathbf{c}$	**d** $5\mathbf{c}$
e $5\mathbf{c} + \mathbf{d}$	**f** $5\mathbf{c} + 2\mathbf{d}$	**g** $2\mathbf{c} + 5\mathbf{d}$	**h** $2\mathbf{c} - \mathbf{d}$
i $\lvert\mathbf{d} - 2\mathbf{c}\rvert$	**j** $\lvert\mathbf{c}\rvert + \lvert\mathbf{d}\rvert$	**k** $\lvert\mathbf{c} + \mathbf{d}\rvert$	**l** $\lvert\mathbf{c} - \mathbf{d}\rvert$

14 If $\mathbf{a} = \langle 5, 4 \rangle$ and $\mathbf{b} = \langle 2, -3 \rangle$, find

a $\mathbf{a} + \mathbf{b}$	**b** $\mathbf{a} - \mathbf{b}$	**c** $2\mathbf{a}$	**d** $3\mathbf{a} + \mathbf{b}$
e $2\mathbf{b} - \mathbf{a}$	**f** $\lvert\mathbf{a}\rvert$	**g** $\lvert\mathbf{a} + \mathbf{b}\rvert$	**h** $\lvert\mathbf{a}\rvert + \lvert\mathbf{b}\rvert$

15 If $\mathbf{c} = \begin{pmatrix} 3 \\ 4 \end{pmatrix}$ and $\mathbf{d} = \begin{pmatrix} -1 \\ 0 \end{pmatrix}$, find

a $\mathbf{c} + \mathbf{d}$	**b** $\mathbf{c} - \mathbf{d}$	**c** $\mathbf{d} - \mathbf{c}$	**d** $2\mathbf{c} + \mathbf{d}$
e $\mathbf{c} + 2\mathbf{d}$	**f** $\mathbf{c} - 2\mathbf{d}$	**g** $\lvert\mathbf{c} - 2\mathbf{d}\rvert$	**h** $\lvert 2\mathbf{d} - \mathbf{c}\rvert$

16 If $\mathbf{a} = \begin{pmatrix} 2 \\ 7 \end{pmatrix}$ and $\mathbf{b} = \begin{pmatrix} -2 \\ 3 \end{pmatrix}$, find the exact magnitudes of

a $\mathbf{a}$	**b** $\mathbf{b}$	**c** $2\mathbf{a}$	**d** $\mathbf{a} + \mathbf{b}$	**e** $\mathbf{a} - \mathbf{b}$

17 The wing of an aircraft in flight experiences a force of 4000 N acting upwards at 20° to the vertical. Find

   a   the magnitude of the vertical component of this force (called the lift),

   b   the magnitude of the horizontal component of this force (called the drag).

Find the resultant for each set of vectors shown in numbers **18** to **23** giving your answers in the form $a\mathbf{i} + b\mathbf{j}$, where **i** is a unit vector due east and **j** is a unit vector due north. (Give a and b correct to one decimal place).

18

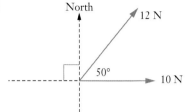

19

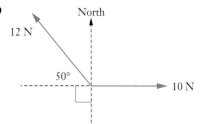

20

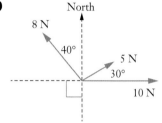

21

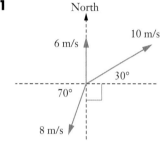

22

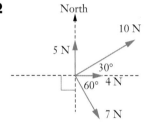

23

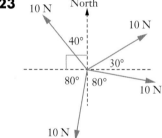

24 Forces $\mathbf{F}_1$, $\mathbf{F}_2$ and $\mathbf{F}_3$ act on a body. If $\mathbf{F}_1 = (2\mathbf{i} + 3\mathbf{j})$ N, $\mathbf{F}_2 = (4\mathbf{i} + 3\mathbf{j})$ N and $\mathbf{F}_3 = (2\mathbf{i} - 4\mathbf{j})$ N, find the magnitude of the resultant force acting on the body.

25 Find **a** and **b** if $\mathbf{a} + \mathbf{b} = 3\mathbf{i} + \mathbf{j}$ and $\mathbf{a} - \mathbf{b} = \mathbf{i} - 7\mathbf{j}$.

26 Find **c** and **d** if $2\mathbf{c} + \mathbf{d} = -\mathbf{i} + 6\mathbf{j}$ and $2(\mathbf{c} + \mathbf{d}) = 2\mathbf{i} - 10\mathbf{j}$.

Further examples

EXAMPLE 2

If $\mathbf{a} = 2\mathbf{i} + 3\mathbf{j}$, $\mathbf{b} = 3\mathbf{i} - 4\mathbf{j}$ and $\mathbf{c} = x\mathbf{i} + \mathbf{j}$ find.

a a vector in the same direction as **a** but twice the magnitude of **a**,

b a unit vector in the same direction as **a**,

c a vector in the same direction as **a** but the same magnitude as **b**,

d the possible values of x if $|\mathbf{c}| = |\mathbf{a}|$.

Solution

a Any vector in the same direction as **a** will be a positive scalar multiple of **a**.

To be twice the magnitude the scalar multiple must be 2.

Thus the required vector is 2**a**, i.e. $4\mathbf{i} + 6\mathbf{j}$.

b **a** has magnitude $\sqrt{2^2 + 3^2}$

$$= \sqrt{13} \text{ units}$$

Thus the unit vector, in the same direction as **a**, would be $\dfrac{1}{\sqrt{13}}(2\mathbf{i} + 3\mathbf{j})$

i.e. $\dfrac{2}{\sqrt{13}}\mathbf{i} + \dfrac{3}{\sqrt{13}}\mathbf{j}$

or, with rationalised denominators

$$\dfrac{2\sqrt{13}}{13}\mathbf{i} + \dfrac{3\sqrt{13}}{13}\mathbf{j}$$

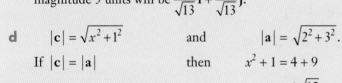

$$\text{unitV}\left(\begin{bmatrix} 2 \\ 3 \end{bmatrix}\right)$$

$$\begin{bmatrix} \dfrac{2 \cdot \sqrt{13}}{13} \\ \dfrac{3 \cdot \sqrt{13}}{13} \end{bmatrix}$$

c $|\mathbf{b}| = \sqrt{3^2 + (-4)^2}$

$$= 5 \text{ units.}$$

Thus, using our answer for part **b**, the vector in the same direction as **a** but of

magnitude 5 units will be $\dfrac{10}{\sqrt{13}}\mathbf{i} + \dfrac{15}{\sqrt{13}}\mathbf{j}$.

d $|\mathbf{c}| = \sqrt{x^2 + 1^2}$ and $|\mathbf{a}| = \sqrt{2^2 + 3^2}$.

If $|\mathbf{c}| = |\mathbf{a}|$ then $x^2 + 1 = 4 + 9$

$$\therefore \qquad x = \pm\sqrt{12}.$$

The possible values of x are $\pm 2\sqrt{3}$.

Note: A **unit vector** in the direction of vector **a** is sometimes written $\hat{\mathbf{a}}$.

Thus $\hat{\mathbf{a}} = \dfrac{\mathbf{a}}{|\mathbf{a}|}$. Similarly $\hat{\mathbf{b}} = \dfrac{\mathbf{b}}{|\mathbf{b}|}$, $\hat{\mathbf{c}} = \dfrac{\mathbf{c}}{|\mathbf{c}|}$ etc.

WS

Unit vectors

EXAMPLE 3

A body is moving with velocity $(7\mathbf{i} + 24\mathbf{j})$ m/s. How far will it travel in twenty seconds?

Solution

If the velocity $= (7\mathbf{i} + 24\mathbf{j})$ m/s

$$\text{speed} = |\text{velocity}|$$
$$= |7\mathbf{i} + 24\mathbf{j}|$$
$$= 25 \text{ m/s}.$$

Thus in twenty seconds the body will travel 500 metres.

EXAMPLE 4

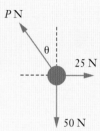

The forces acting on a body are as shown in the diagram. If the body is in equilibrium find P and θ.

Solution

If the body is in equilibrium there can be no 'surplus' force in any direction because, if there was, the body would tend to move in that direction.

The horizontal forces must balance. $\qquad \therefore P\sin\theta = 25 \qquad$ [1]

The vertical forces must balance. $\qquad \therefore P\cos\theta = 50 \qquad$ [2]

Dividing [1] by [2] gives $\qquad\qquad \tan\theta = 0.5$

$$\theta \approx 26.6°$$

Substituting for θ into [1] gives $\qquad\qquad P \approx 55.9$

If the body is in equilibrium then $P \approx 56$ and $\theta \approx 27°$.

Alternatively, we could approach example 4 as follows:

The resultant of the 25 N force and the 50 N force is a vector $\mathbf{a}$, see diagram on the right.

Thus the force of P N must be $-\mathbf{a}$ to counteract the effect of $\mathbf{a}$ and reduce the system to equilibrium.

By Pythagoras: $\qquad\qquad P = \sqrt{25^2 + 50^2}$
i.e. $\qquad\qquad\qquad P \approx 56$

By trigonometry: $\qquad \tan\theta = \dfrac{25}{50}$

i.e. $\qquad\qquad\qquad \theta \approx 27°$

If the body is in equilibrium then $P \approx 56$ and $\theta \approx 27°$, as before.

EXAMPLE 5

Airports A and B are such that $\overrightarrow{AB} = (600\mathbf{i} + 200\mathbf{j})$ km. An aircraft is to be flown directly from A to B. The aircraft can maintain a steady speed of 390 km/h in still air. There is a wind blowing with velocity $(30\mathbf{i} - 20\mathbf{j})$ km/h.

Find, in the form $a\mathbf{i} + b\mathbf{j}$, the velocity vector the pilot should set so that this velocity, together with the wind, causes the plane to travel directly from A to B.

Solution

The resultant of the planes velocity due to its engines, $(a\mathbf{i} + b\mathbf{j})$ km/h, and the wind, $(30\mathbf{i} - 20\mathbf{j})$ km/h, must be along $\overrightarrow{AB}$. i.e. the resultant must be a positive scalar multiple of $600\mathbf{i} + 200\mathbf{j}$.

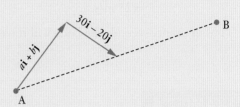

Thus	$a\mathbf{i} + b\mathbf{j} + 30\mathbf{i} - 20\mathbf{j} = \lambda(600\mathbf{i} + 200\mathbf{j})$ $\lambda > 0$	
Equating the $\mathbf{i}$ components	$a + 30 = 600\lambda$	[1]
Equating the $\mathbf{j}$ components	$b - 20 = 200\lambda$	[2]
Dividing [1] by [2]	$\dfrac{a+30}{b-20} = \dfrac{600}{200}$	
i.e.	$a = 3b - 90$	[3]
We also know that	$|a\mathbf{i} + b\mathbf{j}| = 390$ i.e. $a^2 + b^2 = 390^2$	[4]
From [3] and [4]	$(3b - 90)^2 + b^2 = 390^2$	
Using a calculator to solve this equation gives	$b = 150$ or $b = -96$	

But λ must be positive so, from [2], $b \neq -96$.

(Note: $b = -96$ gives a resultant in direction $\overrightarrow{BA}$, not $\overrightarrow{AB}$ as required.)

If $b = 150$, $a = 360$ i.e.   velocity $= 360\mathbf{i} + 150\mathbf{j}$.

The required velocity is $(360\mathbf{i} + 150\mathbf{j})$ km/h.

Note:
- The above example could be solved by instead changing the given vectors to magnitude and direction form and then solving as in chapter 3.

- To avoid some of the algebra we could use the ability of some calculators to solve equations [1], [2] and [4] simultaneously to determine a, b and λ.

- The justification for 'equating the $\mathbf{i}$ components' and 'equating the $\mathbf{j}$ components' is given below:

  If $a\mathbf{i} + b\mathbf{j} = c\mathbf{i} + d\mathbf{j}$ then $a\mathbf{i} - c\mathbf{i} = d\mathbf{j} - b\mathbf{j}$
  i.e. $(a - c)\mathbf{i} = (d - b)\mathbf{j}$

  As $\mathbf{i}$ and $\mathbf{j}$ are not parallel it follows that $a - c = 0$ and $d - b = 0$.
  i.e. $a = c$ and $d = b$.

We usually choose to express vectors in terms of the horizontal unit vector, **i**, and vertical unit vector, **j**, because these are convenient directions. However any two non-parallel vectors can be chosen as base vectors if horizontal and vertical are not convenient.

EXAMPLE 6

Using $\mathbf{a} = 2\mathbf{i} + 3\mathbf{j}$ and $\mathbf{b} = 4\mathbf{i} - \mathbf{j}$ as base vectors, express each of the following in the form $\lambda\mathbf{a} + \mu\mathbf{b}$.

a $5\mathbf{i} + 3\mathbf{j}$

b $6\mathbf{i} - 4\mathbf{j}$

Solution

a Let

$5\mathbf{i} + 3\mathbf{j} = \lambda\mathbf{a} + \mu\mathbf{b}$

i.e.

$5\mathbf{i} + 3\mathbf{j} = \lambda(2\mathbf{i} + 3\mathbf{j}) + \mu(4\mathbf{i} - \mathbf{j})$

Equating the **i** components

$5 = 2\lambda + 4\mu$ [1]

Equating the **j** components

$3 = 3\lambda - \mu$ [2]

Solving [1] and [2] simultaneously:

$\lambda = \dfrac{17}{14}$ and $\mu = \dfrac{9}{14}$

Thus

$5\mathbf{i} + 3\mathbf{j} = \dfrac{17}{14}\mathbf{a} + \dfrac{9}{14}\mathbf{b}$

b Let

$6\mathbf{i} - 4\mathbf{j} = \lambda\mathbf{a} + \mu\mathbf{b}$

i.e.

$6\mathbf{i} - 4\mathbf{j} = \lambda(2\mathbf{i} + 3\mathbf{j}) + \mu(4\mathbf{i} - \mathbf{j})$

Equating the **i** components

$6 = 2\lambda + 4\mu$ [1]

Equating the **j** components

$-4 = 3\lambda - \mu$ [2]

Solving simultaneously

$\lambda = -\dfrac{5}{7}$ and $\mu = \dfrac{13}{7}$

Thus

$6\mathbf{i} - 4\mathbf{j} = -\dfrac{5}{7}\mathbf{a} + \dfrac{13}{7}\mathbf{b}$

Exercise 4B

1 For *each* of the vectors **a** to **d** shown below, write in the form $h\mathbf{i} + k\mathbf{j}$

  i   the vector itself,

  ii   a vector in the same direction as the given vector but twice as long,

  iii a unit vector in the same direction as the given vector,

  iv   a vector in the same direction as the given vector but of length 2 units.

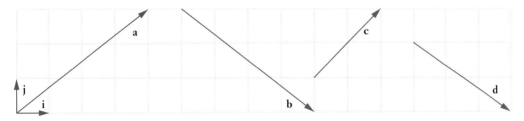

2 Given that $\mathbf{a} = -3\mathbf{i} + 4\mathbf{j}$, $\mathbf{b} = 2\mathbf{i} + \mathbf{j}$ and $\mathbf{c} = 3\mathbf{i} - 2\mathbf{j}$ find

   a   a unit vector in the same direction as $\mathbf{b}$,

   b   a vector in the same direction as $\mathbf{b}$ but equal in magnitude to $\mathbf{a}$,

   c   a vector in the same direction as $\mathbf{a}$ but equal in magnitude to $\mathbf{c}$,

   d   a vector in the same direction as the resultant of $\mathbf{a}$, $\mathbf{b}$ and $\mathbf{c}$ but equal in magnitude to $\mathbf{a}$.

3 For this question $\mathbf{a} = 2\mathbf{i} - 4\mathbf{j}$, $\mathbf{b} = 4\mathbf{i} + 2\mathbf{j}$, $\mathbf{c} = \mathbf{i} - 8\mathbf{j}$, $\mathbf{d} = \mathbf{i} - 2\mathbf{j}$ and $\mathbf{e} = 4\mathbf{i} - 2\mathbf{j}$.

   a   Which of these vectors are parallel to each other?

   b   Find the resultant of $\mathbf{a}$, $\mathbf{b}$, $\mathbf{c}$, $\mathbf{d}$ and $\mathbf{e}$.

   c   Find the magnitude of this resultant.

   d   Taking the direction of the unit vector $\mathbf{j}$ as due North and $\mathbf{i}$ as due East, express the direction of the resultant as a bearing, measured clockwise from North and to the nearest degree.

4 For this question $\mathbf{a} = w\mathbf{i} + 3\mathbf{j}$, $\mathbf{b} = -\mathbf{i} + x\mathbf{j}$, $\mathbf{c} = 0.5\mathbf{i} + y\mathbf{j}$ and $\mathbf{d} = -\mathbf{i} - z\mathbf{j}$.

Find the possible values of w, x, y and z given that all of the following are true:

- $\mathbf{a}$ is of magnitude 5 units and w is negative,
- $\mathbf{b}$ is parallel to $\mathbf{a}$,
- $\mathbf{c}$ is a unit vector,
- the resultant of $\mathbf{a}$ and $\mathbf{d}$ has magnitude 13 units.

5 For this question $\mathbf{p} = 0.6\mathbf{i} - a\mathbf{j}$, $\mathbf{q} = b\mathbf{i} + c\mathbf{j}$, $\mathbf{r} = d\mathbf{i} + e\mathbf{j}$ and $\mathbf{s} = f\mathbf{i} + g\mathbf{j}$.

Find the possible values of a, b, c, d, e, f and g given that all of the following are true:

- $\mathbf{p}$ is a unit vector and a is a positive constant,
- $\mathbf{q}$ is in the same direction as $\mathbf{p}$ and five times the magnitude,
- $\mathbf{r} + 2\mathbf{q} = 11\mathbf{i} - 20\mathbf{j}$,
- $\mathbf{s}$ is in the same direction as $\mathbf{r}$ but equal in magnitude to $\mathbf{q}$.

6 Find, correct to one decimal place, the magnitude of $\mathbf{R}$ given that $\mathbf{R}$ is the resultant of $\mathbf{a}$, $\mathbf{b}$, $\mathbf{c}$ and $\mathbf{d}$ shown below.

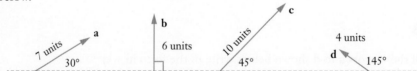

Express in the form $\lambda\mathbf{i} + \mu\mathbf{j}$ the vector $\mathbf{e}$ such that $\mathbf{a} + \mathbf{b} + \mathbf{c} + \mathbf{d} + \mathbf{e} = 0$. (Give λ and μ correct to one decimal place.)

Find P, correct to one decimal place, and θ, to the nearest degree, in each of the following if in each case the body is in equilibrium.

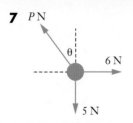

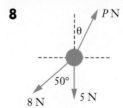

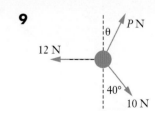

Each of the following diagrams shows a body of weight 100 N, supported in equilibrium by two wires. Find the magnitudes of T_1 and T_2, the tensions in the wires.

10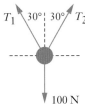

11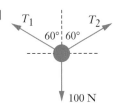

12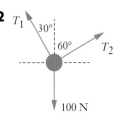

13 Two particles, A and B, have velocities of $(21\mathbf{i} + 17\mathbf{j})$ m/s and $(26\mathbf{i} - 2\mathbf{j})$ m/s respectively. Which particle is moving the fastest?

14 How far will a body moving with a constant velocity of $(5\mathbf{i} - 2\mathbf{j})$ m/s travel in one minute?

15 A helicopter can fly at 75 m/s in still air. The pilot wishes to fly from airport A to a second airport B, 300 km due North of A. If $\mathbf{i}$ is a unit vector due East and $\mathbf{j}$ a unit vector due North, find (in the form $a\mathbf{i} + b\mathbf{j}$) the velocity vector that the pilot should set and the time the journey will take if

a there is no wind blowing,

b there is a wind of $(21\mathbf{i} + 10\mathbf{j})$ m/s blowing.

16 The helicopter of **question 15** now wishes to return from B to A. If the wind of $(21\mathbf{i} + 10\mathbf{j})$ m/s still blows what velocity vector should the pilot now set and how long will the journey take?

17 An engineer finds that for a lot of her work she uses isometric paper. For her, the usual base vectors → **i** and ↑ **j** are not so useful. Instead she expresses vectors in terms of two base vectors that she calls **p** and **q** (see the diagram on the right).

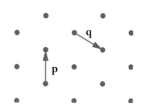

Express each of the vectors shown below in terms of **p** and/or **q**.

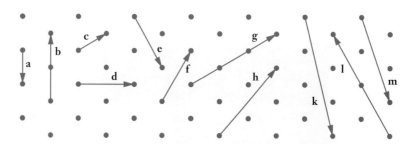

18 The diagram shows a particle of weight 10 N on an inclined plane that is angled at 30° to the horizontal. Express this weight as a vector $(a\mathbf{i} + b\mathbf{j})$ N where **i** and **j** are unit vectors down and perpendicular to the slope respectively (see diagram).

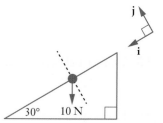

19 An unconventional mathematician decides not to use the horizontal and vertical unit vectors **i** and **j** as base vectors but instead wishes to express all vectors in terms of new base vectors **a** and **b**, where:

$$\mathbf{a} = 2\mathbf{i} + 3\mathbf{j} \qquad \text{and} \qquad \mathbf{b} = \mathbf{i} - \mathbf{j}.$$

Express each of the following in the form $x\mathbf{a} + y\mathbf{b}$.

a $3\mathbf{i} + 2\mathbf{j}$ **b** $5\mathbf{i} + 5\mathbf{j}$ **c** $\mathbf{i} + 9\mathbf{j}$

d $4\mathbf{i} + 7\mathbf{j}$ **e** $3\mathbf{i} - \mathbf{j}$ **f** $3\mathbf{i} + 7\mathbf{j}$

20 Airports A and B are such that $\overrightarrow{AB} = (-250\mathbf{i} + 750\mathbf{j})$ km.

An aircraft is to be flown directly from A to B. In still air the aircraft can maintain a steady speed of 400 km/h. There is a wind blowing with velocity $(-13\mathbf{i} - 9\mathbf{j})$ km/h.

a Find, in the form $a\mathbf{i} + b\mathbf{j}$, the velocity vector the pilot should set so that this velocity, together with the wind, causes the plane to travel directly from A to B.

b If the wind remains unchanged find, in the form $a\mathbf{i} + b\mathbf{j}$, the velocity vector the pilot should now set to return directly from B to A.

ISBN 9780170390477

Position vectors

Consider the points O, A, B, C, D and P shown in the diagram.

The vectors $\overrightarrow{OP}$, $\overrightarrow{AB}$ and $\overrightarrow{CD}$ are each $3\mathbf{i} + \mathbf{j}$.

However, with O as the origin, only point P has a **position vector** of $3\mathbf{i} + \mathbf{j}$.

Point A has position vector $\mathbf{i} + 2\mathbf{j}$,

Point B has position vector $4\mathbf{i} + 3\mathbf{j}$,

Point C has position vector $4\mathbf{j}$,

Point D has position vector $3\mathbf{i} + 5\mathbf{j}$.

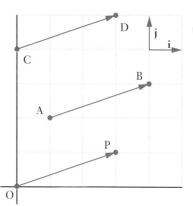

Displacement vectors

Addition of vectors using components

EXAMPLE 7

Points A and B have position vectors $2\mathbf{i} + 3\mathbf{j}$ and $5\mathbf{i} - \mathbf{j}$ respectively. Find $\overrightarrow{AB}$.

Solution

Initially draw a rough sketch of the situation:

From the diagram
$$\overrightarrow{AB} = \overrightarrow{AO} + \overrightarrow{OB}$$
$$= -\overrightarrow{OA} + \overrightarrow{OB}$$
$$= -(2\mathbf{i} + 3\mathbf{j}) + (5\mathbf{i} - \mathbf{j})$$
$$= 3\mathbf{i} - 4\mathbf{j}$$

Thus $\overrightarrow{AB} = 3\mathbf{i} - 4\mathbf{j}$

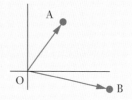

EXAMPLE 8

At 1 p.m. a ship is at a location A, position vector $(2\mathbf{i} + 8\mathbf{j})$ km, and is moving with velocity $(5\mathbf{i} - \mathbf{j})$ km/h. If the ship continues with this velocity what will be its position vector at 4 p.m.?

Solution

Suppose the ship is at point B at 4 p.m. (see diagram).

Then
$$\overrightarrow{OB} = \overrightarrow{OA} + \overrightarrow{AB}$$
$$= (2\mathbf{i} + 8\mathbf{j}) + 3(5\mathbf{i} - \mathbf{j})$$
$$= 17\mathbf{i} + 5\mathbf{j}$$

By 4 p.m. the ship will be at the point with position vector $(17\mathbf{i} + 5\mathbf{j})$.

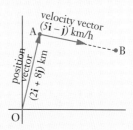

Points A and B have position vectors $\mathbf{i} + 7\mathbf{j}$ and $10\mathbf{i} + 4\mathbf{j}$ respectively. Find the position vector of the point that divides AB internally in the ratio 1 : 2.

(Note: If a point P divides AB in the ratio 1 : 2 then AP : PB = 1 : 2.)

Solution

We require the position vector of the point P (see diagram) where

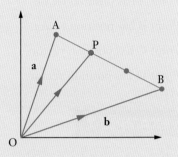

$$AP : PB = 1 : 2$$

$$\overrightarrow{OP} = \overrightarrow{OA} + \overrightarrow{AP}$$

$$= \overrightarrow{OA} + \frac{1}{3}\overrightarrow{AB}$$

$$= \mathbf{a} + \frac{1}{3}(-\mathbf{a} + \mathbf{b})$$

$$= \frac{2}{3}\mathbf{a} + \frac{1}{3}\mathbf{b}$$

$$= \frac{2}{3}(\mathbf{i} + 7\mathbf{j}) + \frac{1}{3}(10\mathbf{i} + 4\mathbf{j})$$

$$= 4\mathbf{i} + 6\mathbf{j}$$

The point dividing AB internally in the ratio 1 : 2 has position vector $4\mathbf{i} + 6\mathbf{j}$.

Exercise 4C

1 Points A, B, C and D have cartesian coordinates (2, 5), (−3, 6), (0, −5) and (3, 8) respectively. Give the position vectors of each point in the form $a\mathbf{i} + b\mathbf{j}$ where $\mathbf{i}$ is a unit vector in the direction of the positive x-axis and $\mathbf{j}$ is a unit vector in the direction of the positive y-axis.

2 Points A and B have position vectors $3\mathbf{i} + \mathbf{j}$ and $2\mathbf{i} − \mathbf{j}$ respectively.

Express in the form $a\mathbf{i} + b\mathbf{j}$.   **a** $\overrightarrow{AB}$ **b** $\overrightarrow{BA}$

3 Points A, B and C have position vectors $−\mathbf{i} + 4\mathbf{j}$, $2\mathbf{i} − 3\mathbf{j}$ and $\mathbf{i} + 5\mathbf{j}$ respectively.

Express in the form $a\mathbf{i} + b\mathbf{j}$.   **a** $\overrightarrow{AB}$ **b** $\overrightarrow{BC}$ **c** $\overrightarrow{CA}$

4 Points A, B, C and D have position vectors $\mathbf{i} + 2\mathbf{j}$, $4\mathbf{i} − 2\mathbf{j}$, $−\mathbf{i} + 11\mathbf{j}$ and $6\mathbf{i} − 13\mathbf{j}$ respectively.

Find   **a** $\overrightarrow{AB}$ **b** $\overrightarrow{BC}$ **c** $\overrightarrow{CD}$

       d   a vector in the same direction as $\overrightarrow{AB}$ but equal in magnitude to $\overrightarrow{CD}$.

5 With respect to an origin O, points A and B have position vectors $3\mathbf{i} + 7\mathbf{j}$ and $−2\mathbf{i} + \mathbf{j}$ respectively.

Find   **a** $|\overrightarrow{OA}|$ **b** $|\overrightarrow{OB}|$ **c** $|\overrightarrow{AB}|$

6 Points A, B and C have position vectors $2\mathbf{i} + 3\mathbf{j}$, $5\mathbf{i} − \mathbf{j}$ and $3\mathbf{i} + 7\mathbf{j}$ respectively.

Find   **a** $|\overrightarrow{AB}|$ **b** $|\overrightarrow{BA}|$ **c** $|\overrightarrow{AC}|$ **d** $|\overrightarrow{BC}|$

7 Points A and B have position vectors $-\mathbf{i} + 6\mathbf{j}$ and $5\mathbf{i} + 3\mathbf{j}$ respectively.

How far is   **a**   A from the origin?   **b**   B from the origin?   **c**   A from B?

8 With respect to an origin O, points A, B, C and D have position vectors $2\mathbf{i} - 3\mathbf{j}$, $\mathbf{i} + 2\mathbf{j}$, $9\mathbf{i} + 21\mathbf{j}$ and $6\mathbf{i} - 2\mathbf{j}$ respectively.

Find   **a** $\overrightarrow{AB}$ in component form,   **b** $\overrightarrow{BC}$ in component form,

       c $\overrightarrow{CD}$ in component form,   **d** $|\overrightarrow{AC}|$,

       e $\overrightarrow{OA} + \overrightarrow{AB}$ in component form,   **f** $\overrightarrow{OA} + 2\overrightarrow{AC}$, in component form.

9 If point A has position vector $3\mathbf{i} + 4\mathbf{j}$ and $\overrightarrow{AB} = 7\mathbf{i} - \mathbf{j}$ find the position vector of point B.

10 Point A has position vector $-\mathbf{i} + 7\mathbf{j}$, $\overrightarrow{AB} = 2\mathbf{i} + 3\mathbf{j}$ and $\overrightarrow{AC} = 4\mathbf{i} - 3\mathbf{j}$.

Find   **a**   the position vector of point B,   **b**   the position vector of point C,

       c $\overrightarrow{BC}$.

11 Point A has position vector $-\mathbf{i} + 9\mathbf{j}$. Point C has position vector $7\mathbf{i} - \mathbf{j}$.

Points B and D are such that $\overrightarrow{BC} = 4\mathbf{i} - 6\mathbf{j}$ and $\overrightarrow{DC} = 3\mathbf{i} + 2\mathbf{j}$.

Find   **a**   the position vector of point B,   **b**   the position vector of point D,

       c $\overrightarrow{BD}$,   **d** $|\overrightarrow{AD}|$.

12 A particle has an initial position vector of $(2\mathbf{i} + 9\mathbf{j})$ m with respect to an origin O. If the particle moves with a constant velocity of $(2\mathbf{i} - 5\mathbf{j})$ m/s what will be its position vector after

  a   1 second?   **b**   2 seconds?   **c**   10 seconds?

How far is the particle from the origin after five seconds?

13 A particle has an initial position vector of $(5\mathbf{i} - 6\mathbf{j})$ m with respect to an origin O. If the particle moves with a constant velocity of $(\mathbf{i} + 6\mathbf{j})$ m/s,

  a   what will be its position vector after

       i   2 seconds?   **ii**   3 seconds?   **iii**   7 seconds?

  b   How far is the particle from the origin after five seconds?

  c   After how many seconds will the particle be 50 metres from the origin?

14 Points A, B and C have position vectors $3\mathbf{i} - \mathbf{j}$, $-\mathbf{i} + 15\mathbf{j}$ and $9\mathbf{i} - 25\mathbf{j}$ respectively. Use vectors to prove that A, B and C are collinear.

15 Points D, E and F have position vectors $9\mathbf{i} - 7\mathbf{j}$, $-11\mathbf{i} + 8\mathbf{j}$ and $25\mathbf{i} - 19\mathbf{j}$ respectively. Use vectors to prove that D, E and F are collinear.

16 Points A and B have position vectors $2\mathbf{i} + 5\mathbf{j}$ and $12\mathbf{i} + 10\mathbf{j}$ respectively. Find the position vector of the point dividing AB internally in the ratio $4:1$.

17 Points A and B have position vectors $-2\mathbf{i} + 2\mathbf{j}$ and $10\mathbf{i} - \mathbf{j}$ respectively. Find the position vector of the point that divides AB internally in the ratio $1:2$.

18 Points A and B have position vectors $\mathbf{i} + 8\mathbf{j}$ and $19\mathbf{i} + 2\mathbf{j}$ respectively. Find the position vector of the point that divides AB internally in the ratio $1:4$.

Miscellaneous exercise four

This miscellaneous exercise may include questions involving the work of this chapter, the work of any previous chapters, and the ideas mentioned in the Preliminary work section at the beginning of the book.

1 If $\mathbf{a} = 2\mathbf{i} + 3\mathbf{j}$, $\mathbf{b} = 3\mathbf{i} - 4\mathbf{j}$ and $\mathbf{c} = 2\mathbf{i} + \mathbf{j}$ find λ and μ such that $\mathbf{c} = \lambda\mathbf{a} + \mu\mathbf{b}$.

2 Two or three letter codes are to be formed using the letters A, B, C, D, E, F.

Each code must not use the same letter more than once. How many such codes are possible?

3 If a positive whole number ends in a five, then the number is a multiple of five.

Write the converse statement and the contrapositive statement and in each case state whether it is true or false.

4 A careers awareness form requires a student to choose five careers from a list of twelve and write them in order of preference, putting most preferred first to least preferred last.
How many different ordered lists are possible?

5 From the triangles shown below, find four pairs that must be congruent, stating the reason each time. (Diagrams not necessarily drawn to scale.)

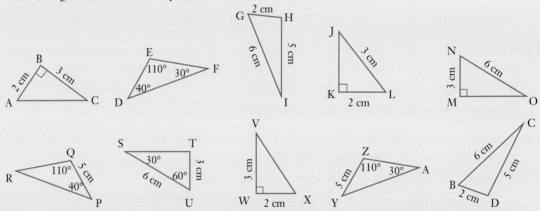

6 Points A and B have position vectors $a\mathbf{i} - 15\mathbf{j}$ and $10\mathbf{i} + b\mathbf{j}$ respectively. The point C, position vector $4\mathbf{i} - 3\mathbf{j}$, lies on AB, between A and B, and is such that $AC:CB = 2:3$. Find the values of a and b.

7 How many four or five digit numbers greater than 5000 can be made using some or all of the digits 1, 2, 3, 4 and 5 if

a a digit cannot appear more than once in a number?

b a digit can appear more than once in a number?

8 In a cricket match the batsman is at a point O and he hits the ball with a velocity of $(7\mathbf{i} + 24\mathbf{j})$ m/s. A fielder at point A, position vector $2\mathbf{i} + 3\mathbf{j}$ relative to O, does not move at all as the ball passes by. Assuming the ball suffers no change in velocity:

a how long will the ball take to reach the boundary, 60 metres from O?

b find the least distance between the fielder at point A and the ball during its journey to the boundary.

5.

Geometric proofs

- Proof
- Definitions, axioms and theorems
- Circle properties
- Angles in circles
- Tangents and secants
- Miscellaneous exercise five

Proof

An attempt to prove a statement to be true involves establishing and presenting sufficient evidence to convince others of the fact that the statement is true. In a courtroom situation the prosecuting team attempts to present sufficient evidence to convince the members of the jury that the defendant is guilty. The defence team, on the other hand, attempts to cause the jury members to doubt the claim that the defendant is guilty.

Similarly, in mathematics if we wish to prove a statement true, for example that the angles of a triangle add up to 180°, we must present evidence to convince others of the truth of the statement. The *Preliminary work*, and some questions in *Miscellaneous exercises one* and *two* involved the use of similar triangles and congruent triangles to prove various things to be true.

Definition, axioms and theorems

Some statements do not need to be proved because they are true *by definition*.

For example, if we define 1 centimetre (1 cm) to be one 100th part of a metre it follows from this definition that $100 \text{ cm} = 1$ metre.

If we define 1 degree (1°) to be one 90th part of a right angle it follows from this definition that $90° = 1$ right angle.

From basic definitions other true statements can follow. For example, if we define the angle of a straight line to equal two right angles it follows that with a and b as shown in the diagram on the right

$$a + b = \text{two right angles}$$

and so
$$a + b = 2 \times 90°$$
$$= 180°$$

There are some statements that are simply accepted as being true without the need for proof. Such statements are called **axioms**.

For example we accept the statement below as being true without proof.

> *There is only one straight line that can be drawn to join two specific points in space*

The statement is an axiom. It is simply accepted as fact.

Theorems are statements that can be proved to be true using accepted definitions, axioms and other (proven) theorems. The truth of the theorem is arrived at by reasoning from other accepted truths. It is deduced. In this **proof by deduction** the validity of the proof depends upon the correctness of the axioms and theorems used to deduce it.

For example, suppose we accept the following two statements as fact:

Angles that together form a straight line have a sum of 180°.

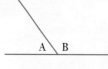

$$A + B = 180°$$

When a transversal cuts parallel lines, corresponding angles are equal.

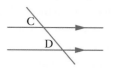

$$C = D$$

it is then possible to prove the following statements true, for example:

- When two straight lines intersect, the vertically opposite angles are equal.

  I.e., in the diagram on the right, A = B.

- When a transversal cuts parallel lines, alternate angles are equal.

  I.e., in the diagram on the right, C = D.

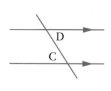

The proofs of these two statements, given the bold statements at the top of the page, follow as examples 1 and 2.

EXAMPLE 1

Proof of: When two straight lines intersect, the vertically opposite angles are equal.

Solution

Given: Two straight lines AB and CD intersecting at E.

To prove: Vertically opposite angles are equal.

I.e., in the diagram on the right,

$$\angle AED = \angle BEC$$

and $$\angle AEC = \angle DEB.$$

Proof: $\angle AED + \angle AEC = 180°$ (Angle sum of straight line DC.)

∴ $\angle AED = 180° - \angle AEC$

$\angle BEC + \angle AEC = 180°$ (Angle sum of a straight line AB.)

∴ $\angle BEC = 180° - \angle AEC$

Hence $\angle AED = \angle BEC$ as required.   (Each equal to $180° - \angle AEC$.)

Also $\angle AEC = 180° - \angle AED$ (Angle sum of straight line DC.)

$\angle DEB = 180° - \angle AED$ (Angle sum of straight line AB.)

Hence $\angle AEC = \angle DEB$ as required.   (Each equal to $180° - \angle AED$.)

ISBN 9780170390477

Notice that as was the case with the proofs given in the *Preliminary work*, the previous proof features

- a clear statement of what is 'given'
- a statement of what it is that we are attempting 'to prove'
- a clear diagram
- reference to known truths in order to justify a statement.

Note also that as was mentioned in the *Preliminary work* section, these justifications do not need to be essays but are instead brief statements that clearly indicate which truth justifies the statement.

EXAMPLE 2

Proof of: When a transversal cuts parallel lines, alternate angles are equal.

Solution

Given: A transversal cutting a pair of parallel lines with angles A, B, C, D, E, F, G and H as shown in the diagram on the right.

To prove: Alternate angles are equal.

I.e., in the diagram on the right, C = E

and D = F.

Proof:

		C = G	(Corresponding angles are equal.)
But		E = G	(Vertically opposite angles are equal, just proved.)
∴		C = E,	as required.
Similarly		D = H	(Corresponding angles are equal.)
But		F = H	(Vertically opposite angles are equal, just proved.)
∴		D = F,	as required.

Some proofs may end with the abbreviation QED rather than the 'as required' statement used above. QED stands for

quad erat demonstrandum

which is a Latin phrase meaning 'which was to be demonstrated'.

Parts of a circle

Circle properties

This chapter concentrates on proving and using various circle properties. In the examples and exercises that follow, the properties of straight lines and triangles mentioned or proven in earlier pages may be stated as fact, without proof.

In particular, see the following list.

- Angles that together form a straight line have a sum of 180°.
  (And, conversely, if angles have a sum of 180° then they together form a straight line.)

- When a transversal cuts parallel lines, corresponding angles are equal.
  (And conversely, if corresponding angles are equal then we have parallel lines.)

- When two straight lines intersect the vertically opposite angles are equal.
  (And conversely, if vertically opposite angles are equal then the intersecting lines are straight lines.)

- When a transversal cuts parallel lines, alternate angles are equal.
  (And conversely, if alternate angles are equal then we have parallel lines.)

- When a transversal cuts parallel lines, co-interior angles are supplementary.
  (And conversely, if co-interior angles are supplementary then we have parallel lines.)

- The angles of a triangle add up to 180°.
  (And conversely, if the angles of a polygon sum to 180° then the polygon is a triangle.)

- The angles of a quadrilateral add up to 360°.
  (And conversely, if the angles of a polygon sum to 360° then the polygon is a quadrilateral.)

- If a triangle has two sides of the same length then the angles opposite these sides are of equal size. I.e. in the isosceles triangle shown on the right, with $XY = XZ$ then the base angles, $\angle XYZ$ and $\angle XZY$, are equal.
  (And conversely, if a triangle has two angles of the same size then the sides opposite these angles are of equal length.)

Also, any fact proved in a previous question can be used as fact to prove any later question.

Don't forget that the concepts of similar triangles and congruent triangles can be useful for some proofs.

Angles in circles

Angles at the centre and circumference

The next two examples, and the exercise that follows, considers (and uses) some geometrical results concerning angles in circles.

First though, note the use of the word *subtends*:

Angles in semicircles and subtended by equal chords

> We say that the arc AB shown in the diagram on the right **subtends** the angle ACB at C.

I.e. straight lines drawn from the extremities of the arc, to the point C, form the angle ACB.

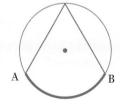

A page of circles

Examples 3 and 4 involve the following result:

The angle an arc subtends at the centre of a circle is twice the angle the same arc subtends at the circumference.

This result is referred to as:

Angle at the centre is twice the angle at the circumference.

EXAMPLE 3

Proof of the above statement.

Solution

Given:	Points A, B and C lying on the circumference of a circle centre O, as shown in the diagram.
To prove:	In the given diagram $\angle AOC = 2 \times \angle ABC$.
Construction:	Draw a line from B to pass through O to some point D. Let $\angle ABO = x°$ and $\angle CBO = y°$.
Proof:	$OA = OB$ (Radii)

$\therefore \triangle OAB$ is isosceles.

Thus $\angle OAB = \angle OBA = x°$ (Base angles of isosceles triangle)

$\therefore$ $\angle AOB = 180° - 2x°$ (Angle sum of a triangle)

and so $\angle AOD = 2x°$ (Angle of straight line BD)

Similar reasoning for $\triangle OCB$ gives

$\angle COD = 2y°$

Now $\angle AOC = \angle AOD + \angle COD$

$= 2x° + 2y°$

$= 2(x° + y°)$

$= 2 \times \angle ABC$, as required.

Note: Situations involving this rule may not always look quite like the diagram above. See the diagrams below for examples.

EXAMPLE 4

In the diagram on the right, point O is the centre of the circle and points A, B, C and D lie on the circle.

$$\angle CBO = 70°,$$
$$\angle COD = 90°$$
and $\qquad \angle BAD = x°.$

Prove that $\qquad x = 65.$

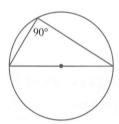

Solution

To prove: That for the given diagram, $x = 65$.

Proof: $\qquad\qquad\qquad$ OB = OC $\qquad$ (Radii)

$\therefore \triangle OBC$ is isosceles.

Thus $\qquad \angle OCB = 70°$ $\qquad$ (Base angles of isosceles triangle)

and $\qquad \angle BOC = 40°$ $\qquad$ (Angle sum of a triangle)

Now $\qquad \angle BOD = \angle BOC + \angle COD$

$$= 40° + 90°$$
$$= 130°$$

Hence $\qquad \angle BAD = 65°$ $\qquad$ (Angle at centre is twice angle at circumference)

Thus $\qquad x = 65$, as required.

Exercise 5A

1 Prove the following result:

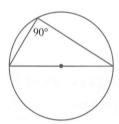

The angle a diameter subtends at the circumference is a right angle.

Or, as we tend to remember it:

Angles in a semicircle are right angles.

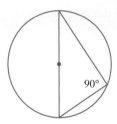 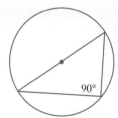 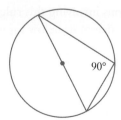

2 Prove the following result:

Angles that the minor arc AB subtends at points on the major arc AB are equal (see diagram I below) and, similarly, angles that the major arc AB subtends at points on the minor arc AB are equal (see diagram II below).

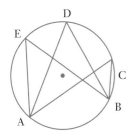

Diagram I

∠AEB = ∠ADB = ∠ACB

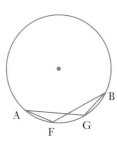

Diagram II

∠AFB = ∠AGB

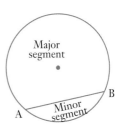

Diagram III

With the chord AB splitting the circle into a minor segment and a major segment (see diagram III above) we tend to refer to this fact as

Angles in the same segment are equal.

3 If the four vertices of a quadrilateral lie on the circumference of a circle we call the quadrilateral a **cyclic quadrilateral**.

Prove that:

The opposite angles of a cyclic quadrilateral add up to 180°.

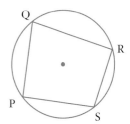

You will find the facts proved in example 3 and questions 1 to 3 of this exercise useful for questions 4 to 9. The facts are summarised below.

> The angle at the centre of a circle is twice the angle at the circumference.
>
> Angles in a semicircle are right angles.
>
> Angles in the same segment are equal.
>
> The opposite angles of a cyclic quadrilateral are supplementary.

4 In the diagram on the right points A, B and C lie on the circle centre O.

Given that ∠ABC = 65°

and ∠OAC = x°

prove that $x = 25$.

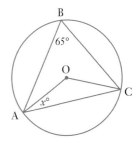

5 In the diagram on the right, points A, B, C and D lie on the circle centre O and AC is a diameter of the circle.

Given that $\angle ACB = 65°$

and $\angle BDC = x°$

prove that $x = 25$.

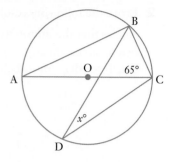

6 In the diagram on the right, points A, B C and D lie on the circle centre O.

E is a point outside the circle and ADE is a straight line.

Given that $BA = BC$

               $\angle BAC = 58°$

and $\angle CDE = x°$

prove that $x = 64$.

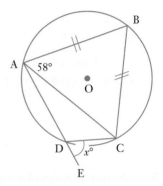

7 In the diagram on the right, points A, B and D lie on the circle centre O.

Point C lies inside the circle as shown in the diagram.

Given that $\angle DAB = 46°$

            $\angle ODC = 60°$

            $\angle OBC = 50°$

and $\angle DCB = x°$

prove that $x = 158$.

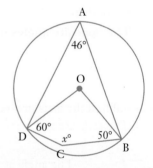

8 In the diagram on the right, points B, C, D and E lie on the circle centre O.

FCG is a straight line and $\angle FCB = 90°$.

CB produced (continued) meets DE produced at point A.

Given that $\angle BAE = 52°$

           $\angle CBE = 118°$

and $\angle DCG = x°$

prove that $x = 24$.

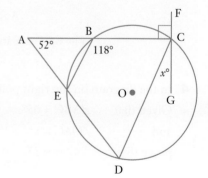

ISBN 9780170390477

9 In the diagram on the right, points A, B, C and D lie on the circle centre O, with BD as a diameter.

F is a point on AD produced (continued).

E is a point on CD produced.

AB is parallel to OC.

Given that $\angle ODC = 20°$

and $\angle EDF = x°$

prove that $x = 70$.

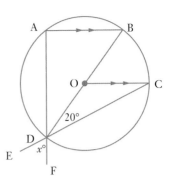

10 Prove that:

A line drawn from the centre of a circle to the midpoint of any chord of the circle meets that chord at right angles.

I.e., for the diagram shown on the right prove that
$$\angle OBA = \angle OBC = 90°.$$

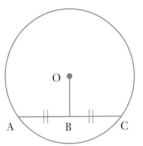

11 Prove that:

If a line is drawn perpendicular to any chord of a circle, and passing through the centre of the circle, then the line will bisect the chord.

I.e., for the diagram shown on the right prove that
$$DE = EF.$$

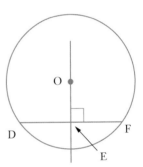

12 Prove that:

The perpendicular bisector of any chord of a circle passes through the centre of that circle.

I.e., for the diagram shown on the right prove that the line JK consists of points equidistant from G and I and one of these points must be the centre of the circle.

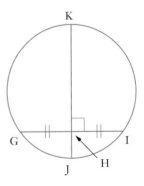

13 Prove that:

Chords of equal length subtend equal angles at the centre.

14 Prove that:

Chords that subtend equal angles at the centre of a circle are equal in length.

15 Prove that:

If any two chords of a circle, that we will call AB and CD, intersect at some point, that we will call E, then

$$AE \times EB = DE \times EC.$$

Hence determine x and y in the following:

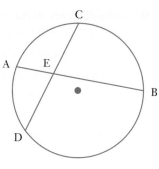

AB = 4 cm
AC = 10 cm
BE = 8 cm
DB = x cm

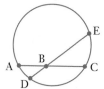

ST = 5 cm
TQ = 6 cm
PR = y cm
PT = 10 cm

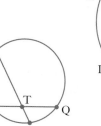

Tangents and secants

In each of the diagrams below, the straight line is a *tangent* to the circle. (From the Latin *tangere*, to touch.) Each tangent just *touches* the circle.

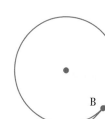

Tangent touches the circle at A.

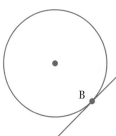

Tangent touches the circle at B.

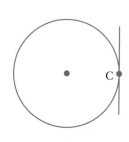

Tangent touches the circle at C.

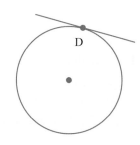

Tangent touches the circle at D.

The angle between a tangent and the radius drawn at the point of contact is a right angle.

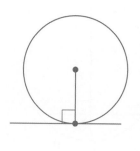

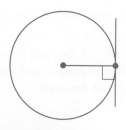

If a straight line cuts a circle at two distinct points it is a **secant** to the circle. (From the Latin *secare*, to cut.)

Thus in the diagram on the right the line AB is a secant of the circle shown, cutting it at points C and D.

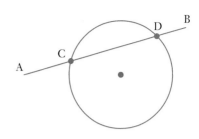

The 'angle between tangent and radius' statement on the previous page can be used in any questions that follow.

You may also quote, without proof, any of the 'circle facts' encountered earlier in this chapter, for example

$$\text{angle at centre} = 2 \times \text{angle at circumference}$$

(though you should be able to prove this statement if required).

 EXAMPLE 5

In the diagram on the right, AB is a tangent to the circle centre O, with C the point of contact. ED is a diameter of the circle.

Given that	$\angle EOC = 50°$
and	$\angle DCB = x°$
prove that	$x = 65$.

Solution

Given:	Diagram as shown.
To prove:	$x = 65$

Proof:

	$\angle ODC = 25°$	(Angle at centre = 2 × angle at circumference)
$\triangle OCD$ is isosceles.		(OD and OC are radii)
$\therefore$	$\angle OCD = 25°$	($\angle OCD = \angle ODC$, base angles of isosceles triangle)
But	$\angle OCD + \angle DCB = 90°$	(Angle between tangent and radius)
$\therefore$	$25 + x = 90$	
and so	$x = 65$, as required.	

Exercise 5B

1 In the diagram on the right, AB is a tangent to the circle, centre O, with C the point of contact. Point D lies on the circle. Point E lies outside the circle with ODE a straight line.

Given that $\angle DCB = 65°$

              $\angle EDC = x°$

prove that $x = 155$.

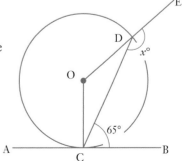

2 In the diagram on the right, AB is a tangent to the circle centre O, point C being the point of contact. Points D and E lie on the circle as shown.

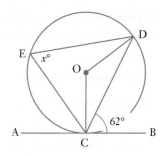

Given that $\angle DCB = 62°$

and $\angle CED = x°$

prove that $x = 62$.

3 The diagram on the right shows point P lying outside a circle centre C with the two tangents from P to the circle shown.

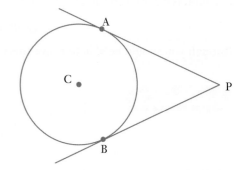

Prove that:

a   the two tangents are of equal length
(i.e. that PA = PB),

b   PC bisects $\angle APB$ (i.e. that $\angle CPA = \angle CPB$).

4 (The general case of situation shown in **question 2**.)

Prove that: *The angle between a tangent and a chord drawn at the point of contact is equal to the angle in the alternate segment.*

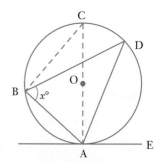

I.e. for the diagram shown on the right, prove that $\angle DAE = x°$.

Note: If you are not sure what is meant by the phrase *the alternate segment* see below:

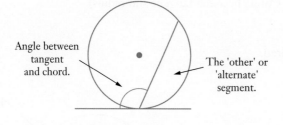

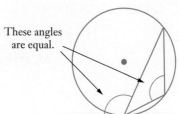

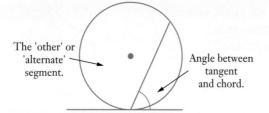

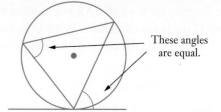

In the remaining questions of this exercise you may use any of the following facts, without proof, in order to justify statements you make (in addition to the standard facts about parallel lines, straight lines, triangles etc.):

The angle at the centre is twice the angle at the circumference.

Angles in a semicircle are right angles.

Angles in the same segment are equal.

The opposite angles of a cyclic quadrilateral are supplementary.

The angle between a tangent and a radius is a right angle.

The angle between tangent and chord equals angle in alternate segment.

The two tangents drawn from a point to a circle are of equal length.

5 In the diagram on the right, AD and AE are tangents to a circle centre O, with points of contact B and C respectively.

Given that $\angle DAE = 70°$

and $\angle BCE = x°$

prove that $x = 125$.

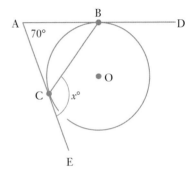

6 In the diagram on the right, AC and AE are tangents to a circle centre O, with points of contact B and D respectively. Points F and G lie on the circle.

Given that $\angle DAB = 80°$

                $\angle GBC = 60°$

and $\angle DFG = x°$

prove that $x = 110$.

7 In the diagram on the right, ABC and ADE are tangents to the circle centre O and points, B, D and F lie on the circle.

Given that $\angle DFB = 70°$

and $\angle DAB = x°$

prove that $x = 40$.

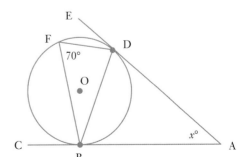

8 In the diagram on the right, points C, E and F lie on the circle centre O.

ACB is a tangent to the circle and the perpendicular from E to AB meets AB at D.

If $\angle DEC = 18°$ and $\angle EFC = x°$ prove that $x = 72$.

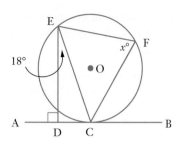

9 The diagram shows two circles. The larger circle has centre O and the smaller has centre P.

Points B, E and F lie on the larger circle and points C, G and H lie on the smaller circle.

ABCD is a tangent to both circles.

$\angle GHC = 50°$ and $\angle BEF = x°$.

Prove that $x = 40$.

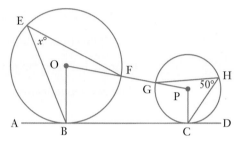

10 The straight lines ABC and DEF are tangents to a circle with points B and E respectively the points of contact with the circle, as shown in the diagram.

A point H lies outside the circle and HE produced (continued) meets the circle again at G.

Given that $\angle BGE = 70°$

               $\angle GBA = 85°$

and $\angle FEH = x°$

prove that $x = 25$.

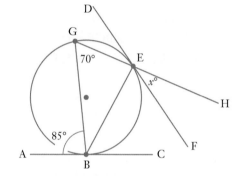

11 The diagram on the right shows a secant cutting a circle at points A and B and a tangent touching the circle at T, both drawn from some external point M.

Prove that $MT^2 = MA \times MB$.

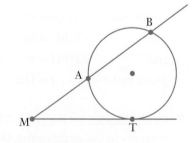

Hence determine x and y in the following:

PQ is a tangent to the circle with Q as the point of contact.

PQ = x cm
PR = 8 cm
RS = 10 cm

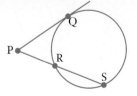

AD is a tangent to the circle with D as the point of contact.

AB = y cm
BC = 30 cm
AD = 20 cm

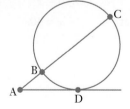

ISBN 9780170390477

12 The diagram on the right shows two secants to a circle, each drawn from the point M, with one cutting the circle at points A and B and the other cutting the circle at points C and D.

Prove that:

$$MA \times MB = MC \times MD.$$

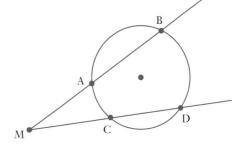

Hence determine x and y in the following:

AB = 40 cm
BC = 60 cm
AD = 50 cm
DE = x cm

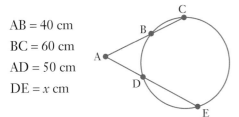

PQ = 6 cm
PR = 15 cm
ST = 13 cm
PS = y cm

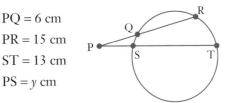

Miscellaneous exercise five

This miscellaneous exercise may include questions involving the work of this chapter, the work of any previous chapters, and the ideas mentioned in the Preliminary work section at the beginning of the book.

1 A band records 13 new songs and wishes to select 8 of the 13 for release on a forthcoming CD. How many different groups of eight songs can be chosen from the 13?

Once the selection of the eight songs has been made in how many different orders can these eight occur on the CD?

2 Set A = {a, b, c, d, e, f, g, h, i}. How many different subsets of A are there that consist of six different letters, two of which are vowels and the other four are consonants?

3 How many combinations of six university courses can be chosen from ten courses if two of the ten are compulsory?

Suppose now that the two compulsory courses are put in a list A, along with three other courses, and the remaining five courses form list B. How many combinations of six courses are there now if the two compulsory courses are still compulsory and three of the six courses must come from List A and three from list B?

4 In still air, an aircraft can maintain a speed of 350 km/h. In what direction should the aircraft be pointing if it is to travel on a bearing 170° and a wind of 50 km/h is blowing from 020°?

5 The diagram on the right shows a vehicle parked on a slope, angle of inclination 20°. The weight of the vehicle acts as a downward force of 10 000 N.

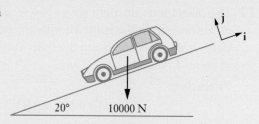

a With the unit vectors **i** and **j** as shown in the diagram express this 10 000 N force in the form $(a\mathbf{i} + b\mathbf{j})$ N with a and b given to the nearest hundred.

b What is the magnitude of the resistance force the brakes must be exerting to prevent the vehicle moving down the slope? (Give your answer to the nearest 100 N.)

6 A teacher asked the class to prove the following:

If in triangle ABC the perpendicular bisector of AC passes through B then ABC is an isosceles triangle.

One student presented the following proof:

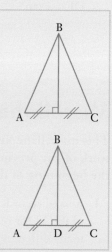

Given:	△ABC with the perpendicular bisector of AC passing through B.
To prove:	That △ABC is an isosceles triangle.
Proof:	Let the midpoint of AC be D (see diagram).

In triangles ABD and BDC

AD = DC   (D is midpoint of AC)

BDA = BDC   (Each angle equals 90°)

BD = BD   (Common side)

Thus △ABD is congruent to △BDC (RHS).

Therefore △ABC is an isosceles triangle.

If the answer were to be marked out of ten, how many marks would you award the above response and why?

7 The diagram shows the parallelogram ABDE, with $\overrightarrow{EA} = \mathbf{a}$, and $\overrightarrow{ED} = \mathbf{b}$. AB is produced to C where $\overrightarrow{BC} = \overrightarrow{AB}$. Point F lies on BC and is such that $\overrightarrow{BF} = \dfrac{1}{3}\overrightarrow{BC}$. G is the midpoint of DC.

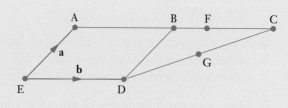

Find the following vectors in terms of **a** and/or **b**.

a $\overrightarrow{AC}$ **b** $\overrightarrow{AF}$ **c** $\overrightarrow{DC}$

d $\overrightarrow{EC}$ **e** $\overrightarrow{EG}$ **f** $\overrightarrow{GF}$

If GF and EC intersect at H with $\overrightarrow{GH} = h\overrightarrow{GF}$ and $\overrightarrow{HC} = k\overrightarrow{EC}$ determine h and k.

ISBN 9780170390477

6.

Relative
displacement
and relative
velocity

- Relative displacement
- Relative velocity
- Miscellaneous exercise six

Situation

The diagram below shows 'you' sitting in a chair watching a sports car approach at 60 km/h. Quite clearly what you see is the sports car approaching at 60 km/h.

Suppose now that as well as the sports car approaching you at 60 km/h, you are approaching the sports car at 40 km/h. This situation is shown below.

What you would see in this case is the same as you would see if you were not moving and the sports car were approaching you at 100 km/h. We say that the situation **relative to you** is that the sports car is approaching at 100 km/h.

Consider now the situation shown below.

In this case the situation relative to you is that the sports car is moving away from you at 20 km/h.

In each of the situations on the next page, describe the motion of the sports car relative to 'you'.

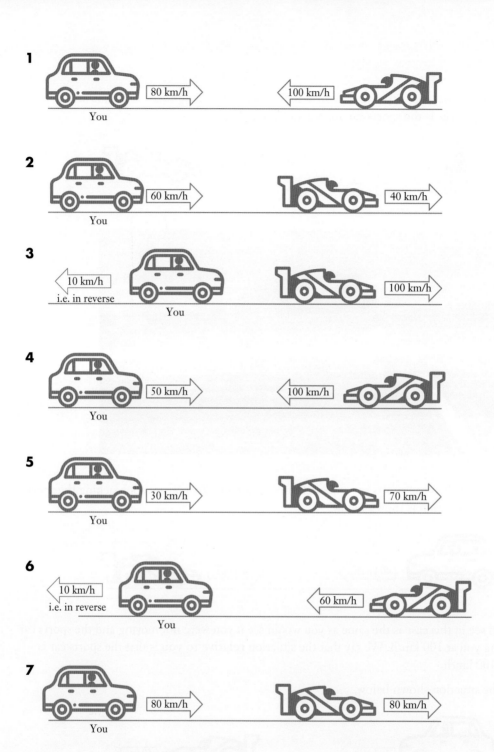

1. You 80 km/h — ← 100 km/h

2. You 60 km/h — 40 km/h →

3. ← 10 km/h i.e. in reverse — You — 100 km/h →

4. You 50 km/h — ← 100 km/h

5. You 30 km/h — 70 km/h →

6. ← 10 km/h i.e. in reverse — You — ← 60 km/h

7. You 80 km/h — 80 km/h →

Relative displacement

Consider the points O, A, B, C and D shown on the right.

With respect to the origin, O, the position vectors of A, B, C and D are:

$$\mathbf{r}_A = \overrightarrow{OA} = 4\mathbf{i} + \mathbf{j},$$

$$\mathbf{r}_B = \overrightarrow{OB} = 5\mathbf{i} + 3\mathbf{j},$$

$$\mathbf{r}_C = \overrightarrow{OC} = 3\mathbf{i} + 3\mathbf{j},$$

$$\mathbf{r}_D = \overrightarrow{OD} = 0\mathbf{i} + 2\mathbf{j}.$$

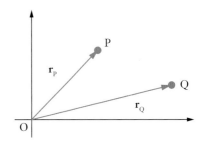

These vectors give the **displacements** of each point from O.

However, if we were situated at A it may be more useful to know the displacement of each of the other points relative to our own position, A. Writing the position vector of B relative to A (or the displacement of B relative to A) as $_B\mathbf{r}_A$ it follows that:

$$_B\mathbf{r}_A = \overrightarrow{AB} = \mathbf{i} + 2\mathbf{j},$$

$$_C\mathbf{r}_A = \overrightarrow{AC} = -\mathbf{i} + 2\mathbf{j},$$

$$_D\mathbf{r}_A = \overrightarrow{AD} = -4\mathbf{i} + \mathbf{j},$$

$$_O\mathbf{r}_A = \overrightarrow{AO} = -4\mathbf{i} - \mathbf{j}.$$

It follows that $\quad _P\mathbf{r}_Q = \overrightarrow{QP}$

$$= -\,\mathbf{r}_Q + \mathbf{r}_P$$

i.e. $\qquad \boxed{_P\mathbf{r}_Q = \mathbf{r}_P - \mathbf{r}_Q}$

Though this result is shown 'boxed', in most cases it is better to think about what is being asked for rather than to simply substitute into the rule.

Note: We could write the position vector of A relative to O as $_A\mathbf{r}_O$ but, because it is usual to give position vectors relative to the origin, we tend to simply write this as $\mathbf{r}_A$, as we did at the top of this page.

If we are asked for the position vector of a point then this should be taken as being relative to the origin, position vector $0\mathbf{i} + 0\mathbf{j}$, unless stated otherwise.

EXAMPLE 1

The diagram on the right shows the origin O and the points A, B, C and D. Find each of the following in the form $a\mathbf{i} + b\mathbf{j}$.

a The position vector of A.

b The position vector of A relative to B.

c The displacement of B relative to A.

d The displacement of C relative to A.

e The position vector of A relative to C.

f $_D\mathbf{r}_A$

g $_A\mathbf{r}_D$

h $_C\mathbf{r}_D$

Solution

a $\mathbf{r}_A = \mathbf{i} + 3\mathbf{j}$

b $_A\mathbf{r}_B = \overrightarrow{BA}$
$= -2\mathbf{i} + \mathbf{j}$

c $_B\mathbf{r}_A = \overrightarrow{AB}$
$= 2\mathbf{i} - \mathbf{j}$

d $_C\mathbf{r}_A = \overrightarrow{AC}$
$= \mathbf{i} - 2\mathbf{j}$

e $_A\mathbf{r}_C = \overrightarrow{CA}$
$= -\mathbf{i} + 2\mathbf{j}$

f $_D\mathbf{r}_A = \overrightarrow{AD}$
$= 3\mathbf{i} + \mathbf{j}$

g $_A\mathbf{r}_D = \overrightarrow{DA}$
$= -3\mathbf{i} - \mathbf{j}$

h $_C\mathbf{r}_D = \overrightarrow{DC}$
$= -2\mathbf{i} - 3\mathbf{j}$

EXAMPLE 2

Ship A has position vector $(20\mathbf{i} + 10\mathbf{j})$ km. Relative to an observer on A, a second ship, B, has a position vector $(3\mathbf{i} - 8\mathbf{j})$ km. Find the position vector of the second ship.

Solution

If A is $(20\mathbf{i} + 10\mathbf{j})$ km from the origin and B is a further $(3\mathbf{i} - 8\mathbf{j})$ km from A, it follows that B is $(20\mathbf{i} + 10\mathbf{j}) + (3\mathbf{i} - 8\mathbf{j})$ from the origin. Thus ship B has position vector $(23\mathbf{i} + 2\mathbf{j})$ km.

Alternatively, the answer could be obtained using the formula $_P\mathbf{r}_Q = \mathbf{r}_P - \mathbf{r}_Q$:

We are told that $\mathbf{r}_A = 20\mathbf{i} + 10\mathbf{j}$ and $_B\mathbf{r}_A = 3\mathbf{i} - 8\mathbf{j}$.

Using $_B\mathbf{r}_A = \mathbf{r}_B - \mathbf{r}_A$

$3\mathbf{i} - 8\mathbf{j} = \mathbf{r}_B - (20\mathbf{i} + 10\mathbf{j})$

Thus $\mathbf{r}_B = (3\mathbf{i} - 8\mathbf{j}) + (20\mathbf{i} + 10\mathbf{j})$

$= (23\mathbf{i} + 2\mathbf{j})$

Thus ship B has position vector $(23\mathbf{i} + 2\mathbf{j})$ km, as before.

Exercise 6A

1 The diagram on the right shows the origin O and the points A, B, C and D. Find each of the following in the form $a\mathbf{i} + b\mathbf{j}$.

   a The position vector of A.

   b The position vector of B.

   c The position vector of C.

   d The position vector of A relative to B.

   e The displacement of A relative to C.

   f The position vector of A relative to D.

   g The displacement of D relative to A.

   h The position vector of B relative to C.

   i The position vector of C relative to D.

   j The displacement of D relative to C.

2 On a copy of the grid shown on the right, show points A to L given the following information.

   $\mathbf{r}_A = 2\mathbf{i} + 4\mathbf{j}$ $\mathbf{r}_B = 4\mathbf{i} + 3\mathbf{j}$

   $_C\mathbf{r}_A = -3\mathbf{j}$ $_D\mathbf{r}_A = -\mathbf{i} - \mathbf{j}$

   $\mathbf{r}_E = -2\mathbf{i} + 4\mathbf{j}$ $_F\mathbf{r}_E = \mathbf{i} + \mathbf{j}$

   $_G\mathbf{r}_D = -4\mathbf{i} - \mathbf{j}$ $_H\mathbf{r}_G = \mathbf{i} - \mathbf{j}$

   $_I\mathbf{r}_A = 3\mathbf{i} + \mathbf{j}$ $_I\mathbf{r}_J = 3\mathbf{j}$

   $_H\mathbf{r}_K = -5\mathbf{i} + \mathbf{j}$ $_I\mathbf{r}_L = 5\mathbf{i} + \mathbf{j}$

3 Ship A has position vector $(7\mathbf{i} + 11\mathbf{j})$ km. Relative to an observer on A a second ship, B, has position vector $(-12\mathbf{i} - 8\mathbf{j})$ km. Find the position vector of ship B.

4 Is $\overrightarrow{AB}$ the displacement of A relative to B or is it the displacement of B relative to A?

5 Ship A has position vector $(11\mathbf{i} - 3\mathbf{j})$ km. Relative to an observer on a second ship B, ship A has position vector $(-8\mathbf{i} - 8\mathbf{j})$ km. How far is ship B from the origin?

   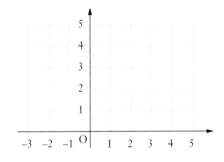

6 Ship A has position vector $(5\mathbf{i} + 2\mathbf{j})$ km. Relative to an observer on a second ship B, ship A has position vector $(8\mathbf{i} + 3\mathbf{j})$ km. Find the distance B is from the origin.

7 Point A has position vector $2\mathbf{i} + 4\mathbf{j}$. Points B and C are such that the position vector of B relative to C is $7\mathbf{i} - 4\mathbf{j}$ and the position vector of C relative to A is $-4\mathbf{i} - \mathbf{j}$. Find the position vector of B.

8 Prove that $_D\mathbf{r}_E + _E\mathbf{r}_F = _D\mathbf{r}_F$.

Relative velocity

The result $_B\mathbf{r}_A = \mathbf{r}_B - \mathbf{r}_A$ can be extended to velocity vectors to determine the velocity of one body relative to another.

The velocity of A relative to B is given by $_A\mathbf{v}_B$ where $_A\mathbf{v}_B = \mathbf{v}_A - \mathbf{v}_B$.

By the phrase 'the velocity of A relative to B' we mean the velocity of A as seen by an observer on B.

Suppose A and B have velocities $\mathbf{v}_A$ and $\mathbf{v}_B$ respectively. To consider the situation from the point of view of an observer on B we need to imagine a velocity of $-\mathbf{v}_B$ is imposed on the whole system. It will then seem as though B is reduced to rest ($\mathbf{v}_B - \mathbf{v}_B = 0$) and the velocity A has in this system ($= \mathbf{v}_A - \mathbf{v}_B$) will equal the velocity of A as seen by an observer on B. i.e. $_A\mathbf{v}_B = \mathbf{v}_A - \mathbf{v}_B$.

Note: All velocities are relative to something. When we write $\mathbf{v}_A$ we usually mean the velocity of A relative to the Earth. $\mathbf{v}_A$ could be written $_A\mathbf{v}_{Earth}$ and $_A\mathbf{v}_B = \mathbf{v}_A - \mathbf{v}_B$ then becomes

$$_A\mathbf{v}_B = {_A\mathbf{v}_{Earth}} - {_B\mathbf{v}_{Earth}}.$$

This 'velocity of A as seen by an observer on B' was the sort of thing we considered at the beginning of this chapter. Consider again the situation:

If we take $\mathbf{i}$ as a unit vector to the right then:

$$\mathbf{v}_{you} = 40\mathbf{i} \text{ km/h} \qquad \text{and} \qquad \mathbf{v}_{sports} = -60\mathbf{i} \text{ km/h}.$$

The velocity of the sports car as it appears to 'you' is given by $_{sports}\mathbf{v}_{you}$ where

$$_{sports}\mathbf{v}_{you} = \mathbf{v}_{sports} - \mathbf{v}_{you}$$
$$= -60\mathbf{i} - 40\mathbf{i}$$
$$= -100\mathbf{i}$$

Thus, with the sports car positioned 'to the right' of you and moving 'left', what you see is the sports car approaching at 100 km/h, as we said when considering this situation intuitively at the beginning of this chapter. This then raises the following question:

Question: We were able to determine the motion of the sports car as seen by 'you' before we had met the result $_{sports}\mathbf{v}_{you} = \mathbf{v}_{sports} - \mathbf{v}_{you}$ so what is the point of this result?

Answer: The benefit of a result like $_A\mathbf{v}_B = \mathbf{v}_A - \mathbf{v}_B$ is that it can be used to find the velocity of one body relative to another body when the directions involved make an intuitive solution difficult. (See **example 3**).

ISBN 9780170390477

EXAMPLE 3

Ship A is travelling due North at 10 km/h.

Ship B is travelling on a bearing 030° at 15 km/h.

What is the velocity of ship B relative to ship A? (i.e. What is the velocity of B as seen by an observer on A?)

Solution

We are given: $\qquad \mathbf{v}_A = \Big\uparrow 10$ km/h $\qquad$ and $\qquad \mathbf{v}_B =$ (30°, 15 km/h)

and we require: $_B\mathbf{v}_A$.

Now $\qquad\qquad _B\mathbf{v}_A = \mathbf{v}_B - \mathbf{v}_A$

$\qquad\qquad\qquad\qquad = \mathbf{v}_B + (-\mathbf{v}_A) \qquad$ (See diagram).

By the cosine rule: $\qquad |_B\mathbf{v}_A|^2 = 10^2 + 15^2 - 2\,(10)\,(15)\cos 30°$

$\qquad\qquad\qquad\qquad |_B\mathbf{v}_A| \approx 8.07$ km/h

By the sine rule: $\qquad \dfrac{|_B\mathbf{v}_A|}{\sin 30°} = \dfrac{|-\mathbf{v}_A|}{\sin \theta}$

i.e. $\qquad\qquad \dfrac{|_B\mathbf{v}_A|}{\sin 30°} = \dfrac{10}{\sin \theta}$

giving $\qquad\qquad\qquad \theta \approx 38.3°$

Note: The obtuse solution to the above equation, i.e 141.7°, can be ignored in this case because θ is opposite one of the smaller sides of the triangle.

Relative to A, ship B is travelling with speed 8.1 km/hr on a bearing 068°.

EXAMPLE 4

If $\mathbf{v}_A = 7\mathbf{i} - 2\mathbf{j}$ and $\mathbf{v}_B = 6\mathbf{i} + \mathbf{j}$, find $_A\mathbf{v}_B$.

Solution

$\qquad _A\mathbf{v}_B = \mathbf{v}_A - \mathbf{v}_B$

$\qquad\qquad = (7\mathbf{i} - 2\mathbf{j}) - (6\mathbf{i} + \mathbf{j})$

$\qquad\qquad = \mathbf{i} - 3\mathbf{j}$

Thus $_A\mathbf{v}_B$, the velocity of A relative to B, is $\mathbf{i} - 3\mathbf{j}$.

EXAMPLE 5

To a person on a ship travelling with velocity $(15\mathbf{i} + 2\mathbf{j})$ km/h the wind appears to have velocity $(-9\mathbf{i} + 2\mathbf{j})$ km/h. Find the true velocity of the wind.

Solution

We are given that $\qquad \mathbf{v}_{\text{Ship}} = 15\mathbf{i} + 2\mathbf{j} \qquad$ and $\qquad {}_{\text{Wind}}\mathbf{v}_{\text{Ship}} = -9\mathbf{i} + 2\mathbf{j}$

Now $\qquad\qquad {}_{\text{Wind}}\mathbf{v}_{\text{Ship}} = \mathbf{v}_{\text{Wind}} - \mathbf{v}_{\text{Ship}}$

$\therefore \qquad\qquad\qquad \mathbf{v}_{\text{Wind}} = {}_{\text{Wind}}\mathbf{v}_{\text{Ship}} + \mathbf{v}_{\text{Ship}}$

$$= (-9\mathbf{i} + 2\mathbf{j}) + (15\mathbf{i} + 2\mathbf{j})$$

$$= 6\mathbf{i} + 4\mathbf{j}$$

The true velocity of the wind is $(6\mathbf{i} + 4\mathbf{j})$ km/h.

EXAMPLE 6

To a person on a ship moving at 18 km/h on a bearing 150°, the wind appears to come from the South with speed 10 km/h. Find the true velocity of the wind.

Solution

We are given: $\qquad \mathbf{v}_{\text{ship}} = $ 150° 18 km/h $\qquad {}_{\text{wind}}\mathbf{v}_{\text{ship}} = $ 10 km/h

and we require: $\qquad \mathbf{v}_{\text{wind}}$

Now $\qquad\qquad {}_{\text{wind}}\mathbf{v}_{\text{ship}} = \mathbf{v}_{\text{wind}} - \mathbf{v}_{\text{ship}}$

$\therefore \qquad\qquad\qquad \mathbf{v}_{\text{wind}} = {}_{\text{wind}}\mathbf{v}_{\text{ship}} + \mathbf{v}_{\text{ship}}$

By the cosine rule $\quad |\mathbf{v}_{\text{wind}}|^2 = 10^2 + 18^2 - 2\,(10)\,(18)\cos 30°$

$\therefore \qquad\qquad |\mathbf{v}_{\text{wind}}| \approx 10.6$ km/h

By the sine rule $\qquad \dfrac{|\mathbf{v}_{\text{wind}}|}{\sin 30°} = \dfrac{10}{\sin \theta} \qquad$ from which $\qquad \theta \approx 28.2°$

The true velocity of the wind is 10.6 km/h from 302°.

(Again the obtuse solution to the sine rule could be ignored because we chose to determine an angle opposite one of the smaller sides of the triangle.)

EXAMPLE 7

To an observer on ship A travelling North-East at 18 km/h, the wind appears to come from the North. To an observer on ship B travelling at 22 km/h in direction S 70° E, the wind appears to come from S 40° E. Use a scale drawing to determine the true magnitude and direction of the wind.

Solution

We are given:

$$\mathbf{v}_A = \qquad {}_w\mathbf{v}_A = \qquad \mathbf{v}_B = \qquad {}_w\mathbf{v}_B =$$

Now $${}_w\mathbf{v}_A = \mathbf{v}_w - \mathbf{v}_A$$ and $${}_w\mathbf{v}_B = \mathbf{v}_w - \mathbf{v}_B$$

∴ $$\mathbf{v}_w = {}_w\mathbf{v}_A + \mathbf{v}_A$$ and $$\mathbf{v}_w = {}_w\mathbf{v}_B + \mathbf{v}_B$$

With $\mathbf{v}_w$ common to both triangles, a single diagram allows $\mathbf{v}_w$ to be determined:

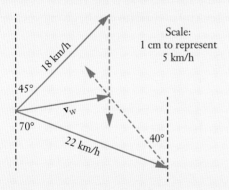

Scale:
1 cm to represent
5 km/h

or:

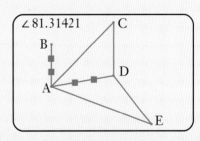

The true magnitude and direction of the wind is 13 km/h from S 81° W.

Exercise 6B

In each of the following $\mathbf{v}_A$ is the velocity of A relative to the Earth and $\mathbf{v}_B$ is the velocity of B relative to the Earth. Find $_A\mathbf{v}_B$, the velocity of A relative to B, giving your answers in the form $a\mathbf{i} + b\mathbf{j}$.

1 $\mathbf{v}_A = 2\mathbf{i} - 3\mathbf{j}$, $\quad$ $\mathbf{v}_B = -4\mathbf{i} + 7\mathbf{j}$.

2 $\mathbf{v}_A = 4\mathbf{i} + 2\mathbf{j}$, $\quad$ $\mathbf{v}_B = 7\mathbf{i} - \mathbf{j}$.

3 $\mathbf{v}_A = 3\mathbf{i} - 2\mathbf{j}$, $\quad$ $\mathbf{v}_B = 4\mathbf{i} + 7\mathbf{j}$.

4 $\mathbf{v}_A = 6\mathbf{i} + 2\mathbf{j}$, $\quad$ $\mathbf{v}_B = 3\mathbf{i} - \mathbf{j}$.

In each of the following $\mathbf{v}_A$ is the velocity of A relative to the Earth and $\mathbf{v}_B$ is the velocity of B relative to the Earth. Use trigonometry to determine $_A\mathbf{v}_B$, the velocity of A relative to B, giving your answer as magnitude and direction.

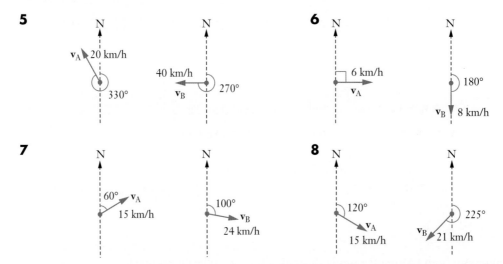

In each of the following $\mathbf{v}_A$ is the velocity of A relative to the Earth and $\mathbf{v}_B$ is the velocity of B relative to the Earth. Use a scale diagram to determine $_A\mathbf{v}_B$, the velocity of A relative to B, giving your answer in magnitude and direction.

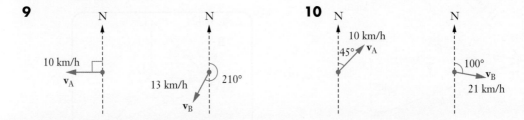

11 Ship A has velocity $(7\mathbf{i} - 10\mathbf{j})$ km/h and ship B has velocity $(2\mathbf{i} + 20\mathbf{j})$ km/h.
Find:

a the velocity of ship A as seen by an observer on B,

b the velocity of ship B as seen by an observer on A.

12 Ship A is travelling due West with speed 12 km/h and ship B is travelling on a bearing 060° with speed 15 km/h.

Find:

a the velocity of ship A as seen by an observer on B,

b the velocity of ship B as seen by an observer on A.

13 A car is travelling at 100 km/h on a bearing 120° and a motorcycle is travelling due West at 80 km/h. Find the velocity of the motorcyclist relative to the car.

14 If $\mathbf{v}_A = 3\mathbf{i} - 7\mathbf{j}$ and $_A\mathbf{v}_B = 2\mathbf{i} + 5\mathbf{j}$, find $\mathbf{v}_B$.

15 If $\mathbf{v}_A = -4\mathbf{i} + 3\mathbf{j}$ and $_B\mathbf{v}_A = 2\mathbf{i} + 3\mathbf{j}$, find $\mathbf{v}_B$.

16 Prove that $_A\mathbf{v}_B + {_B\mathbf{v}_C} = {_A\mathbf{v}_C}$.

17 If $\mathbf{v}_A = 20$ km/h due North and $_A\mathbf{v}_B = 30$ km/h due North, find $\mathbf{v}_B$.

18 If $\mathbf{v}_A = 80$ km/h due South and $_A\mathbf{v}_B = 60$ km/h due South, find $\mathbf{v}_B$.

19 If $\mathbf{v}_A = 100$ km/h in a direction 060° and $_B\mathbf{v}_A = 70$ km/h in direction 100°, find the magnitude and direction of $\mathbf{v}_B$.

20 To a person on a ship moving with velocity $(7\mathbf{i} + 2\mathbf{j})$ km/h the wind seems to have velocity $(6\mathbf{i} - \mathbf{j})$ km/h. Find the true velocity of the wind.

21 To a person walking with a velocity of $(3\mathbf{i} - 4\mathbf{j})$ km/h the wind seems to have a velocity of $(\mathbf{i} + 2\mathbf{j})$ km/h. Find the true velocity of the wind.

22 To a person on a ship moving South-East at 25 km/h the wind seems to come from the South with speed 15 km/h. Find the true velocity of the wind.

23 To a person walking due North at 5 km/h the wind seems to come from the West with speed 3 km/h. Find the true velocity of the wind.

24 Ship A is travelling due North at 10 km/h. The radar operator on this ship monitors the motion of three other ships B, C and D shown as dots on the radar screen. The movement of the dots give the impression that B is stationary, C is moving South at 3 km/h and D is moving North at 5 km/h but these velocities are all relative to A's motion. Find the true velocities of B, C and D.

25 To a bird flying at 14 km/h on a bearing 160° the wind seems to be from the South at 22 km/h. Find the true velocity of the wind.

26 Ship E is travelling at 15 km/h in direction 040°. The radar operator on this ship monitors the motion of three other ships F, G and H shown as dots on the radar screen. The movement of the dots give the impression that F is moving at 15 km/h in direction 220°, G is moving at 5 km/h due North and H is moving at 30 km/h in direction 300°. However, these velocities are all relative to E's motion.

Find the true velocities of F, G and H.

27 The velocity of ship A, as seen by an observer on ship B, is $7\mathbf{i} - 10\mathbf{j}$.

The velocity of ship A, as seen by an observer on ship C, is $13\mathbf{i} - 2\mathbf{j}$.

Find the velocity of ship B as seen by an observer on C.

28 When a cyclist travels East at 20 km/h, the wind seems to come from the North-East. When the cyclist travels South at 25 km/h, the same wind seems to come from S 30° W. Use a scale diagram to determine the true magnitude and direction of the wind.

29 To a person walking due North at 6 km/h, the wind seems to come from N 70° W. To a second person walking due East at 8 km/h, the same wind seems to come from S 40° E. Use a scale diagram to determine the true magnitude and direction of the wind.

Miscellaneous exercise six

This miscellaneous exercise may include questions involving the work of this chapter, the work of any previous chapters, and the ideas mentioned in the Preliminary work section at the beginning of the book.

1 If **i** is a unit vector due East and **j** is a unit vector due North, express each of the following vectors in the form $a\mathbf{i} + b\mathbf{j}$.

p: 6 units on a bearing 030°.

q: $8\sqrt{2}$ units North-West.

r: 10 units on a bearing 330°.

s: 8 units on a bearing S 60° E.

2 Vectors **a**, **b** and **c** are such that $\mathbf{a} = 2\mathbf{i} + 2\mathbf{j}$, $\mathbf{b} = x\mathbf{i} + 5\mathbf{j}$ and $\mathbf{c} = 7\mathbf{i} + y\mathbf{j}$.

If **a** and **b** are parallel and **b** and **c** have equal magnitudes, determine the values of x and y.

3 A school assembly is to be addressed by eight speakers, A, B, C, D, E, F, G and H. How many orders of speakers are possible in each of the following situations?

a   A must be first.

b   A must be first and C must be last.

c   A must be first, C must be last and D must immediately follow F.

Alamy Stock Photo/Ian Shaw

ISBN 9780170390477

4 A teacher asked the class to prove the following:

All cyclic quadrilaterals have opposite angles that are supplementary.

One student presented the following proof:

Given: The cyclic quadrilateral ABCD with AC a diameter of the circle. (See diagram.)

To prove: That $\angle ABC + \angle ADC = 180°$

and $\angle BAD + \angle BCD = 180°$

Proof: $\angle ABC = 90°$ (Angle in a semi-circle)

$\angle ADC = 90°$ (Angle in a semi-circle)

$\therefore$ $\angle ABC + \angle ADC = 180°$, as required. [1]

$\angle ABC + \angle ADC + \angle BAD + \angle BCD = 360°$ (Quadrilateral angle sum)

Thus $180° + \angle BAD + \angle BCD = 360°$ (Using [1])

$\therefore$ $\angle BAD + \angle BCD = 180°$, as required.

If the answer were to be marked out of ten, how many marks would you award the above response and why?

5 In the diagram on the right points A, B and D lie on the circle centre O with BD as a diameter.

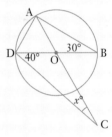

C is a point outside of the circle with A, O and C collinear and $\angle ACD = x°$.

Given that $\angle ABD = 30°$

and $\angle BDC = 40°$

prove that $x = 20$.

6. Point P has position vector $-7\mathbf{i} + 13\mathbf{j}$ and point Q has position vector $13\mathbf{i} - 2\mathbf{j}$.

Point R lies on the straight line joining P to Q and is such that PR : PQ = 3 : 5.

Find:

a how far R is from Q,

b the position vector of R,

c how far R is from the origin.

7 If a statement 'if P then Q' is true then which of the following must also be true:

If Q then P.

If not P then not Q.

If not Q then not P.

8 How many different combinations of five letters, all different, can be formed from the letters of the alphabet if each combination must contain two vowels and three consonants?

How many of these combinations contain an a or an e?

(Remember in mathematics we interpret 'or' to mean 'one or the other or both', i.e we interpret it to mean 'at least one of'.)

9 Five places exist on a course for teachers. The five are to be chosen from the 11 teachers who have expressed interest in attending. These 11 come from three education districts with 6 from District A, 4 from District B and 1 from District C.

How many different groups of five are there?

How many different groups of five are there if each group must contain at least one teacher from each of the three districts?

10 Three soccer team managers are to be invited to attend a TV talk program.

Eight managers are shortlisted: A, B, C, D, E, F, G, H.

   a How many different combinations of three managers are possible?

   b Bitter rivalry exists between managers C and D so three options are considered:

   Option I: Invite both C and D as lively debate should ensue.

   Option II: Avoid conflict and, to be fair, invite neither C nor D.

   Option III: Avoid the 'invite both C and D' situation but allow any other group of 3 that is decided upon.

   For each of these three options how many different combinations of three managers are possible?

A B Ɔ ꓷ Ǝ F G H

Proofs using vectors

- Miscellaneous exercise seven

A vector approach can be used to prove certain geometrical facts, as the next example demonstrates. When using such an approach our accepted facts or axioms include the basic ideas that follow from our understanding of vectors, and the results that follow from these basic ideas. For example:

- Equal vectors have the same magnitude and the same direction.
- If $\mathbf{a} = \lambda\mathbf{b}$,   then if $\lambda > 0$, $\mathbf{a}$ and $\mathbf{b}$ are like parallel vectors

                         and if $\lambda < 0$, $\mathbf{a}$ and $\mathbf{b}$ are unlike parallel vectors.

- Vectors can be added (or subtracted) using a triangle of vectors or the parallelogram law.
- If $h\mathbf{a} = k\mathbf{b}$ then either $\mathbf{a}$ and $\mathbf{b}$ are parallel vectors

                       or $h = k = 0$.

EXAMPLE 1

To prove: The line from the midpoint of one side of a triangle to the midpoint of a second side is parallel to, and half as long as, the third side.

Solution

Consider triangle OAB with C the midpoint of OA and D the midpoint of AB.

We have to prove that $\overrightarrow{CD} = \frac{1}{2}\overrightarrow{OB}$.

Let $\overrightarrow{OA} = \mathbf{a}$ and $\overrightarrow{OB} = \mathbf{b}$.

Now $\overrightarrow{CD} = \overrightarrow{CA} + \overrightarrow{AD}$

$$= \frac{1}{2}\overrightarrow{OA} + \frac{1}{2}\overrightarrow{AB} \qquad \text{(C and D are midpoints.)}$$

$$= \frac{1}{2}\mathbf{a} + \frac{1}{2}(-\mathbf{a} + \mathbf{b})$$

$$= \frac{1}{2}\mathbf{b}$$

$$= \frac{1}{2}\overrightarrow{OB} \text{ as required.}$$

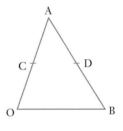

Thus the line from the midpoint of one side of a triangle to the midpoint of a second side is parallel to, and half as long as, the third side.

Exercise 7A

1 To prove: The line drawn from the point that divides one side of a triangle in a certain ratio, to the point that divides a second side in the same ratio is parallel to the third side.

In triangle OAB, $\overrightarrow{OA} = \mathbf{a}$, and $\overrightarrow{OB} = \mathbf{b}$.

C and D are points on OA and OB respectively such that $\overrightarrow{OC} = h\overrightarrow{OA}$ and $\overrightarrow{OD} = h\overrightarrow{OB}$.

Prove that CD is parallel to AB.

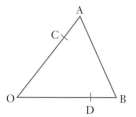

2 To prove: The midpoints of the sides of a quadrilateral form a parallelogram.

Consider the quadrilateral OABC with $\overrightarrow{OA} = \mathbf{a}$, $\overrightarrow{OB} = \mathbf{b}$ and $\overrightarrow{OC} = \mathbf{c}$.
P, Q, R and S are the midpoints of OA, AB, BC and OC respectively.

Find vector expressions for each of $\overrightarrow{PQ}$, $\overrightarrow{QR}$, $\overrightarrow{SR}$ and $\overrightarrow{PS}$ and hence prove that the midpoints of the sides of a quadrilateral form a parallelogram.

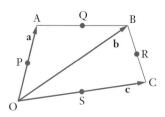

3 To prove: The diagonals of a parallelogram bisect each other. (Method 1.)

OABC is a parallelogram with $\overrightarrow{OA} = \mathbf{a}$, and $\overrightarrow{OC} = \mathbf{c}$.
M is the midpoint of the diagonal OB.

Find $\overrightarrow{CM}$ and $\overrightarrow{CA}$ in terms of $\mathbf{a}$ and $\mathbf{c}$ and hence show that M lies on CA and is the midpoint of CA.

4 To prove: The diagonals of a parallelogram bisect each other.
(Method 2.)

OABC is a parallelogram with

$\overrightarrow{OA} = \mathbf{a}$ and $\overrightarrow{OC} = \mathbf{c}$.

The diagonals OB and AC meet at M.

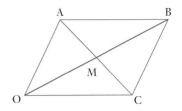

If $\overrightarrow{AM} = h\overrightarrow{AC}$ and $\overrightarrow{OM} = k\overrightarrow{OB}$ use the fact that $\overrightarrow{OM} = \overrightarrow{OA} + \overrightarrow{AM}$ to show that $h = k = \dfrac{1}{2}$.

5 a   To prove: If a quadrilateral is such that its diagonals bisect each other then the quadrilateral is a parallelogram.

OABC is a quadrilateral. If M is the midpoint of both diagonals prove that

$$\overrightarrow{AB} = \overrightarrow{OC} \text{ and } \overrightarrow{OA} = \overrightarrow{CB}.$$

i.e. prove that the quadrilateral is a parallelogram.

b   Combine the result proved in this question with the result of the previous question into one statement using the symbol $\Leftrightarrow$ and also write the statement using the 'if and only if' statement. (See page 7 if need be.)

6 To prove: The medians of a triangle intersect at a point two-thirds of the way along their length measured from the vertex. (Method 1.)

In triangle OAB, C, D and E are the midpoints of OB, AB and

OA respectively. M is a point on OD such that $\overrightarrow{OM} = \dfrac{2}{3}\overrightarrow{OD}$.

$\overrightarrow{OA} = \mathbf{a}$ and $\overrightarrow{OB} = \mathbf{b}$.

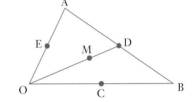

a   Find the following in terms of $\mathbf{a}$ and/or $\mathbf{b}$.

  i $\overrightarrow{AB}$ **ii** $\overrightarrow{AC}$ **iii** $\overrightarrow{AD}$

  iv $\overrightarrow{OD}$ **v** $\overrightarrow{OM}$ **vi** $\overrightarrow{AM}$

b   Prove that M lies on AC and is such that AM : MC = 2 : 1.

c   Prove that M lies on BE and is such that BM : ME = 2 : 1.

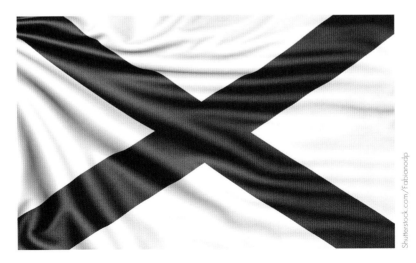

7. Proofs using vectors ●●●●●●●●●●●●●

7 To prove: In a triangle, if a line is drawn from a point that divides one side in a given ratio, parallel to a second side, then it divides the third side in the same ratio.

In triangle ABC, $\overrightarrow{AB} = \mathbf{a}$, $\overrightarrow{AC} = \mathbf{b}$. D is a point on AB such that $\overrightarrow{AD} = h\overrightarrow{AB}$. A line through D, parallel to AC, cuts CB at point E. Prove that $\overrightarrow{CE} = h\overrightarrow{CB}$.

(Hint: Let $\overrightarrow{CE} = k\overrightarrow{CB}$ and then prove $k = h$.)

8 To prove: The medians of a triangle intersect at a point two thirds of the way along their length measured from the vertex. (Method 2.)

In triangle OAB, C, D and E are the midpoints of OB, AB and OA respectively.

OD and AC meet at M.

$\overrightarrow{OA} = \mathbf{a}$, $\overrightarrow{OB} = \mathbf{b}$, $\overrightarrow{OM} = h\overrightarrow{OD}$ and $\overrightarrow{AM} = k\overrightarrow{AC}$.

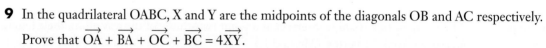

Use the fact that $\overrightarrow{OA} + \overrightarrow{AM} = \overrightarrow{OM}$ to determine h and k.

Show that M also lies on BE and, if $\overrightarrow{BM} = \lambda\overrightarrow{BE}$, find λ.

9 In the quadrilateral OABC, X and Y are the midpoints of the diagonals OB and AC respectively. Prove that $\overrightarrow{OA} + \overrightarrow{BA} + \overrightarrow{OC} + \overrightarrow{BC} = 4\overrightarrow{XY}$.

10 In triangle OAB, $\overrightarrow{OA} = \mathbf{a}$, $\overrightarrow{OB} = \mathbf{b}$ and C is the midpoint of AB. D and E are points on OA and OB respectively and DE cuts OC at F.

$\overrightarrow{OD} = h\overrightarrow{OA}$, $\overrightarrow{OE} = k\overrightarrow{OB}$ and $\overrightarrow{OF} = m\overrightarrow{OC}$.

a Express $\overrightarrow{DF}$ in terms of h, m, $\mathbf{a}$ and $\mathbf{b}$.

b Express $\overrightarrow{FE}$ in terms of k, m, $\mathbf{a}$ and $\mathbf{b}$.

If $\overrightarrow{DF} = \overrightarrow{FE}$ prove that

c $h = k = m$.

d DE is parallel to AB.

Miscellaneous exercise seven

This miscellaneous exercise may include questions involving the work of this chapter, the work of any previous chapters, and the ideas mentioned in the Preliminary work section at the beginning of the book.

1 Discuss the correctness or otherwise of each of the following 'if and only if' statements.

  a A triangle is scalene if and only if it has three different length sides.

  b A positive whole number is a multiple of 5 if and only if it ends with a zero.

2 In how many ways can the three positions of Chairman, Secretary and Treasurer be chosen from a committee of 8 people if each position must be held by a different person?

3 a How many different subcommittees of six people could be selected from a full committee of 15 people?

  b In how many ways can a particular subcommittee of six be arranged in a line for a photograph?

4 How many different 6 letter arrangements can be made each consisting of 6 different letters of the alphabet, with exactly one of the 6 being a vowel?

5 In the diagram on the right, points A, B, C and D lie on a circle centre O.

  Given that $\angle BOA = 80°$

                 $\angle BCD = 64°$

  and $\angle ADO = x°$

  prove that $x = 66$.

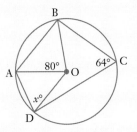

6 Points A and B have position vectors $\mathbf{i} + 5\mathbf{j}$ and $7\mathbf{i} - \mathbf{j}$ respectively.

  Find the position vector of

  a point P, on AB, such that $\overrightarrow{AP} : \overrightarrow{PB} = 4 : 1$,

  b point Q on AB produced, such that $\overrightarrow{AQ} : \overrightarrow{QB} = 4 : -1$.

7 Point A has position vector $-2\mathbf{i} + 7\mathbf{j}$.

  Relative to point A a second point, B, has position vector $8\mathbf{i} + 3\mathbf{j}$.

  I.e. $_B\mathbf{r}_A = 8\mathbf{i} + 3\mathbf{j}$. All units are in metres.

  When timing commences an object moving with constant velocity of

$$(3\mathbf{i} - 2\mathbf{j}) \text{ m/sec}$$

  is at point B. *Exactly* how far is this object from the origin 2 seconds later?

8 Vectors **a**, **b** and **c** are such that
$$\mathbf{b} = -7\mathbf{i} + 24\mathbf{j},$$
$$\mathbf{c} = 3\mathbf{i} - 4\mathbf{j},$$
a and **b** have the same magnitude,

a and **c** have exactly the same direction.

Find the exact magnitude of (**a** + **b**).

9 Airfield B is 600 km south-east of airfield A.

An aeroplane that can fly at 300 km/h in still air is to make the journey from A to B with a wind of 40 km/h blowing from the west.

If the wind remains the same throughout determine the time taken (to the nearest minute) for the aircraft to fly from

a A to B,

b B to A.

10 In the diagram $\overrightarrow{OA} = \mathbf{a}$, $\overrightarrow{AB} = \mathbf{b}$ and $\overrightarrow{OC} = 3\mathbf{b}$.

D is a point on BC such that BD : DC = 1 : 2.

Express each of the following vectors in terms of **a** and **b**.

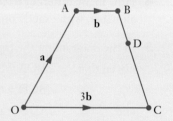

a $\overrightarrow{BC}$ **b** $\overrightarrow{BD}$ **c** $\overrightarrow{OD}$

Now suppose that OD continued meets AB continued at E and that:

$$\overrightarrow{BE} = h\mathbf{b} \text{ and } \overrightarrow{DE} = k\overrightarrow{OD}.$$

Find h and k.

11 A company wishes to give each of the products it sells a code using the letters of the alphabet, i.e. A, B, C, … Z. In each code the order of the letters is significant. Thus whilst one product might have code ABCD, the code ABDC is different and would indicate a different product.

How many different codes are possible if each code consists of

a 4 different letters?

b 4 letters with multiple use of letters in a code permitted?

c 6 different letters?

d 6 letters with multiple use of letters in a code permitted?

If the company wants

• all the codes to have the same number of letters,

• all the codes to start with the letters AR, in that order,

• no code to feature a letter more than once,

• to have at least 12 500 different codes,

what is the least number of letters it should have in each code?

Scalar product

- Scalar product
- Algebraic properties of the scalar product
- The scalar product from the components
- Proofs using the scalar product
- Miscellaneous exercise eight

So far in our consideration of vectors we have not developed any way of testing whether two vectors, $a\mathbf{i} + b\mathbf{j}$ and $c\mathbf{i} + d\mathbf{j}$, are perpendicular.

This chapter considers a concept called the **scalar product** of two vectors and this does give such a test for perpendicularity between two vectors.

The idea of forming a product of two vectors may seem rather confusing initially. How do we multiply together quantities which have magnitude and direction? Well we could define what we mean by vector multiplication in all sorts of ways but there are two methods of performing vector multiplication that prove to be useful. One method of vector multiplication gives an answer that is a scalar. We call this the **scalar product** and will consider the definition of this product in a moment. A second method gives an answer that is a vector. We call this the **vector product**, a concept that is beyond the scope of this unit.

Scalar product

We define the scalar product of two vectors $\mathbf{a}$ and $\mathbf{b}$ to be the magnitude of $\mathbf{a}$ multiplied by the magnitude of $\mathbf{b}$ multiplied by the cosine of the angle between $\mathbf{a}$ and $\mathbf{b}$. We write this product as $\mathbf{a.b}$ and say this as 'a dot b'. For this reason the scalar product is also referred to as the 'dot product'.

Component form of the dot product

Parallel and perpendicular vectors

Thus

$$\mathbf{a.b} = |\mathbf{a}|\,|\mathbf{b}|\cos\theta \qquad \text{where } \theta \text{ is the angle between } \mathbf{a} \text{ and } \mathbf{b}.$$

Note

- It follows from the above definition that for two perpendicular vectors,

  $$\mathbf{a.b} = 0.$$

- '$|\mathbf{a}|\cos\theta$' is the **scalar projection** of $\mathbf{a}$ onto $\mathbf{b}$ or the **resolved part** of $\mathbf{a}$ in the direction of $\mathbf{b}$.

  The **vector projection** of $\mathbf{a}$ onto $\mathbf{b}$ is

  $$|\mathbf{a}|\cos\theta\,\hat{\mathbf{b}}$$

  (Remember, $\hat{\mathbf{b}}$ is a unit vector in the direction of $\mathbf{b}$.)

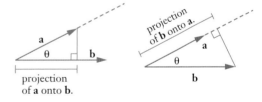

  Similarly '$|\mathbf{b}|\cos\theta$' can be referred to as the scalar projection of $\mathbf{b}$ onto $\mathbf{a}$ and the vector projection of $\mathbf{b}$ onto $\mathbf{a}$ is $|\mathbf{b}|\cos\theta\,\hat{\mathbf{a}}$.

- Remember that the angle between two vectors refers to the angle between the vectors when they are either both directed away from a point or both directed towards it.

EXAMPLE 1

With **a** and **b** as defined in the diagram determine **a.b**.

Solution

$$\mathbf{a.b} = |\mathbf{a}|\ |\mathbf{b}|\ \cos\theta$$
$$= (6)\ (4)\ \cos 60°$$
$$= 12$$

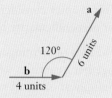

EXAMPLE 2

Given that $|\mathbf{a}| = 7$, $|\mathbf{b}| = 2$ and $\mathbf{a.b} = 10$, find the angle between **a** and **b**, correct to the nearest degree.

Solution

$$\mathbf{a.b} = |\mathbf{a}|\ |\mathbf{b}|\ \cos\theta$$
$$\therefore \quad 10 = (7)\ (2)\ \cos\theta$$
$$\cos\theta \approx 0.7143$$
$$\therefore \quad \theta \approx 44°$$

The angle between **a** and **b** is 44°, correct to the nearest degree.

Properties of the
dot product

Algebraic properties of the scalar product

A number of algebraic properties of the scalar product follow as a consequence of the way we define the scalar product, **a.b**.

It follows that:

$$\mathbf{a.b} = \mathbf{b.a}$$

Proof: $\mathbf{a.b} = |\mathbf{a}|\ |\mathbf{b}|\ \cos\theta$
$$= |\mathbf{b}|\ |\mathbf{a}|\ \cos\theta$$
$$= \mathbf{b.a}$$

It also follows that:

$$\mathbf{a}.(\lambda\mathbf{b}) = \lambda(\mathbf{a.b})$$
$$= (\lambda\mathbf{a}).\mathbf{b}$$

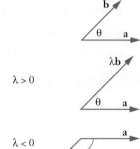

Proof: If $\lambda > 0$

$$\mathbf{a}.(\lambda\mathbf{b}) = |\mathbf{a}|\ |\lambda\mathbf{b}|\ \cos\theta$$
$$= |\mathbf{a}|\ |\lambda|\ |\mathbf{b}|\ \cos\theta$$
$$= \lambda\ |\mathbf{a}|\ |\mathbf{b}|\ \cos\theta$$
$$= \lambda(\mathbf{a.b})$$

$\lambda > 0$

If $\lambda < 0$

$$\mathbf{a}.(\lambda\mathbf{b}) = |\mathbf{a}|\ |\lambda\mathbf{b}|\ \cos(180° - \theta)$$
$$= -|\mathbf{a}|\ |\lambda|\ |\mathbf{b}|\ \cos\theta$$
$$= -|\lambda|\ |\mathbf{a}|\ |\mathbf{b}|\ \cos\theta$$
$$= \lambda\ |\mathbf{a}|\ |\mathbf{b}|\ \cos\theta$$
$$= \lambda(\mathbf{a.b})$$

$\lambda < 0$

ISBN 9780170390477

Another property is: $\qquad$ $\mathbf{a}.\mathbf{a} = |\mathbf{a}|^2$

If we write $|\mathbf{a}|$ simply as a this last result can be written as follows:

$$\mathbf{a}.\mathbf{a} = a^2$$

Proof:
$$\begin{aligned}
\mathbf{a}.\mathbf{a} &= |\mathbf{a}|\,|\mathbf{a}|\cos 0° \\
&= |\mathbf{a}|\,|\mathbf{a}|\,(1) \\
&= |\mathbf{a}|^2 \\
&= a^2
\end{aligned}$$

An important result is: $\qquad$ $\mathbf{a}.(\mathbf{b} + \mathbf{c}) = \mathbf{a}.\mathbf{b} + \mathbf{a}.\mathbf{c}$

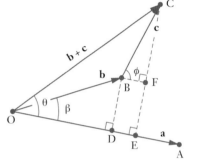

Proof: The diagram on the right shows $\mathbf{a}\,(= \overrightarrow{OA})$, $\mathbf{b}\,(= \overrightarrow{OB})$ and $\mathbf{c}\,(= \overrightarrow{BC})$. Thus $\overrightarrow{OC} = \mathbf{b} + \mathbf{c}$.

The angle between $\mathbf{a}$ and $\mathbf{b}$ is β, between $\mathbf{a}$ and $\mathbf{c}$ is ϕ and between $\mathbf{a}$ and $(\mathbf{b} + \mathbf{c})$ is θ.

$$\begin{aligned}
\mathbf{a}.(\mathbf{b} + \mathbf{c}) &= |\mathbf{a}|\,|\mathbf{b} + \mathbf{c}|\cos\theta \\
&= |\mathbf{a}|\,(OC)\cos\theta \\
&= |\mathbf{a}|\,(OE) \\
&= |\mathbf{a}|\,(OD + DE) \\
&= |\mathbf{a}|\,(OD) + |\mathbf{a}|\,(DE)
\end{aligned}$$

But $OD = |\mathbf{b}|\cos\beta$ and $DE = BF = |\mathbf{c}|\cos\phi$

$\therefore \qquad \begin{aligned}[t]
\mathbf{a}.(\mathbf{b} + \mathbf{c}) &= |\mathbf{a}|\,|\mathbf{b}|\cos\beta + |\mathbf{a}|\,|\mathbf{c}|\cos\phi \\
&= \mathbf{a}.\mathbf{b} + \mathbf{a}.\mathbf{c}
\end{aligned}$

Similarly:
$$(\mathbf{a} + \mathbf{b}).\mathbf{c} = \mathbf{a}.\mathbf{c} + \mathbf{b}.\mathbf{c}$$

and so:
$$(\mathbf{a} + \mathbf{b}).(\mathbf{c} + \mathbf{d}) = \mathbf{a}.\mathbf{c} + \mathbf{a}.\mathbf{d} + \mathbf{b}.\mathbf{c} + \mathbf{b}.\mathbf{d}$$

Summary:
$$\mathbf{a}.\mathbf{b} = \mathbf{b}.\mathbf{a}$$
$$\mathbf{a}.(\lambda\mathbf{b}) = (\lambda\mathbf{a}).\mathbf{b} = \lambda(\mathbf{a}.\mathbf{b})$$
$$\mathbf{a}.\mathbf{a} = a^2$$
$$\mathbf{a}.(\mathbf{b} + \mathbf{c}) = \mathbf{a}.\mathbf{b} + \mathbf{a}.\mathbf{c}$$
$$(\mathbf{a} + \mathbf{b}).\mathbf{c} = \mathbf{a}.\mathbf{c} + \mathbf{b}.\mathbf{c}$$
$$(\mathbf{a} + \mathbf{b}).(\mathbf{c} + \mathbf{d}) = \mathbf{a}.\mathbf{c} + \mathbf{a}.\mathbf{d} + \mathbf{b}.\mathbf{c} + \mathbf{b}.\mathbf{d}$$

EXAMPLE 3

a   Expand and simplify $(2\mathbf{p} + \mathbf{q}).(5\mathbf{p} - 2\mathbf{q})$.

b   Simplify $(\mathbf{u} + 2\mathbf{v}).(\mathbf{u} - \mathbf{v})$ given that $\mathbf{u}$ is perpendicular to $\mathbf{v}$.

Solution

a $(2\mathbf{p} + \mathbf{q}).(5\mathbf{p} - 2\mathbf{q}) = 2\mathbf{p}.5\mathbf{p} + 2\mathbf{p}.(-2\mathbf{q}) + \mathbf{q}.5\mathbf{p} + \mathbf{q}.(-2\mathbf{q})$

$$= 10\mathbf{p}.\mathbf{p} - 4\mathbf{p}.\mathbf{q} + 5\mathbf{q}.\mathbf{p} - 2\mathbf{q}.\mathbf{q}$$

$$= 10p^2 - 4\mathbf{p}.\mathbf{q} + 5\mathbf{p}.\mathbf{q} - 2q^2$$

$$= 10p^2 + \mathbf{p}.\mathbf{q} - 2q^2$$

b $(\mathbf{u} + 2\mathbf{v}).(\mathbf{u} - \mathbf{v}) = \mathbf{u}.\mathbf{u} - \mathbf{u}.\mathbf{v} + 2\mathbf{v}.\mathbf{u} - 2\mathbf{v}.\mathbf{v}$

$$= u^2 + \mathbf{u}.\mathbf{v} - 2v^2$$

$$= u^2 - 2v^2 \text{ (because } \mathbf{u} \text{ is perpendicular to } \mathbf{v} \text{ and hence } \mathbf{u}.\mathbf{v} = 0).$$

Exercise 8A

In questions **1** to **6** evaluate the given scalar product where **a**, **b**, **c** and **d** are as shown in the diagram. (Leave answers in exact form.)

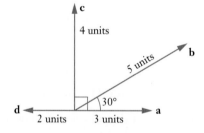

1   a.b   **2**   a.c   **3**   a.d

4   b.c   **5**   b.d   **6**   c.d

In questions **7** to **12** evaluate the given scalar product where **e**, **f** and **g** are as shown in the diagram. (Leave answers in exact form.)

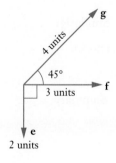

7   e.f   **8**   f.e   **9**   e.e

10   f.f   **11**   f.g   **12**   g.f

In questions **13** to **18** find the scalar product of the two vectors shown. (Leave answers in exact form).

13
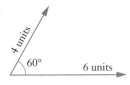

14
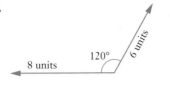

15
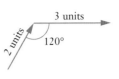

16

17

18
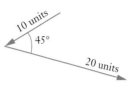

19 For each of the following, state whether scalar or vector.

   a $|\mathbf{a}|$ **b** $\mathbf{a.b}$ **c** $\mathbf{a} + \mathbf{b}$ **d** $\mathbf{a} - \mathbf{b}$

   e $\mathbf{a} + 2\mathbf{b}$ **f** $\mathbf{a}.(2\mathbf{b})$ **g** $(\mathbf{a} + \mathbf{b}).(\mathbf{c} + \mathbf{d})$ **h** $|\mathbf{a} + \mathbf{b}|$

   i $\mathbf{a} + \lambda\mathbf{b}$ **j** $\mathbf{a}.(\lambda\mathbf{b})$

20 With **i** and **j** representing horizontal and vertical unit vectors respectively, find

   a $\mathbf{i.i}$ **b** $\mathbf{i.j}$ **c** $\mathbf{j.j}$

21 Expand and simplify each of the following. (Writing $|\mathbf{a}|$ as a and $|\mathbf{b}|$ as b.)

   a $(\mathbf{a} + \mathbf{b}).(\mathbf{a} - \mathbf{b})$ **b** $(\mathbf{a} + \mathbf{b}).(\mathbf{a} + \mathbf{b})$ **c** $(\mathbf{a} - \mathbf{b}).(\mathbf{a} - \mathbf{b})$

   d $(2\mathbf{a} + \mathbf{b}).(2\mathbf{a} - \mathbf{b})$ **e** $(\mathbf{a} + 3\mathbf{b}).(\mathbf{a} - 2\mathbf{b})$ **f** $\mathbf{a}.(\mathbf{a} - \mathbf{b}) + \mathbf{a.b}$

22 If **a** and **b** are perpendicular vectors prove that $(\mathbf{a} + \mathbf{b}).(\mathbf{a} - 2\mathbf{b}) = a^2 - 2b^2$, where a is $|\mathbf{a}|$ and b is $|\mathbf{b}|$.

23 If **a** and **b** are perpendicular vectors, and writing $|\mathbf{a}|$ as a and $|\mathbf{b}|$ as b, which of the following statements are necessarily true?

   a $a = b$ **b** $\mathbf{a.b} = 0$ **c** $ab = 0$ **d** $\mathbf{a}.(\mathbf{a} + \mathbf{b}) = a^2$

24 If **a** is perpendicular to $(\mathbf{b} - \mathbf{c})$ which of the following statements are necessarily true?

   a $\mathbf{a}.(\mathbf{b} - \mathbf{c}) = 0$ **b** $\mathbf{a.b} = \mathbf{a.c}$

   c $\mathbf{b} = \mathbf{c}$ **d**   **a** is perpendicular to $(\mathbf{c} - \mathbf{b})$

25 Find an expression for **a.b** given that $\mathbf{a} = x_1\mathbf{i} + y_1\mathbf{j}$, $\mathbf{b} = x_2\mathbf{i} + y_2\mathbf{j}$ and **i** and **j** are perpendicular vectors each of unit length.

26 Find the scalar projection of **a** onto **b** given that
$$\mathbf{a} \cdot \mathbf{b} = 14 \text{ and } |\mathbf{b}| = 5.$$

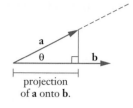

projection of **a** onto **b**.

27 Find the scalar projection of **b** onto **a** given that
$$\mathbf{a} \cdot \mathbf{b} = 18 \text{ and } |\mathbf{a}| = 25.$$

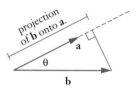

28 If **a** and **b** are perpendicular vectors which of the following statements are necessarily true?

a $\mathbf{a} \cdot (\mathbf{a} - \mathbf{b}) = 0$ **b** $(\mathbf{a} + \mathbf{b}) \cdot (\mathbf{a} - \mathbf{b}) = 0$

c $(\mathbf{a} + \mathbf{b}) \cdot (\mathbf{a} + \mathbf{b}) = a^2 + b^2$

29 Given that $|\mathbf{a}| = 5$, $|\mathbf{b}| = 3$ and $\mathbf{a} \cdot \mathbf{b} = 7$, find:

a   the angle between **a** and **b**, correct to the nearest degree,

b $\mathbf{a} \cdot \mathbf{a}$ **c** $\mathbf{b} \cdot \mathbf{b}$ **d** $(\mathbf{a} - \mathbf{b}) \cdot (\mathbf{a} - \mathbf{b})$ **e** $|\mathbf{a} - \mathbf{b}|$

30 Evaluate $\mathbf{p} \cdot \mathbf{q}$ given that $\mathbf{p} = 3\mathbf{a} + 2\mathbf{b}$, $\mathbf{q} = 4\mathbf{a} - \mathbf{b}$, $|\mathbf{a}| = 3$, $|\mathbf{b}| = 2$ and **a** and **b** are perpendicular to each other.

31 Explain why with three vectors **a**, **b** and **c**, we can talk of $\mathbf{a} \cdot (\mathbf{b} + \mathbf{c})$ and also of $\mathbf{a} \cdot (\mathbf{b} - \mathbf{c})$ but $\mathbf{a} \cdot (\mathbf{b} \cdot \mathbf{c})$ has no meaning.

32 a   Prove that $|\mathbf{a} \cdot \mathbf{b}| \le |\mathbf{a}| \, |\mathbf{b}|$

b   Expand $(\mathbf{a} + \mathbf{b}) \cdot (\mathbf{a} + \mathbf{b})$ and use this, together with the inequality from **a** to prove the triangle inequality: $|\mathbf{a} + \mathbf{b}| \le |\mathbf{a}| + |\mathbf{b}|$

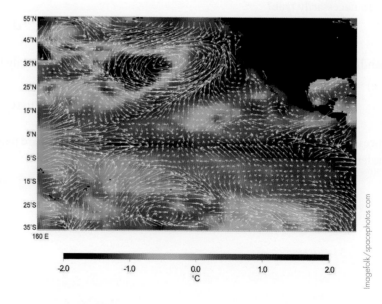

The scalar product from the components

If you managed question **25** of the previous exercise then you would have obtained an important result that allows **a.b** to be determined quickly when **a** and **b** are given in component form. Your working for question **25** should have followed the same reasoning as is shown below.

Let $\mathbf{a} = x_1\mathbf{i} + y_1\mathbf{j}$ and $\mathbf{b} = x_2\mathbf{i} + y_2\mathbf{j}$.

Then

$$\mathbf{a.b} = (x_1\mathbf{i} + y_1\mathbf{j}).(x_2\mathbf{i} + y_2\mathbf{j})$$
$$= (x_1\mathbf{i}).(x_2\mathbf{i}) + (x_1\mathbf{i}).(y_2\mathbf{j}) + (y_1\mathbf{j}).(x_2\mathbf{i}) + (y_1\mathbf{j}).(y_2\mathbf{j})$$
$$= x_1x_2(\mathbf{i.i}) + x_1y_2(\mathbf{i.j}) + y_1x_2(\mathbf{j.i}) + y_1y_2(\mathbf{j.j}) \quad [1]$$

But

$\mathbf{i.i} = (1)(1)\cos 0°$ $\qquad$ $\mathbf{i.j} = \mathbf{j.i}$ $\qquad$ $\mathbf{j.j} = (1)(1)\cos 0°$

$\quad = 1$ $\qquad\qquad\qquad$ $= (1)(1)\cos 90°$ $\qquad$ $= 1$

$\qquad\qquad\qquad\qquad\qquad = 0$

$\therefore$ from [1] $\mathbf{a.b} = x_1x_2 + y_1y_2$

If $\mathbf{a} = x_1\mathbf{i} + y_1\mathbf{j}$ and $\mathbf{b} = x_2\mathbf{i} + y_2\mathbf{j}$ then $\mathbf{a.b} = x_1x_2 + y_1y_2$

and, by definition, $\mathbf{a.b} = |\mathbf{a}|\,|\mathbf{b}|\cos\theta$

EXAMPLE 4

Given that $\mathbf{u} = 5\mathbf{i} - 3\mathbf{j}$, $\mathbf{v} = 2\mathbf{i} + \mathbf{j}$ and $\mathbf{w} = -4\mathbf{i} + 3\mathbf{j}$, find

a $\quad$ **u.v** $\qquad\qquad\qquad$ **b** $\quad$ **u.w** $\qquad\qquad\qquad$ **c** $\quad$ **u.(2v − w)**

Solution

a $\qquad\qquad \mathbf{u.v} = (5\mathbf{i} - 3\mathbf{j}).(2\mathbf{i} + \mathbf{j})$
$$= (5)(2) + (-3)(1)$$
$$= 7$$

b $\qquad\qquad \mathbf{u.w} = (5\mathbf{i} - 3\mathbf{j}).(-4\mathbf{i} + 3\mathbf{j})$
$$= (5)(-4) + (-3)(3)$$
$$= -29$$

c $\qquad\quad 2\mathbf{v} - \mathbf{w} = 2(2\mathbf{i} + \mathbf{j}) - (-4\mathbf{i} + 3\mathbf{j})$
$$= 8\mathbf{i} - \mathbf{j}$$

$\therefore$ $\quad \mathbf{u.(2v − w)} = (5\mathbf{i} - 3\mathbf{j}).(8\mathbf{i} - \mathbf{j})$
$$= (5)(8) + (-3)(-1)$$
$$= 43$$

Or, using parts **a** and **b**: $\qquad \mathbf{u.(2v − w)} = 2\mathbf{u.v} - \mathbf{u.w}$
$$= 2 \times 7 - (-29)$$
$$= 14 + 29$$
$$= 43, \text{ as before.}$$

EXAMPLE 5

Find the value of λ given that $\mathbf{a}$ $(= 2\mathbf{i} + 3\mathbf{j})$ is perpendicular to $\mathbf{b}$ $(= \lambda\mathbf{i} - 5\mathbf{j})$.

Solution

$\mathbf{a}$ and $\mathbf{b}$ are perpendicular. Thus $\quad \mathbf{a}.\mathbf{b} = 0$

$\therefore \qquad\qquad (2\mathbf{i} + 3\mathbf{j}).(\lambda\mathbf{i} - 5\mathbf{j}) = 0$

i.e. $\qquad\qquad\qquad 2\lambda - 15 = 0$

giving $\qquad\qquad\qquad\qquad \lambda = 7.5$

With $\mathbf{a}$ and $\mathbf{b}$ perpendicular λ must equal 7.5.

EXAMPLE 6

Find the angle between $\mathbf{a}$ and $\mathbf{b}$ given that $\mathbf{a} = 5\mathbf{i} + 2\mathbf{j}$ and $\mathbf{b} = -3\mathbf{i} + \mathbf{j}$. (Give your answer correct to the nearest degree.)

Solution

$\mathbf{a}.\mathbf{b} = (5\mathbf{i} + 2\mathbf{j}).(-3\mathbf{i} + \mathbf{j})$ $\qquad$ $|\mathbf{a}| = \sqrt{5^2 + 2^2}$ $\qquad$ $|\mathbf{b}| = \sqrt{(-3)^2 + 1^2}$

$\qquad = (5)(-3) + (2)(1)$ $\qquad\qquad = \sqrt{29}$ $\qquad\qquad\qquad = \sqrt{10}$

$\qquad = -13$

But $\qquad \mathbf{a}.\mathbf{b} = |\mathbf{a}|\,|\mathbf{b}|\cos\theta$

$\therefore \qquad -13 = \sqrt{29}\sqrt{10}\cos\theta$

giving $\qquad \theta \approx 139.8°.$

The angle between $\mathbf{a}$ and $\mathbf{b}$ is 140°, correct to the nearest degree.

TECHNOLOGY

Some calculators and computer programs are able to determine the scalar product of two vectors directly, as well as give other information about vectors, as suggested by the display.

Explore your calculator in this regard.

However, make sure that you can show how to obtain the scalar product and the angle between vectors algebraically if required to do so.

$\text{dotP}\left(\begin{bmatrix} 5 \\ 2 \end{bmatrix}, \begin{bmatrix} -3 \\ 1 \end{bmatrix}\right)$

$\qquad\qquad\qquad -13$

$\text{angle}\left(\begin{bmatrix} 5 \\ 2 \end{bmatrix}, \begin{bmatrix} -3 \\ 1 \end{bmatrix}\right)$

$\qquad\qquad 139.7636417$

$\text{solve}\left(\text{dotP}\left(\begin{bmatrix} 2 \\ 3 \end{bmatrix}, \begin{bmatrix} \lambda \\ -5 \end{bmatrix}\right) = 0, \lambda\right)$

$\qquad\qquad\qquad \lambda = 7.5$

Exercise 8B

1 Given that $\mathbf{a} = 3\mathbf{i} - 2\mathbf{j}$, $\mathbf{b} = 5\mathbf{i} + 6\mathbf{j}$ and $\mathbf{c} = 2\mathbf{i} - \mathbf{j}$ evaluate the following.

a $\mathbf{a}.\mathbf{b}$ **b** $\mathbf{b}.\mathbf{a}$ **c** $\mathbf{a}.\mathbf{c}$ **d** $\mathbf{b}.\mathbf{c}$

2 Given that $\mathbf{x} = \begin{pmatrix} 2 \\ 3 \end{pmatrix}$, $\mathbf{y} = \begin{pmatrix} 5 \\ -1 \end{pmatrix}$ and $\mathbf{z} = \begin{pmatrix} 4 \\ 2 \end{pmatrix}$ evaluate the following.

a $\mathbf{x}.\mathbf{y}$ **b** $\mathbf{x}.\mathbf{z}$ **c** $\mathbf{z}.\mathbf{x}$ **d** $\mathbf{y}.\mathbf{z}$

3 Given that $\mathbf{p} = <3, 1>$, $\mathbf{q} = <2, -1>$ and $\mathbf{r} = <5, 2>$ evaluate the following.

a $\mathbf{q}.\mathbf{r}$ **b** $2\mathbf{q}.3\mathbf{r}$ **c** $\mathbf{p}.(\mathbf{q} + \mathbf{r})$ **d** $\mathbf{p}.(\mathbf{q} - \mathbf{r})$

4 State whether each pair of vectors are perpendicular vectors or not.

a $2\mathbf{i} + 3\mathbf{j}$ and $4\mathbf{i} - 2\mathbf{j}$ **b** $-2\mathbf{i} + \mathbf{j}$ and $4\mathbf{i} - 2\mathbf{j}$

c $3\mathbf{i} - \mathbf{j}$ and $2\mathbf{i} + 6\mathbf{j}$ **d** $<12, -3>$ and $<1, 4>$

e $\begin{pmatrix} 5 \\ 2 \end{pmatrix}$ and $\begin{pmatrix} -3 \\ 7 \end{pmatrix}$ **f** $\begin{pmatrix} 14 \\ 8 \end{pmatrix}$ and $\begin{pmatrix} -4 \\ 7 \end{pmatrix}$

5 Given that $\mathbf{d} = 3\mathbf{i} + \mathbf{j}$, $\mathbf{e} = 2\mathbf{i} + 4\mathbf{j}$ and $\mathbf{f} = -2\mathbf{i} - 3\mathbf{j}$, evaluate the following.

a $\mathbf{d}.\mathbf{e}$ **b** $\mathbf{e}.\mathbf{f}$ **c** $\mathbf{d}.(\mathbf{e} + \mathbf{f})$ **d** $(\mathbf{d} + \mathbf{e}).\mathbf{f}$

6 Given that $\mathbf{a} = 2\mathbf{i} - \mathbf{j}$, $\mathbf{b} = 3\mathbf{i} + 2\mathbf{j}$ and $\mathbf{c} = 4\mathbf{i} - 3\mathbf{j}$, evaluate the following.

a $\mathbf{a}.(\mathbf{b} + \mathbf{c})$ **b** $(\mathbf{a} + \mathbf{b}).\mathbf{c}$ **c** $\mathbf{b}.(\mathbf{a} + \mathbf{c})$ **d** $(\mathbf{a} - \mathbf{b}).(\mathbf{b} - \mathbf{c})$

7 If $\mathbf{a} = 2\mathbf{i} + 3\mathbf{j}$, $\mathbf{b} = \mathbf{i} - 4\mathbf{j}$ and $\mathbf{c} = -4\mathbf{i} + 5\mathbf{j}$ evaluate

a $\mathbf{a}.\mathbf{b}$ **b** $\mathbf{a}.\mathbf{c}$ **c** $\mathbf{b} + \mathbf{c}$

Use your answer to part **c** to determine $\mathbf{a}.(\mathbf{b} + \mathbf{c})$ and hence verify that

$$\mathbf{a}.(\mathbf{b} + \mathbf{c}) = \mathbf{a}.\mathbf{b} + \mathbf{a}.\mathbf{c}$$

8 If $\mathbf{p} = 3\mathbf{i} + 4\mathbf{j}$ and $\mathbf{q} = 5\mathbf{i} - 12\mathbf{j}$ find

a $|\mathbf{p}|$ **b** $|\mathbf{q}|$ **c** $\mathbf{p}.\mathbf{q}$

Hence find the angle between $\mathbf{p}$ and $\mathbf{q}$, to the nearest degree.

9 If $\mathbf{c} = 7\mathbf{i} + 7\mathbf{j}$ and $\mathbf{d} = 15\mathbf{i} - 8\mathbf{j}$ find

a $|\mathbf{c}|$ **b** $|\mathbf{d}|$ **c** $\mathbf{c}.\mathbf{d}$

Hence find the angle between $\mathbf{c}$ and $\mathbf{d}$, to the nearest degree.

10 If $\mathbf{a} = 2\mathbf{i} - 3\mathbf{j}$, $\mathbf{b} = 4\mathbf{i} + 5\mathbf{j}$ and $\mathbf{c} = 2\mathbf{i} - \mathbf{j}$, prove that $\mathbf{a}$ is perpendicular to $(\mathbf{b} + \mathbf{c})$.

11 If $\mathbf{a} = -2\mathbf{i} + 2\mathbf{j}$, $\mathbf{b} = 5\mathbf{i} + 2\mathbf{j}$ and $\mathbf{c} = 4\mathbf{i} - \mathbf{j}$, prove that $(\mathbf{a} + 2\mathbf{b})$ and $(\mathbf{b} - 2\mathbf{c})$ are perpendicular vectors.

For each of numbers **12** to **17** find:

a the scalar product of the two vectors,

b the angle between the two vectors (to the nearest degree).

12 $\mathbf{a} = 3\mathbf{i} + 4\mathbf{j}$, $\mathbf{b} = 4\mathbf{i} + 3\mathbf{j}$.   **13** $\mathbf{c} = 24\mathbf{i} + 7\mathbf{j}$, $\mathbf{d} = 5\mathbf{i} + 12\mathbf{j}$.

14 $\mathbf{e} = 2\mathbf{i} + \mathbf{j}$, $\mathbf{f} = 3\mathbf{i} - 2\mathbf{j}$.   **15** $\mathbf{g} = 2\mathbf{i} + \mathbf{j}$, $\mathbf{h} = -4\mathbf{i} + 8\mathbf{j}$.

16 $\mathbf{m} = -3\mathbf{i} + 4\mathbf{j}$, $\mathbf{n} = 9\mathbf{i} - 12\mathbf{j}$.   **17** $\mathbf{p} = \mathbf{i}$, $\mathbf{q} = 12\mathbf{i} - 5\mathbf{j}$.

18 $\mathbf{a} = 2\mathbf{i} + 3\mathbf{j}$, $\mathbf{b} = \lambda\mathbf{i} + 12\mathbf{j}$ and $\mathbf{c} = \mu\mathbf{i} - 7\mathbf{j}$.

Find λ given that $\mathbf{a}$ and $\mathbf{b}$ are parallel.

Find μ given that $\mathbf{a}$ and $\mathbf{c}$ are perpendicular.

19 $\mathbf{d} = w\mathbf{i} + \mathbf{j}$, $\mathbf{e} = -\mathbf{i} + 7\mathbf{j}$ and $\mathbf{f} = x\mathbf{i} + 5\mathbf{j}$.

Find w given that $\mathbf{d}$ and $\mathbf{e}$ are perpendicular.

Find x given that it is a negative number and that $|\mathbf{d}| = |\mathbf{f}|$.

20 The scalar projection of $\mathbf{a}$ onto $\mathbf{b}$ is $|\mathbf{a}| \cos \theta$,

$$\text{i.e. } \frac{\mathbf{a}.\mathbf{b}}{|\mathbf{b}|}$$

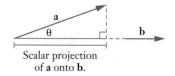

Scalar projection
of $\mathbf{a}$ onto $\mathbf{b}$.

The vector projection of $\mathbf{a}$ onto $\mathbf{b}$ is $|\mathbf{a}| \cos \theta \, \hat{\mathbf{b}}$,

$$\text{i.e. } \frac{\mathbf{a}.\mathbf{b}}{\mathbf{b}.\mathbf{b}} \mathbf{b}$$

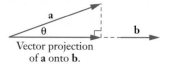

Vector projection
of $\mathbf{a}$ onto $\mathbf{b}$.

Find

a the scalar projection of $3\mathbf{i} + 4\mathbf{j}$ onto $2\mathbf{i} + \mathbf{j}$,

b the vector projection of $3\mathbf{i} + 4\mathbf{j}$ onto $2\mathbf{i} + \mathbf{j}$,

c the scalar projection of $2\mathbf{i} + \mathbf{j}$ onto $3\mathbf{i} + 4\mathbf{j}$,

d the vector projection of $2\mathbf{i} + \mathbf{j}$ onto $3\mathbf{i} + 4\mathbf{j}$.

21 Find the two vectors in the $\mathbf{i} - \mathbf{j}$ plane that are perpendicular to $3\mathbf{i} - 4\mathbf{j}$ and have magnitude of 25 units.

22 Find the two unit vectors in the $\mathbf{i} - \mathbf{j}$ plane that are perpendicular to $2\mathbf{i} + \mathbf{j}$.

23 Points A, B, C and D have position vectors $2\mathbf{i} + 4\mathbf{j}$, $6\mathbf{i} + 6\mathbf{j}$, $7\mathbf{i} + 2\mathbf{j}$ and $4\mathbf{i} + \mathbf{j}$ respectively.

Find $\overrightarrow{AC}$ and $\overrightarrow{BD}$ and hence prove that AC is perpendicular to BD.

24 The position vectors of points A, B and C are $4\mathbf{i} + 7\mathbf{j}$, $6\mathbf{i} + 2\mathbf{j}$ and $8\mathbf{i} + 9\mathbf{j}$ respectively.

Find   **a** $\overrightarrow{AC}$ **b** $\overrightarrow{AB}$ **c** $\overrightarrow{AC}.\overrightarrow{AB}$

d the size of angle CAB correct to the nearest degree.

25 The angle between $\mathbf{a}$ and $\mathbf{b}$ is $45°$. If $\mathbf{a} = 3\mathbf{i} - \mathbf{j}$ and $\mathbf{b} = 4\mathbf{i} + y\mathbf{j}$ find the two possible values of y.

ISBN 9780170390477

Proofs using the scalar product

In the previous chapter, we used our understanding of vectors to prove various geometrical facts. We can now add the various properties of the scalar product to our toolbox of facts that we can use in such proofs. Useful facts include the following:

- If, for non zero vectors **a** and **b**, **a.b** = 0, then **a** and **b** are perpendicular vectors. (And conversely, if **a** and **b** are perpendicular vectors, **a.b** = 0.)

- The various properties of the scalar product listed on page 139, for example:
$$(a + b).(c + d) = a.c + a.d + b.c + b.d$$

EXAMPLE 7

To prove: The diagonals of a rhombus are perpendicular to each other.

The diagram shows a rhombus OABC. $\overrightarrow{OA}$ = **a** and $\overrightarrow{OC}$ = **c**.

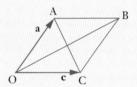

a   Find $\overrightarrow{OB}$ and $\overrightarrow{AC}$ in terms of **a** and **c**.

b   With OABC a rhombus it follows that $|a| = |c|$.

Use this fact to prove that the diagonals of a rhombus are perpendicular to each other.

Solution

a $\overrightarrow{OB} = \overrightarrow{OA} + \overrightarrow{AB}$ $\qquad\qquad$ $\overrightarrow{AC} = \overrightarrow{AO} + \overrightarrow{OC}$
$\qquad\quad$ = **a** + **c** $\qquad\qquad\qquad\qquad$ = –**a** + **c**

b $\overrightarrow{OB}.\overrightarrow{AC} = (a + c).(–a + c)$

$\qquad\qquad\quad = -a.a + a.c - c.a + c.c$ $\qquad$ Standard expansion.

$\qquad\qquad\quad = -a^2 + a.c - a.c + c^2$ $\qquad\quad$ Because $x.x = x^2$ and $x.y = y.x$.

$\qquad\qquad\quad = c^2 - a^2$

$\qquad\qquad\quad = 0$ $\qquad\qquad\qquad\qquad\qquad$ Because, with OABC a rhombus, $|a| = |c|$ (i.e. a = c).

Thus $\overrightarrow{OB}$ is perpendicular to $\overrightarrow{AC}$ and hence the diagonals of a rhombus are perpendicular to each other.

Exercise 8C

1 To prove: The theorem of Pythagoras.

In the right triangle shown, $\overrightarrow{OA} = \mathbf{a}$ and $\overrightarrow{OB} = \mathbf{b}$.

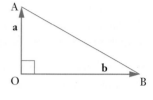

a   Express $\overrightarrow{AB}$ in terms of $\mathbf{a}$ and $\mathbf{b}$.

b   Use the fact that $\overrightarrow{AB}.\overrightarrow{AB} = (AB)^2$ to prove that
$(AB)^2 = (OA)^2 + (OB)^2$.

2 To prove: The diagonals of a rectangle are congruent.

In the rectangle shown, $\overrightarrow{OA} = \mathbf{a}$ and $\overrightarrow{OC} = \mathbf{c}$.

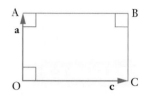

a   Write down the value of $\mathbf{a}.\mathbf{c}$.

b   Express $\overrightarrow{AC}$ and $\overrightarrow{OB}$ in terms of $\mathbf{a}$ and $\mathbf{c}$.

c   Prove that $|\overrightarrow{AC}| = |\overrightarrow{OB}|$.

(Hint: Find $(AC)^2$ by using $\overrightarrow{AC}.\overrightarrow{AC} = (AC)^2$.)

3 To prove: If the diagonals of a parallelogram cut at right angles then the parallelogram is a rhombus.

OABC is a parallelogram with $\overrightarrow{OA} = \mathbf{a}$ and $\overrightarrow{OC} = \mathbf{c}$.

If $\overrightarrow{AC}$ and $\overrightarrow{OB}$ are perpendicular, prove that $|\mathbf{a}| = |\mathbf{c}|$.

4 To prove: In an isosceles triangle, a line drawn from the vertex that is common to the congruent sides, to the midpoint of the third side, is perpendicular to that third side.

ABC is an isosceles triangle with AB = CB.

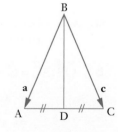

D is the midpoint of AC, $\overrightarrow{BA} = \mathbf{a}$ and $\overrightarrow{BC} = \mathbf{c}$.

a   Find $\overrightarrow{AC}$ and $\overrightarrow{BD}$ in terms of $\mathbf{a}$ and $\mathbf{c}$.

b   Use $\overrightarrow{AC}.\overrightarrow{BD}$ to prove that $\angle BDA$ is a right angle.

5 To prove: The angle in a semicircle is a right angle.

The diagram shows a circle centre O with AB a diameter and C a point on the circumference.

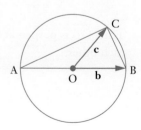

$\overrightarrow{OB} = \mathbf{b}$ and $\overrightarrow{OC} = \mathbf{c}$.

a   Express $\overrightarrow{CB}$, $\overrightarrow{AO}$ and $\overrightarrow{AC}$ in terms of $\mathbf{b}$ and/or $\mathbf{c}$.

b   Prove that $\angle ACB$ is a right angle.

ISBN 9780170390477

6 Prove that the sum of the squares of the lengths of the diagonals of a parallelogram is equal to the sum of the squares of the lengths of the sides.

7 To prove: The altitudes of a triangle are concurrent.

The diagram shows triangle ABC.

The perpendicular from A to BC meets BC at D and the perpendicular from B to AC meets AC at E.

AD and BE intersect at F.

Use the facts that $\mathbf{a}.\overrightarrow{BC} = 0$ and $\mathbf{b}.\overrightarrow{AC} = 0$ to prove that $\overrightarrow{CF}$ is perpendicular to $\overrightarrow{AB}$.

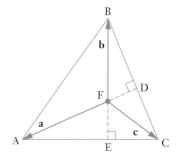

8 To prove: The diagonals of a rhombus bisect the angles of the rhombus.

The diagram shows a rhombus OABC with $\overrightarrow{OA} = \mathbf{a}$ and $\overrightarrow{OC} = \mathbf{c}$.
∠BOA = α and ∠BOC = β.

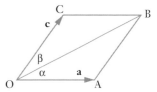

a   Writing $|\mathbf{a}|$ as a and $|\mathbf{c}|$ as c, prove that $\overrightarrow{OB}.\overrightarrow{OA} = a^2 + \mathbf{a}.\mathbf{c}$

and $\overrightarrow{OB}.\overrightarrow{OC} = c^2 + \mathbf{a}.\mathbf{c}$

b   Hence prove that α = β.

(Note: The above proves that BO bisects ∠COA. From the properties of isosceles triangles it follows that BO also bisects ∠CBA. By repeating this process at C the initial statement is proved.)

9 To prove: The perpendicular bisectors of the sides of a triangle are concurrent.

The diagram shows triangle ABC. Points D, E and F are the midpoints of BC, AC and AB respectively. The perpendicular bisector of BC meets the perpendicular bisector of AC at G.

Vectors $\mathbf{a}$, $\mathbf{b}$ and $\mathbf{c}$ are as shown in the diagram, and $|\mathbf{a}| = a$, $|\mathbf{b}| = b$ and $|\mathbf{c}| = c$.

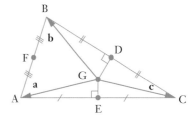

a   Use the fact that $\overrightarrow{GE}.\overrightarrow{AC} = 0$ to prove $a^2 = c^2$.

b   Use the fact that $\overrightarrow{GD}.\overrightarrow{BC} = 0$ to prove $b^2 = c^2$.

c   By determining $\overrightarrow{GF}.\overrightarrow{AB}$, prove that GF is perpendicular to AB.

Miscellaneous exercise eight

This miscellaneous exercise may include questions involving the work of this chapter, the work of any previous chapters, and the ideas mentioned in the Preliminary work section at the beginning of the book.

1 Discuss the validity of each of the following statements.

- Quadrilateral ABCD is a rhombus $\Leftrightarrow$ Quadrilateral ABCD is a parallelogram.

- Diagonals of quadrilateral PQRS cut at right angles $\Leftrightarrow$ PQRS is a rhombus.

- The diagonals of parallelogram WXYZ cut at right angles $\Leftrightarrow$ WXYZ is a rhombus.

2 If **a** is perpendicular to (**b** − **a**), which of the following statements are necessarily true?

   a $\mathbf{a}.(\mathbf{b} - \mathbf{a}) = 0$ **b** $\mathbf{a}.\mathbf{b} = \mathbf{a}.\mathbf{a}$ **c** $\mathbf{a} = \mathbf{b}$ **d** $\mathbf{a}.\mathbf{b} = |\mathbf{a}|^2$

3 Points A and B have position vectors $\mathbf{i} + 3\mathbf{j}$ and $21\mathbf{i} - 7\mathbf{j}$ respectively.

Find the position vector of the point P such that $\overrightarrow{AP} : \overrightarrow{PB} = 3 : 7$.

4 How many different groups of eight letters are there if each group must contain two vowels, six consonants and no letter used more than once?

5 Point A has position vector $-4\mathbf{i} + 6\mathbf{j}$.

Relative to point A a second point, B, has location $6\mathbf{i} - \mathbf{j}$.

Relative to point B a third point, C, has location $4\mathbf{i} + 5\mathbf{j}$.

   a How far is point C from point A? (Give your answer in exact form.)

   b What is the position vector of point C?

   c Find the position vector of the midpoint of the line AC.

6 Three digit numbers are to be made using the digits 1, 2, 3, 4, 5.

How many three digit numbers are possible if

   a each digit can be used more than once in a number?

   b each digit may not be used more than once in a number?

   c multiple use of a digit is not allowed and the number must be even?

   d multiple use of a digit is not allowed and the number must be odd?

   e multiple use of a digit is not allowed and the number must be odd and more than 300?

7 Given that $|\mathbf{c}| = 2$, $|\mathbf{d}| = 3$ and $\mathbf{c}.\mathbf{d} = -5$, find:

   a the angle between **c** and **d**, correct to the nearest degree,

   b $\mathbf{c}.\mathbf{c}$

   c $\mathbf{d}.\mathbf{d}$

   d $(\mathbf{c} + \mathbf{d}).(\mathbf{c} + \mathbf{d})$

   e $|\mathbf{c} + \mathbf{d}|$

ISBN 9780170390477

8 Six files, A, B, C, D, E and F are to be arranged on a shelf.

    a In how many ways can this be done?

In how many of the arrangements

    b is file A at the extreme left?

    c is file A next to file B?

    d are the first three files on the left, A, B and C in that order?

    e are the first three files on the left, A, B and C in any order?

    f are files A, B, C and D together in that order?

    g are files A, B, C and D together in any order?

	A	B	C	D	E	F

9 One section of a river runs from North West to South East with speed 1 m/s.

A person wishes to row a boat from a point A on one bank to a point B on the other bank, B being due North of A.

The person can row the boat with a speed of 2 m/s in still water and the river has constant width of 30 metres.

    a On what bearing should the person row the boat so that their effort, together with the flow of the river, produces the desired result (answer to the nearest degree)?

    b How long would the journey take (to the nearest second)?

10 In how many ways can Amanda, Bridie, Claire, Donelle, Erin, Fran, Gia, Harni and Icolyn be arranged in a row for a photograph if:

    a Donelle is to be in the middle?

    b Donelle is to be in the middle and Erin and Harni are each to be at an end?

    c Donelle is to be in the middle, Erin and Harni must each be at an end and Claire and Icolyn are to stand next to each other?

11 Points A, B and C have position vectors $2\mathbf{p} + \mathbf{q}$, $3\mathbf{p} - \mathbf{q}$ and $6\mathbf{p} - 7\mathbf{q}$.

Prove that A, B and C are collinear and find the ratios $\overrightarrow{AB} : \overrightarrow{BC}$ and $\overrightarrow{AB} : \overrightarrow{AC}$.

12 A committee consists of 13 people: 6 men and 7 women. Five of these thirteen are to be randomly selected to form a sub-committee, and then these chosen five are to be arranged in a line for a photograph.

    a How many different photographs are possible?

    b How many of these consist of two men and three women?

13 The position vectors of points D, E and F are $-3\mathbf{i} + 3\mathbf{j}$, $3\mathbf{i} + 2\mathbf{j}$ and $8\mathbf{i} + 7\mathbf{j}$ respectively.

Find   **a** $\overrightarrow{ED}$ **b** $\overrightarrow{EF}$ **c** $\overrightarrow{ED} . \overrightarrow{EF}$

    d the size of angle DEF correct to the nearest degree.

14 The angle between $\mathbf{a}$ ($= 5\sqrt{3}\mathbf{i} + \mathbf{j}$) and $\mathbf{b}$ ($= 2\sqrt{3}\mathbf{i} + w\mathbf{j}$) is 60°.

With the assistance of a calculator, find the two possible values of w.

15 Vectors **a**, **b** and **c** are such that $\mathbf{a} = 2\mathbf{i} - 3\mathbf{j}$, $\mathbf{b} = x\mathbf{i} + 4\mathbf{j}$ and $\mathbf{c} = 9\mathbf{i} - y\mathbf{j}$.

If **a** and **b** are perpendicular and **a** and **c** are parallel, determine x and y.

16 Find the resultant of forces **F** and **P** where $\quad\mathbf{F} = (6\mathbf{i} + 4\mathbf{j})$ N

$\qquad\qquad\qquad\qquad\qquad\qquad\text{and}\qquad \mathbf{P} = (2\mathbf{i} - 7\mathbf{j})$ N.

Determine the angle between this resultant and whichever of **F** and **P** has the smaller magnitude, giving your answer to the nearest $0.1°$.

17 In the parallelogram OABC, $\overrightarrow{OA} = \mathbf{a}$, and $\overrightarrow{OC} = \mathbf{c}$. D is the midpoint of CB and E is a point on OC such that $\overrightarrow{OE} = \dfrac{1}{3}\overrightarrow{OC}$. The lines OD and AE intersect at M. If $\overrightarrow{AM} = h\overrightarrow{AE}$ and $\overrightarrow{OM} = k\overrightarrow{OD}$ determine h and k.

18 a i How many different six letter 'words' can be formed using letters chosen from the word HARLEQUIN with no letter being used more than once in a word?

ii How many of these six letter words contain at least one vowel?

b i How many different six letter 'words' can be formed using letters chosen from the word PORTHCAWL with no letter being used more than once in a word?

ii How many of these six letter words contain at least one vowel?

19 In how many ways can twelve people be sorted into two groups of six?

(Note: The two groups are not numbered or labelled in any way. Thus the two groups {A, B, C, D, E, F} and {G, H, I, J, K, L} are not considered different to {G, H, I, J, K, L} and {A, B, C, D, E, F}.)

20 In parallelogram OABC, $\overrightarrow{OA} = \mathbf{a}$, and $\overrightarrow{OC} = \mathbf{c}$.

Point D is the mid-point of AC and E is a point on AB such that $\overrightarrow{AE} = h\overrightarrow{AB}$.

The line drawn from E, through D, meets OC at F with $\overrightarrow{CF} = k\overrightarrow{CO}$.

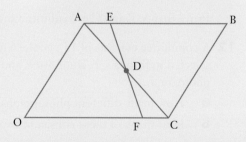

a Obtain an expression for $\overrightarrow{ED}$ in terms of **a**, **c** and h.

b Obtain an expression for $\overrightarrow{DF}$ in terms of **a**, **c** and k.

c If $\overrightarrow{ED} = m\overrightarrow{DF}$ prove that

i $m = 1$ (i.e. $\overrightarrow{ED} = \overrightarrow{DF}$),

ii $h = k$.

ISBN 9780170390477

Mathematics Specialist

Unit Two

UNIT TWO PRELIMINARY WORK

Degrees and radians

Complementary angle properties

Having reached this stage of this book it is naturally assumed that you are already familiar with the work of the previous eight chapters. It is also assumed that you are familiar with the content of Unit One of the *Mathematics Methods* course. The elements of that unit that are of particular relevance to this unit are briefly revised in this section.

Read this 'preliminary work' section and if anything is not familiar to you, and you don't understand the brief mention or explanation given here, you may need to do some further reading to bring your understanding of those concepts up to an appropriate level for this unit.

Radian measure

It is assumed that you are familiar with the idea of using a **radian** as a unit of measurement for angles and with the conversion:

$$\pi \text{ radians} = 180°$$

In this unit, assume the angle measure is in radians unless degrees are clearly indicated.

Unit circle definitions of $y = \sin x$, $y = \cos x$ and $y = \tan x$

With point O as the origin and point A as a point moving around a circle of unit radius and centre at O, we define the sine of the angle that AO makes with the positive x-axis as the y-coordinate of A. From this definition we obtain the graph of the function $y = \sin x$, with degrees as our unit of measure, as shown below right.

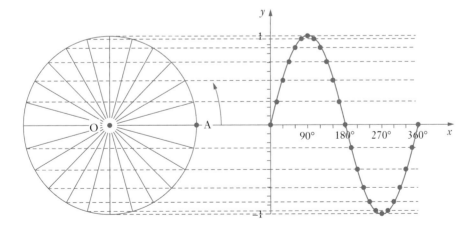

If, having completed one rotation of the circle, we were to continue moving point A around the circle the graph would repeat itself, as shown on the next page for three rotations.

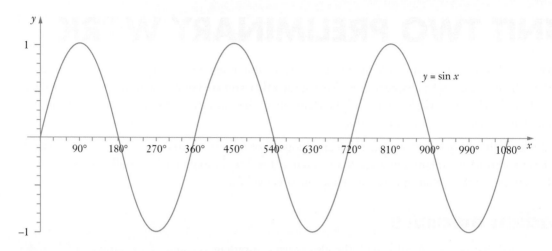

Alternatively angles could be shown in radians and negative angles could also be included, as shown below for $-2\pi \leq x \leq 4\pi$.

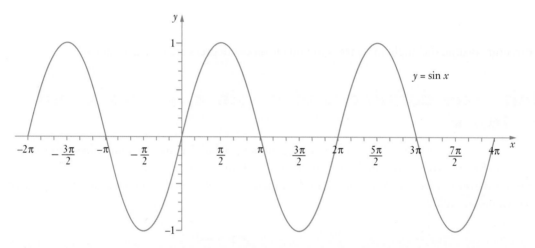

Whilst the graph shown above is for $-2\pi \leq x \leq 4\pi$ this restriction is made purely due to page width limitations. The reader should consider the graph of $y = \sin x$ continuing indefinitely to the left and the right.

Note the following:

- The graph of $y = \sin x$ repeats itself every 2π radians (or 360°).
  We say that the sine function is **periodic**, with **period** 2π.
  Thus $\sin (x \pm 2\pi) = \sin x$.
- We also say that the graph performs one **cycle** each period.
  Thus $y = \sin x$ performs one cycle in 2π radians (or 360°).
- $-1 \leq \sin x \leq 1$.
- If we consider the above graph to have a 'mean' y-coordinate of $y = 0$ then the graph has a maximum value 1 above this mean value and a minimum value 1 below it. We say that $y = \sin x$ has an **amplitude** of 1.
- $\sin (-x) = -\sin x$. (Functions for which $f(-x) = -f(x)$ are called *odd* functions and are unchanged under a 180° rotation about the origin.)

Similar considerations of the **x-coordinate** of point A as it moves around the unit circle gives the graph of

$$y = \cos x$$

shown below for $-2\pi \le x \le 4\pi$.

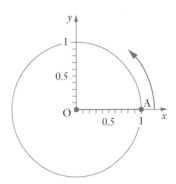

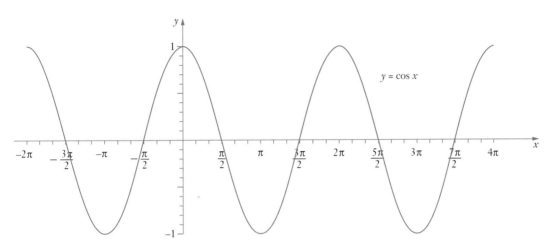

$y = \cos x$

Note the following:

- The cosine function is **periodic**, with **period 2π**. Thus $\cos (x \pm 2\pi) = \cos x$.

- $y = \cos x$ performs one **cycle** in 2π radians (or 360°).

- Note that $-1 \le \cos x \le 1$.

- The graph of $y = \cos x$ has an **amplitude** of 1.

- Note that $\cos (-x) = \cos x$. (Functions for which $f(-x) = f(x)$ are called *even* functions and are unchanged under a reflection in the y-axis.)

- If the above graph of $y = \cos x$ is moved $\dfrac{\pi}{2}$ units right, parallel to the x axis, it would then be the same as the graph of $y = \sin x$. We say that $\sin x$ and $\cos x$ are $\dfrac{\pi}{2}$ out of **phase** with each other.

  It follows that $\cos x = \sin \left(x + \dfrac{\pi}{2} \right)$ and $\sin x = \cos \left(x - \dfrac{\pi}{2} \right)$.

ISBN 9780170390477

Either by considering the relationship $\tan x = \dfrac{\sin x}{\cos x}$, or by defining

the tangent of the angle AOB (shown as a 50° angle in the diagram on the right) as the y-coordinate of the point where OA, continued as necessary, meets the vertical line through B, we obtain the graph of $y = \tan x$ as that shown below, for $0 \le x \le 180°$

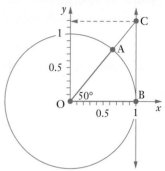

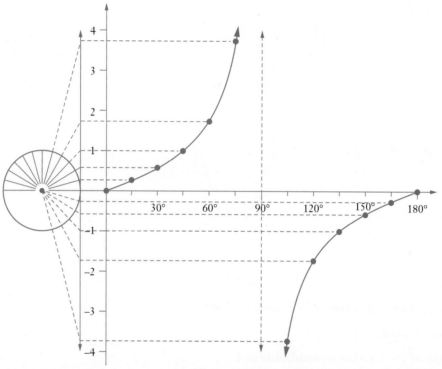

Continuing this graph to left and right we obtain the graph of $y = \tan x$, for $-2\pi \le x \le 4\pi$, as shown below.

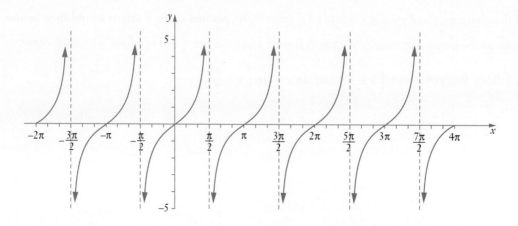

Note the following:

- Though the previous graph is for $-2\pi \le x \le 4\pi$ the reader should consider the graph of $y = \tan x$ continuing indefinitely to the left and right.

- The graph repeats itself every π radians (or 180°).
  The period of the graph is π radians (or 180°). Thus $\tan(x \pm \pi) = \tan x$.
  The graph performs one cycle in π radians (or 180°).

- The term 'amplitude' is meaningless when applied to $y = \tan x$.

- The graph is such that $\tan(-x) = -\tan x$.
  (The tangent function is an *odd* function.)

Transformations of $y = \sin x$ (and of $y = \cos x$ and $y = \tan x$)

The graph of $\quad y = a \sin x \quad$ will be that of $y = \sin x$ dilated $\updownarrow$ scale factor 'a' (and reflected in the x-axis if a is negative).

Hence $y = a \sin x$ will have amplitude $|a|$.

The graph of $\quad y = \sin bx \quad$ will be that of $y = \sin x$, dilated $\leftrightarrow$ scale factor $\dfrac{1}{b}$ (and reflected in the y-axis if b is negative).

Hence whilst $y = \sin x$ performs one cycle in 2π radians, $y = \sin bx$ will perform 'b' cycles in 2π radians.

Thus whilst $y = \sin x$ has a period of 2π radians, $y = \sin bx$ has a period of $\dfrac{2\pi}{b}$ radians.

The graph of $y = \sin[b(x + c)]$ will be that of $y = \sin bx$ moved left 'c' units (a **phase** shift).

The graph of $y = \sin(x) + d$ will be that of $y = \sin x$ moved up 'd' units.

The graph on the right looks like that of $y = \cos x$, but

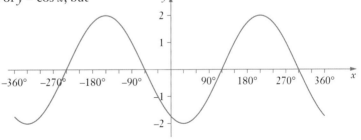

- stretched $\updownarrow$ scale factor 2

- reflected in the x-axis

- not stretched $\leftrightarrow$

- moved $\rightarrow$ 30°

to give its equation as
$y = -2 \cos(x - 30°)$.

Alternatively we could consider it as looking to be that of $y = \sin x$

- stretched $\updownarrow$ scale factor 2

- and moved $\rightarrow$ 120°

to give its equation as $y = 2 \sin(x - 120°)$.

Angle sum and angle difference identities

The fact that on the previous page we obtained two apparently different equations for the one graph should not have troubled you if you remembered the existence of the angle sum and angle difference identities:

$$\sin (A \pm B) = \sin A \cos B \pm \cos A \sin B$$

$$\cos (A \pm B) = \cos A \cos B \mp \sin A \sin B$$

$$\tan (A \pm B) = \frac{\tan A \pm \tan B}{1 \mp \tan A \tan B}$$

Considering again the two equations found for the same graph, i.e. $y = -2 \cos (x - 30°)$
and $y = 2 \sin (x - 120°)$:

$y = -2 \cos (x - 30°)$

$= -2 (\cos x \cos 30° + \sin x \sin 30°)$

$= -2 \left(\cos x \times \dfrac{\sqrt{3}}{2} + \sin x \times \dfrac{1}{2} \right)$

$= -\sqrt{3} \cos x - \sin x$

$y = 2 \sin (x - 120°)$

$= 2 (\sin x \cos 120° - \cos x \sin 120°)$

$= 2 \left(\sin x \times \left(-\dfrac{1}{2} \right) - \cos x \times \dfrac{\sqrt{3}}{2} \right)$

$= -\sin x - \sqrt{3} \cos x,$

$= -\sqrt{3} \cos x - \sin x$ as before.

Note that the above expansions also assume that you are familiar with **exact values**.

The next two pages remind you of the proof of the angle sum and angle difference identities and make use of another identity it will also be assumed you are familiar with, namely the **Pythagorean identity**:

$$\sin^2 \theta + \cos^2 \theta = 1$$

The proofs also assume you are familiar with determining the distance between two points of known coordinates.

ISBN 9780170390477

Proof of angle sum and angle difference identities

Consider the points P and Q lying on the unit circle as shown in the diagram on the right.

From our unit circle definition of sine and cosine the coordinates of P and Q will be as shown.

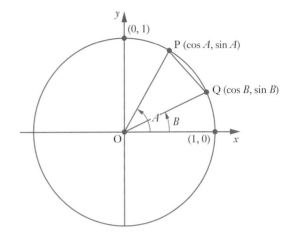

The length of the line joining two points can be found by determining the following:

$$\sqrt{\left(\text{change in the } x\text{-coordinates}\right)^2 + \left(\text{change in the } y\text{-coordinates}\right)^2}$$

Thus
$$
\begin{aligned}
PQ &= \sqrt{\left(\cos A - \cos B\right)^2 + \left(\sin A - \sin B\right)^2} \\
&= \sqrt{\cos^2 A - 2\cos A \cos B + \cos^2 B + \sin^2 A - 2\sin A \sin B + \sin^2 B} \\
&= \sqrt{\cos^2 A + \sin^2 A + \sin^2 B + \cos^2 B - 2\cos A \cos B - 2\sin A \sin B} \\
&= \sqrt{1 + 1 - 2(\cos A \cos B + \sin A \sin B)} \\
&= \sqrt{2 - 2(\cos A \cos B + \sin A \sin B)} \qquad \text{[I]}
\end{aligned}
$$

However, if instead we apply the cosine rule to triangle OPQ:

$$
\begin{aligned}
PQ &= \sqrt{1^2 + 1^2 - 2(1)(1)\cos\left(A - B\right)} \\
&= \sqrt{2 - 2\cos\left(A - B\right)} \qquad \text{[II]}
\end{aligned}
$$

Comparing [I] and [II] we see that

$$\cos\left(A - B\right) = \cos A \cos B + \sin A \sin B \qquad \text{[1]}$$

Replacing B by $(-B)$, and remembering that $\cos(-B) = \cos B$ and $\sin(-B) = -\sin B$, it follows that

$$
\begin{aligned}
\cos\left(A - (-B)\right) &= \cos A \cos(-B) + \sin A \sin(-B) \\
&= \cos A \cos B - \sin A \sin B
\end{aligned}
$$

i.e.
$$\cos\left(A + B\right) = \cos A \cos B - \sin A \sin B \qquad \text{[2]}$$

From [1], $\qquad \cos\left(\dfrac{\pi}{2} - \theta\right) = \cos\dfrac{\pi}{2}\cos\theta + \sin\dfrac{\pi}{2}\sin\theta$

$$= (\,0\,)\cos\theta + (\,1\,)\sin\theta$$

$$= \sin\theta$$

Replacing $\dfrac{\pi}{2} - \theta$ by ϕ (and hence θ by $\dfrac{\pi}{2} - \phi$) it follows that

$$\cos\phi = \sin\left(\dfrac{\pi}{2} - \phi\right)$$

Thus $\qquad \cos\left(\dfrac{\pi}{2} - A\right) = \sin A \qquad$ and $\qquad \sin\left(\dfrac{\pi}{2} - A\right) = \cos A.$

We can now use these facts to determine expansions for $\sin(A + B)$ and for $\sin(A - B)$.

$$\sin(A - B) = \cos[\,90° - (A - B)\,]$$

$$= \cos(90° - A + B)$$

$$= \cos(90° - A)\cos B - \sin(90° - A)\sin B$$

$$= \sin A\cos B - \cos A\sin B$$

i.e. $\qquad \sin(A - B) = \sin A\cos B - \cos A\sin B \qquad$ [3]

Replacing B by $(-B)$ in [3] gives:

$$\sin(A - (-B)) = \sin A\cos(-B) - \cos A\sin(-B)$$

$$= \sin A\cos B + \cos A\sin B$$

i.e. $\qquad \sin(A + B) = \sin A\cos B + \cos A\sin B$

Now: $\qquad \tan(A \pm B) = \dfrac{\sin(A \pm B)}{\cos(A \pm B)}$

$$= \dfrac{\sin A\cos B \pm \cos A\sin B}{\cos A\cos B \mp \sin A\sin B}$$

$$= \dfrac{\dfrac{\sin A\cos B}{\cos A\cos B} \pm \dfrac{\cos A\sin B}{\cos A\cos B}}{\dfrac{\cos A\cos B}{\cos A\cos B} \mp \dfrac{\sin A\sin B}{\cos A\cos B}}$$

Thus $\qquad \tan(A \pm B) = \dfrac{\tan A \pm \tan B}{1 \mp \tan A\tan B}$

Solving trigonometric equations

The diagram on the right summarises the positive and negative nature of sine, cosine and tangent in the various quadrants.

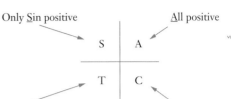

Calculating exact values of trigonometric functions

Using this awareness of 'what's positive where' we can express the sine (or cosine or tangent) of any angle as the sine (or cosine or tangent) of the acute angle made with the x-axis, together with the appropriate sign.

Solving periodic functions

For example, consider sin 200°.

As the diagram on the right shows, an angle of 200° makes 20° with the x-axis and lies in the 3rd quadrant, where the sine function is negative.

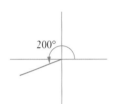

Thus sin 200° = –sin 20°

Now consider cos (300°).

An angle of 300° makes 60° with the x-axis and lies in the 4th quadrant, where the cosine function is positive.

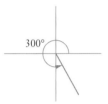

Thus cos 300° = cos 60°
= 0.5

These ideas can help when we have to solve equations involving trigonometric functions.

Suppose we are asked to solve $\sin x = -\dfrac{\sqrt{3}}{2}$ for $0 \le x \le 360°$.

With the sine being negative solutions must lie in the 3rd and 4th quadrants.

From our exact values we know that $\sin 60° = \dfrac{\sqrt{3}}{2}$.

Thus the solutions make 60° with the x-axis as shown diagrammatically on the right.

Using this diagram to obtain solutions in the required interval we have

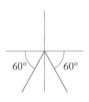

$$x = 240°, 300°.$$

Suppose now that we are asked to solve $\tan x = -\dfrac{1}{\sqrt{3}}$ for $-\pi \le x \le \pi$.

With the tangent being negative solutions must lie in the 2nd and 4th quadrants.

From our exact values we know that $\tan \dfrac{\pi}{6} = \dfrac{1}{\sqrt{3}}$.

Thus the solutions make $\dfrac{\pi}{6}$ radians with the x-axis as shown diagrammatically on the right.

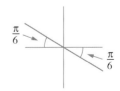

Using this diagram to obtain solutions in the required interval gives:

$$x = -\frac{\pi}{6}, \frac{5\pi}{6}.$$

Trigonometrical identities and equations

- The double angle identities
- $a \cos \theta + b \sin \theta$
- Sec θ, cosec θ and cot θ
- Product to sum and sum to product
- General solutions of trigonometric equations
- Obtaining the rule from the graph
- Modelling periodic motion
- Miscellaneous exercise nine

Situation

Three students, Jennifer, Ravi and Michael, were working on the same mathematics problem but obtained answers that appeared different.

	Jennifer's answer was	$1 - (2 \sin \theta)(\sin \theta - \cos \theta)$,
	Ravi's answer was	$2 \cos^2\theta - 1 + 2 \sin \theta \cos \theta$,
and	Michael's answer was	$\cos \theta + \sin \theta$.

Each of the students checked their working but did not discover any errors so were not prepared to admit to being wrong.

When they checked with the answers in the back of the book they found a different answer again! According to the book the answer was $\cos 2\theta + \sin 2\theta$.

Still unable to discover anything wrong with their own working they decided to evaluate each of the three answers for various values of θ.

- Evaluate Jennifer's answer for $\theta = 30°$.

- Evaluate Ravi's answer for $\theta = 30°$.

- Evaluate Michael's answer for $\theta = 30°$.

- Evaluate the book's answer for $\theta = 30°$.

Repeat this process for other values of θ.

Assuming that the answer in the back of the book was correct, did any of the three students also have the correct answer?

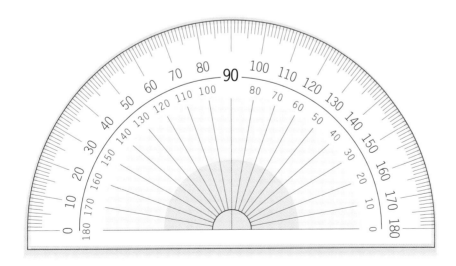

The *Preliminary work* section for this unit reminded us of the fact that two apparently different trigonometrical expressions can actually be the same. In that case it was shown that the rule

$$y = -2 \cos (x - 30°)$$

was the same as

$$y = 2 \sin (x - 120°).$$

Similarly, in the situation on the previous page, whilst all four expressions that were involved may have appeared different, some were just different ways of writing the same thing.

The *Preliminary work* also reminded us of the Pythagorean **identity**:

$$\sin^2 \theta + \cos^2 \theta = 1$$

Remember that this is an *identity* because the left hand side equals the right hand side for **all** values of the variable. I.e. $\sin^2 \theta + \cos^2 \theta = 1$ for **all** values of θ.

Contrast this with the equation $2 \sin \theta = 1$ which is only true for particular values of θ.

Note: In some texts the symbol $\equiv$ is used for an identity.

For example $\sin^2 \theta + \cos^2 \theta \equiv 1$ is an identity,
but $2 \sin \theta = 1$ is an equation.

The Pythagorean identity can be used to prove the truth of other identities, as shown in the next two examples.

EXAMPLE 1

Prove the identity: $\tan \theta - \sin \theta \cos \theta = \sin^2 \theta \tan \theta$

Solution

Left hand side $= \tan \theta - \sin \theta \cos \theta$

$$= \frac{\sin \theta}{\cos \theta} - \sin \theta \cos \theta$$

$$= \frac{\sin \theta - \sin \theta \cos^2 \theta}{\cos \theta}$$

$$= \frac{\sin \theta}{\cos \theta}(1 - \cos^2 \theta)$$

$$= \tan \theta \sin^2 \theta$$

$$= \text{Right hand side}$$

Thus $\tan \theta - \sin \theta \cos \theta = \sin^2 \theta \tan \theta$.

Note

The technique when proving an identity is to work on one side to see if it equals the other side.

ISBN 9780170390477

Or, alternatively:

Right hand side $= \sin^2 \theta \tan \theta$

$$= (1 - \cos^2 \theta) \frac{\sin \theta}{\cos \theta}$$

$$= \frac{\sin \theta}{\cos \theta} - \cos^2 \theta \frac{\sin \theta}{\cos \theta}$$

$$= \tan \theta - \sin \theta \cos \theta$$

$$= \text{Left hand side}$$

Note
- If asked to prove an identity and you are not sure which side to start with then it is usually best to start with 'the more complicated side' and attempt to simplify it. This can be easier than starting with the 'simpler' side and knowing how it should be made more complicated.

- The above example involved $\tan \theta$ which is undefined for $\theta = 90°, 270°, \ldots$. When we say an identity is 'true for all values of θ' this really means 'for all values of θ for which the expression is defined'.

EXAMPLE 2

Prove the identity: $\tan \theta \cos \theta - \sin^3 \theta = \sin \theta \cos^2 \theta$

Solution

Left hand side $= \tan \theta \cos \theta - \sin^3 \theta$

$$= \frac{\sin \theta}{\cos \theta} \cos \theta - \sin^3 \theta$$

$$= \sin \theta - \sin^3 \theta$$

$$= \sin \theta (1 - \sin^2 \theta)$$

$$= \sin \theta \cos^2 \theta$$

$$= \text{Right hand side}$$

Thus $\tan \theta \cos \theta - \sin^3 \theta = \sin \theta \cos^2 \theta$.

> **Note**
>
> The left hand side seems the more complicated.
>
> Hence that is the side we will choose to start with.

Exercise 9A

Prove the following identities.

1 $2 \cos^2 \theta + 3 = 5 - 2 \sin^2 \theta$

2 $\sin \theta - \cos^2 \theta = (\sin \theta)(1 + \sin \theta) - 1$

3 $(\sin \theta + \cos \theta)^2 = 2 \sin \theta \cos \theta + 1$

4 $1 - 2 \sin \theta \cos \theta = (\sin \theta - \cos \theta)^2$

5 $\sin^4 \theta - \cos^4 \theta = 1 - 2 \cos^2 \theta$

6 $\sin^4 \theta - \sin^2 \theta = \cos^4 \theta - \cos^2 \theta$

7 $\sin^2 \theta \tan^2 \theta = \tan^2 \theta - \sin^2 \theta$

8 $(1 + \sin \theta)(1 - \sin \theta) = 1 + (\cos \theta + 1)(\cos \theta - 1)$

9 $\sin \theta \tan \theta + \cos \theta = \dfrac{1}{\cos \theta}$

10 $\dfrac{1}{1 + \tan^2 \theta} = \cos^2 \theta$

11 $\dfrac{\cos^2 \theta + 2 \cos \theta + 1}{\sin^2 \theta} = \dfrac{1 + \cos \theta}{1 - \cos \theta}$

12 $\dfrac{\sin \theta}{1 - \cos \theta} - \dfrac{\cos \theta}{\sin \theta} = \dfrac{1}{\sin \theta}$

13 $\dfrac{1 - \sin \theta \cos \theta - \cos^2 \theta}{\sin^2 \theta + \sin \theta \cos \theta - 1} = \tan \theta$

The *Preliminary work* reminded us of the angle sum and angle difference identities:

$$\sin (A \pm B) = \sin A \cos B \pm \cos A \sin B$$
$$\cos (A \pm B) = \cos A \cos B \mp \sin A \sin B$$
$$\tan (A \pm B) = \dfrac{\tan A \pm \tan B}{1 \mp \tan A \tan B}$$

These too can be used to prove various other identities, as the next example shows.

ISBN 9780170390477

EXAMPLE 3

Prove that $\dfrac{\sin(A+B)}{\sin(A-B)} = \dfrac{\tan A + \tan B}{\tan A - \tan B}$.

Solution

$$\text{Left hand side} = \frac{\sin(A+B)}{\sin(A-B)}$$

$$= \frac{\sin A \cos B + \cos A \sin B}{\sin A \cos B - \cos A \sin B}$$

$$= \frac{\dfrac{\sin A \cos B}{\cos A \cos B} + \dfrac{\cos A \sin B}{\cos A \cos B}}{\dfrac{\sin A \cos B}{\cos A \cos B} - \dfrac{\cos A \sin B}{\cos A \cos B}}$$

$$= \frac{\tan A + \tan B}{\tan A - \tan B}$$

$$= \text{Right hand side}$$

Thus $\dfrac{\sin(A+B)}{\sin(A-B)} = \dfrac{\tan A + \tan B}{\tan A - \tan B}$.

Exercise 9B

Prove the following identities.

1 $\sin(360° + \theta) = \sin \theta$

2 $\cos(360° + \theta) = \cos \theta$

3 $\sin(360° - \theta) = -\sin \theta$

4 $\cos(360° - \theta) = \cos \theta$

5 $\sin(A+B) - \sin(A-B) = 2 \cos A \sin B$

6 $\cos(A-B) + \cos(A+B) = 2 \cos A \cos B$

7 $2 \cos\left(x - \dfrac{\pi}{6}\right) = \sin x + \sqrt{3} \cos x$

8 $\tan\left(\theta + \dfrac{\pi}{4}\right) = \dfrac{1 + \tan \theta}{1 - \tan \theta}$

9 $\dfrac{\cos(A+B)}{\cos(A-B)} = \dfrac{1 - \tan A \tan B}{1 + \tan A \tan B}$

10 $\sqrt{2}(\sin x - \cos x) \sin(x + 45°) = 1 - 2 \cos^2 x$

11 $\tan\left(\theta + \dfrac{\pi}{4}\right) = \dfrac{1 + 2 \sin \theta \cos \theta}{1 - 2 \sin^2 \theta}$

The double angle identities

From the last section we know that $\qquad\qquad \sin(A+B) = \sin A \cos B + \cos A \sin B$

Putting B equal to A we have $\qquad\qquad \sin(A+A) = \sin A \cos A + \cos A \sin A$

i.e. $\qquad\qquad \boxed{\sin 2A = 2 \sin A \cos A}$

Similarly, from $\qquad \cos(A+B) = \cos A \cos B - \sin A \sin B$

$\qquad\qquad\qquad \cos(A+A) = \cos A \cos A - \sin A \sin A$

i.e. $\qquad\qquad\qquad \cos 2A = \cos^2 A - \sin^2 A$

$$\cos 2A = (1 - \sin^2 A) - \sin^2 A \qquad\qquad \cos 2A = \cos^2 A - (1 - \cos^2 A)$$
$$= 1 - 2 \sin^2 A \qquad\qquad\qquad\qquad = 2 \cos^2 A - 1$$

Thus: $\qquad\qquad \boxed{\begin{aligned} \cos 2A &= \cos^2 A - \sin^2 A \\ &= 1 - 2 \sin^2 A \\ &= 2 \cos^2 A - 1 \end{aligned}}$

From $\tan(A+B) = \dfrac{\tan A + \tan B}{1 - \tan A \tan B}$ it follows that $\qquad \boxed{\tan 2A = \dfrac{2 \tan A}{1 - \tan^2 A}}$

EXAMPLE 4

Prove that $\sin 3\theta = 3 \sin\theta - 4 \sin^3 \theta$.

Solution

Left hand side $= \sin 3\theta$

$\qquad\qquad = \sin(\theta + 2\theta)$

$\qquad\qquad = \sin\theta \cos 2\theta + \cos\theta \sin 2\theta$

$\qquad\qquad = \sin\theta (1 - 2\sin^2\theta) + \cos\theta (2\sin\theta \cos\theta)$

$\qquad\qquad = \sin\theta (1 - 2\sin^2\theta) + 2\sin\theta (1 - \sin^2\theta)$

$\qquad\qquad = \sin\theta - 2\sin^3\theta + 2\sin\theta - 2\sin^3\theta$

$\qquad\qquad = 3\sin\theta - 4\sin^3\theta$

$\qquad\qquad =$ Right hand side

Thus $\sin 3\theta = 3\sin\theta - 4\sin^3\theta$.

Note

(Although the left hand side seems less 'complicated' it involves $\sin 3\theta$, which we can attempt to break down into terms involving $\sin\theta$.)

The next examples show the use of the double angle identities in equation-solving.

EXAMPLE 5

Solve $\sin 2x + \sin x = 0$ for $-\pi \leq x \leq \pi$.

Solution

$$\sin 2x + \sin x = 0$$
$\therefore$ $\qquad 2 \sin x \cos x + \sin x = 0$
i.e. $\qquad \sin x\,(2 \cos x + 1) = 0$
$\therefore$ either $\qquad \sin x = 0$ $\qquad\qquad$ or $\qquad\qquad 2 \cos x + 1 = 0$
$\qquad\qquad\qquad\qquad\qquad\qquad\qquad\qquad\qquad\qquad\qquad\qquad \cos x = -0.5$

$x = -\pi,\, 0,\, \pi.$ $\qquad\qquad\qquad\qquad\qquad\qquad x = -\dfrac{2\pi}{3},\, \dfrac{2\pi}{3}.$

Thus for $-\pi \leq x \leq \pi$ the solutions to $\sin 2x + \sin x = 0$ are $-\pi,\, -\dfrac{2\pi}{3},\, 0,\, \dfrac{2\pi}{3},\, \pi.$

EXAMPLE 6

a If $(2 \cos x + 1)(\cos x - 2) = a \cos^2 x + b \cos x + c$, determine a, b and c.

b Solve $\cos 2x = 3 \cos x + 1$ for $-180° \leq x \leq 180°$.

Solution

a Expanding: $(2 \cos x + 1)(\cos x - 2) = 2 \cos^2 x - 4 \cos x + \cos x - 2$
$\qquad\qquad\qquad\qquad\qquad\qquad\qquad\quad = 2 \cos^2 x - 3 \cos x - 2$

Thus $a = 2$, $b = -3$ and $c = -2$.

b We are given the equation $\qquad \cos 2x = 3 \cos x + 1$

If we replace $\cos 2x$ by $(2 \cos^2 x - 1)$ we will obtain a quadratic in $\cos x$.

$\therefore$ $\qquad\qquad\qquad\qquad 2 \cos^2 x - 1 = 3 \cos x + 1$
i.e. $\qquad\qquad\qquad\quad 2 \cos^2 x - 3 \cos x - 2 = 0$
hence, from **a** $\qquad (2 \cos x + 1)(\cos x - 2) = 0$
$\therefore$ either $\qquad 2 \cos x + 1 = 0$ $\qquad$ or $\qquad \cos x - 2 = 0$
i.e. $\qquad\qquad\quad \cos x = -0.5$ $\qquad$ or $\qquad \cos x = 2$
$\qquad\qquad\qquad\qquad\qquad\qquad\qquad\qquad\qquad$ No solutions

$x = \pm120°.$

Thus for $-180° \leq x \leq 180°$ the solutions to $\cos 2x = 3 \cos x + 1$ are $\pm120°$.

Exercise 9C

1 If $\sin A = \dfrac{3}{5}$ and $90° \le A \le 180°$, find exact values for

   a $\sin 2A$ **b** $\cos 2A$ **c** $\tan 2A$

2 If $\tan B = \dfrac{5}{12}$ and $\pi \le B \le \dfrac{3\pi}{2}$, find exact values for

   a $\sin 2B$ **b** $\cos 2B$ **c** $\tan 2B$

3 Express each of the following in the form $a \sin bA$.

   a $6 \sin A \cos A$ **b** $4 \sin 2A \cos 2A$ **c** $\sin \dfrac{A}{2} \cos \dfrac{A}{2}$

4 Express each of the following in the form $a \cos bA$.

   a $2 \cos^2 2A - 2 \sin^2 2A$ **b** $1 - 2 \sin^2 \dfrac{A}{2}$ **c** $2 \cos^2 2A - 1$

5 If θ is obtuse and such that $\cos \theta = -\dfrac{24}{25}$ find exact values for

   a $\sin 2\theta$ **b** $\cos 2\theta$ **c** $\tan 2\theta$

Solve the following equations for the given interval.

- With the occasional help of the information given in the display below right you should be able to solve these equations **without** the assistance of your calculator.

- Give exact answers where possible but if rounding is necessary give answers correct to one decimal place.

6 $4 \sin x \cos x = 1$ for $0 \le x \le 360°$.

7 $\sin 2x + \cos x = 0$ for $-180° \le x \le 180°$.

8 $2 \sin 2x - \sin x = 0$ for $0 \le x \le 360°$.

9 $2 \sin x \cos x = \cos 2x$ for $0 \le x \le 2\pi$.

10 $\cos 2x + 1 - \cos x = 0$ for $0 \le x \le 2\pi$.

11 $\cos 2x + \sin x = 0$ for $-\pi \le x \le \pi$.

12 $2 \sin^2 x + 5 \cos x + \cos 2x = 3$ for $0 \le x \le 540°$.

```
solve(cos(x)=0.25,x) |0 ≤x ≤90°
                    {x=75.52248781}
solve(cos(x)=0.4,x) |0 ≤x ≤90°
                    {x=66.42182152}
factor(2·y²−y−1)
                    (2·y+1)·(y−1)
```

Prove the following identities.

13 $\sin 2\theta \tan \theta = 2 \sin^2 \theta$

14 $\cos \theta \sin 2\theta = 2 \sin \theta - 2 \sin^3 \theta$

15 $\dfrac{1 - \cos 2\theta}{1 + \cos 2\theta} = \tan^2 \theta$

16 $\sin \theta \tan \dfrac{\theta}{2} = 2 - 2 \cos^2 \dfrac{\theta}{2}$

17 $\sin 4\theta = 4 \sin \theta \cos^3 \theta - 4 \sin^3 \theta \cos \theta$

18 $\dfrac{\sin 2\theta - \sin \theta}{1 - \cos \theta + \cos 2\theta} = \tan \theta$

19 $\cos 4\theta = 1 - 8 \cos^2 \theta + 8 \cos^4 \theta$

a cos θ + *b* sin θ

Consider the function $\qquad y = 3 \cos \theta + 4 \sin \theta$

Writing this as $\qquad y = 5 \left(\dfrac{3}{5} \cos \theta + \dfrac{4}{5} \sin \theta \right)$

(The 5 being chosen because $\sqrt{3^2 + 4^2} = 5$.)

it follows that $\qquad y = 5 (\cos \theta \cos \alpha + \sin \theta \sin \alpha)$
$\qquad\qquad\qquad\quad = 5 \cos (\theta - \alpha)$

where $\qquad\qquad \cos \alpha = \dfrac{3}{5}$

and $\qquad\qquad\quad \sin \alpha = \dfrac{4}{5}$

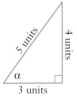

as shown in the diagram on the right.

Using a calculator $\qquad \alpha = 53.1°$ (correct to 1 decimal place).

Thus the initial expression $\qquad y = 3 \cos \theta + 4 \sin \theta$

can be rearranged to $\qquad y = 5 \cos (\theta - 53.1°)$

This means that the graph of $y = 3 \cos x + 4 \sin x$ will be that of $y = \cos x$ but with an amplitude 5 and moved right 53.1°. Confirm this to be the case by viewing the graph of $y = 3 \cos x + 4 \sin x$ on a calculator or computer graphing package.

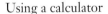

Express $2 \cos \theta - 3 \sin \theta$ in the form $a \cos(\theta + \alpha)$ for known values of a (exact) and α an acute angle (in degrees correct to one decimal place).

Solution

We rearrange to the form $\cos A \cos B - \sin B \sin A$ using $\sqrt{2^2 + 3^2}$, i.e. $\sqrt{13}$.

$$2 \cos \theta - 3 \sin \theta = \sqrt{13}\left(\frac{2}{\sqrt{13}} \cos \theta - \frac{3}{\sqrt{13}} \sin \theta\right)$$

$$= \sqrt{13}(\cos \theta \cos \alpha - \sin \theta \sin \alpha)$$

$$= \sqrt{13} \cos(\theta + \alpha)$$

Using a calculator, $\alpha = 56.3°$ (correct to one decimal place)

Thus $\quad 2 \cos \theta - 3 \sin \theta = \sqrt{13} \cos(\theta + 56.3°)$

This is of the required form with $a = \sqrt{13}$ and $\alpha = 56.3°$.

Again check that the graph of $y = 2 \cos x - 3 \sin x$ is indeed that of $y = \cos x$ stretched vertically, scale factor $\sqrt{13}$, and with an appropriate phase shift.

a Express $5 \sin \theta + 12 \cos \theta$ in the form $R \sin(\theta + \alpha)$ for α an acute angle in degrees, correct to one decimal place.

b Hence determine the maximum value of $5 \sin \theta + 12 \cos \theta$ and the smallest positive value of θ for which it occurs (correct to one decimal place).

Solution

a $\quad 5 \sin \theta + 12 \cos \theta = \sqrt{5^2 + 12^2}\left(\frac{5}{\sqrt{5^2 + 12^2}} \sin \theta + \frac{12}{\sqrt{5^2 + 12^2}} \cos \theta\right)$

$$= 13\left(\frac{5}{13} \sin \theta + \frac{12}{13} \cos \theta\right)$$

$$= 13(\sin \theta \cos \alpha + \cos \theta \sin \alpha)$$

Using a calculator we find that $\alpha = 67.4°$ (correct to 1 decimal place).

Hence $\quad 5 \sin \theta + 12 \cos \theta = 13 \sin(\theta + 67.4°)$.

b $\quad \sin(\theta + 67.4°)$ has a maximum value of 1 when $(\theta + 67.4°) \quad = 90°$

Thus $5 \sin \theta + 12 \cos \theta$ has a maximum value of 13 when $\theta \quad = 90° - 67.4°$

$$= 22.6°$$

In example 7, we expressed $2 \cos \theta - 3 \sin \theta$ in the form $\sqrt{13} \cos(\theta + 56.3°)$. This rearrangement can be useful if we were asked to solve an equation of the form $2 \cos \theta - 3 \sin \theta = c$, as the next example shows.

ISBN 9780170390477

EXAMPLE 9

Use the rearrangement of example 7 to solve $2 \cos x - 3 \sin x = 2.5$ for $0 \le x \le 360°$.

Solution

Given: $\qquad 2 \cos x - 3 \sin x = 2.5$

Hence: $\qquad \sqrt{13} \cos (x + 56.3°) = 2.5$

$$\cos (x + 56.3°) = \frac{5}{2\sqrt{13}}$$

$$(x + 56.3°) = ..., 46.1°, 313.9°, 406.1°, ...$$

$$x = ..., -10.2°, 257.6°, 349.8°, ...$$

Thus for $0 \le x \le 360°$ the solutions to $2 \cos x - 3 \sin x = 2.5$ are $257.6°$ and $349.8°$.

Exercise 9D

Express each of the following in the form $a \cos (\theta + \alpha)$ for α an acute angle in degrees correct to one decimal place.

1 $\ 3 \cos \theta - 4 \sin \theta$

2 $\ 12 \cos \theta - 5 \sin \theta$

Express each of the following in the form $a \cos (\theta - \alpha)$ for α an acute angle in radians correct to two decimal places.

3 $\ 4 \cos \theta + 3 \sin \theta$

4 $\ 7 \cos \theta + 24 \sin \theta$

Express each of the following in the form $a \sin (\theta + \alpha)$ for α an acute angle in degrees correct to one decimal place.

5 $\ 5 \sin \theta + 12 \cos \theta$

6 $\ 7 \sin \theta + 24 \cos \theta$

Express each of the following in the form $a \sin (\theta - \alpha)$ for α an acute angle in radians correct to two decimal places.

7 $\ 4 \sin \theta - 3 \cos \theta$

8 $\ 2 \sin \theta - 3 \cos \theta$

9 Use your answers to questions 7 and 8 to sketch the graphs of $y = 4 \sin x - 3 \cos x$ and $y = 2 \sin x - 3 \cos x$. (Then check your sketches using a graphic calculator.)

10 a Express $\cos \theta + \sin \theta$ in the form $R \cos (\theta - \alpha)$ for α an acute angle in radians.

    b Hence determine the maximum value of $\cos \theta + \sin \theta$ and the smallest positive value of θ for which it occurs. (Give θ in radians.)

11 Solve $3 \cos x + 4 \sin x = 2$ for $0 \le x \le 2\pi$. (Answers correct to 2 decimal places.)

12 Solve $10 \sin x + 5 \cos x = 8$ for $-\pi \le x \le \pi$. (Answers correct to 2 decimal places.)

13 Solve $2 \sin x + 5 \cos x = 3$ for $0 \le x \le 2\pi$. (Answers correct to 2 decimal places.)

Sec θ, cosec θ and cot θ

The reciprocals of sin θ, cos θ and tan θ, i.e. $\dfrac{1}{\sin\theta}$, $\dfrac{1}{\cos\theta}$ and $\dfrac{1}{\tan\theta}$, can occur frequently and are given names of their own:

$$\dfrac{1}{\cos\theta} = \sec\theta \qquad\qquad \dfrac{1}{\sin\theta} = \operatorname{cosec}\theta \qquad\qquad \dfrac{1}{\tan\theta} = \cot\theta \left(= \dfrac{\cos\theta}{\sin\theta}\right)$$

Note: • Sec, cosec and cot are abbreviations for secant, cosecant and cotangent.

• We prefer to define cot θ as $\dfrac{\cos\theta}{\sin\theta}$, then $\cot\left(\dfrac{\pi}{2}\right) = \dfrac{\cos\left(\dfrac{\pi}{2}\right)}{\sin\left(\dfrac{\pi}{2}\right)} = 0$,

thus avoiding $\cot\left(\dfrac{\pi}{2}\right) = \dfrac{1}{\tan\left(\dfrac{\pi}{2}\right)} \left(= \dfrac{1}{\text{undefined}}\right)$.

Pythagorean identities can be established for these reciprocal functions as follows:

$$\sin^2\theta + \cos^2\theta = 1$$

Divide by $\cos^2\theta$

$$\dfrac{\sin^2\theta}{\cos^2\theta} + \dfrac{\cos^2\theta}{\cos^2\theta} = \dfrac{1}{\cos^2\theta}$$

i.e.

$$\tan^2\theta + 1 = \sec^2\theta$$

$$\sin^2\theta + \cos^2\theta = 1$$

Divide by $\sin^2\theta$

$$\dfrac{\sin^2\theta}{\sin^2\theta} + \dfrac{\cos^2\theta}{\sin^2\theta} = \dfrac{1}{\sin^2\theta}$$

i.e.

$$1 + \cot^2\theta = \operatorname{cosec}^2\theta$$

These Pythagorean identities can be used as they are (see example 12) or the question can be re-written in terms of sine, cosine and tangent and then solved as before (see examples 10 and 11).

EXAMPLE 10

Solve $\cot x = 2$ for $-\pi \le x \le \pi$.

Solution

If $\qquad \cot x = 2$

then $\qquad \dfrac{1}{\tan x} = 2$

and so $\qquad \tan x = \dfrac{1}{2}$

Thus for $-\pi \le x \le \pi$ the solutions of $\cot x = 2$ are 0.46 rads and −2.68 rads, correct to 2 decimal places.

EXAMPLE 11

Prove the identity: $\sec \theta - \cos \theta = \sin \theta \tan \theta$

Solution

Left hand side $= \sec \theta - \cos \theta$

$$= \frac{1}{\cos \theta} - \cos \theta$$

$$= \frac{1 - \cos^2 \theta}{\cos \theta}$$

$$= \frac{\sin^2 \theta}{\cos \theta}$$

$$= \frac{\sin \theta}{\cos \theta} \times \sin \theta$$

$$= \tan \theta \sin \theta$$

$$= \text{Right hand side}$$

Thus $\sec \theta - \cos \theta = \sin \theta \tan \theta$.

EXAMPLE 12

Solve $8 \cot^2 \theta = 14 \operatorname{cosec} \theta - 13$ for $-180° \le \theta \le 180°$.

Solution

$$8 \cot^2 \theta = 14 \operatorname{cosec} \theta - 13$$

By substituting for $8 \cot^2 \theta$, we can obtain a quadratic in $\operatorname{cosec} \theta$:

$$8 (\operatorname{cosec}^2 \theta - 1) = 14 \operatorname{cosec} \theta - 13$$

$$8 \operatorname{cosec}^2 \theta - 14 \operatorname{cosec} \theta + 5 = 0$$

$$(4 \operatorname{cosec} \theta - 5)(2 \operatorname{cosec} \theta - 1) = 0$$

Hence either $\quad 4 \operatorname{cosec} \theta - 5 = 0 \quad$ or $\quad 2 \operatorname{cosec} \theta - 1 = 0$

$$\operatorname{cosec} \theta = \frac{5}{4} \quad \text{or} \quad \operatorname{cosec} \theta = \frac{1}{2}$$

$$\sin \theta = \frac{4}{5} \quad \text{or} \quad \sin \theta = 2$$

no solutions

$$\theta = 53.13°, 126.87°$$

Thus for $-180° \le \theta \le 180°$ the solutions to $8 \cot^2 \theta = 14 \operatorname{cosec} \theta - 13$ are $53.13°$ and $126.87°$.

Exercise 9E

Solve the following equations for the given interval.

1 $\sec x = 2$ for $0 \le x \le 2\pi$.

2 $3 \operatorname{cosec}^2 x = 4$ for $-\pi \le x \le \pi$.

3 $\sin x \sec x - 3 \sin x = 0$ for $0 \le x \le 360°$.

4 $(\sec x)(3 - \sec x) = \tan^2 x - 1$ for $-180° \le x \le 180°$.

5 $5 \cos x = \sec x$ for $0 \le x \le 360°$.

6 $\operatorname{cosec}\left(x + \dfrac{\pi}{3}\right) = \sqrt{2}$ for $0 \le x \le 2\pi$.

7 $\sec^2 x + \sec x = 2$ for $0 \le x \le 360°$.

8 $2 \cot^2 x + 5 \operatorname{cosec} x - 1 = 0$ for $0 \le x \le 2\pi$.

Prove the following identities.

9 $1 = \sin^2 \theta \cot^2 \theta + \sin^2 \theta$

10 $(\cot^2 \theta)(1 - \cos^2 \theta) = 1 - \sin^2 \theta$

11 $1 + \cot^2 \theta = \cot^2 \theta \sec^2 \theta$

12 $(\sec \theta - 1)(\operatorname{cosec} \theta + \cot \theta) = \tan \theta$

13 $\tan^4 \theta - 1 = \tan^2 \theta \sec^2 \theta - \sec^2 \theta$

14 $\dfrac{1 + \sin \theta}{1 - \sin \theta} = 2 \tan^2 \theta + 1 + 2 \tan \theta \sec \theta$

15 $\dfrac{1 + \sin \theta}{1 - \sin \theta} = \sec^2 \theta + 2 \tan \theta \sec \theta + \tan^2 \theta$

16 $\dfrac{1 + \sec \theta}{1 - \sec \theta} = 1 - 2 \operatorname{cosec}^2 \theta - 2 \cot \theta \operatorname{cosec} \theta$

17 This question requires you to think about what the graph of $y = \operatorname{cosec} x$ looks like.

The graph below shows $y = \sin x$ for $-2\pi \leq x \leq 4\pi$, which involves 3 cycles.

With $\operatorname{cosec} x$ being $\dfrac{1}{\sin x}$ it follows that the graph of $y = \operatorname{cosec} x$ will be asymptotic for any values

of x for which $\sin x = 0$.

Make a copy of the diagram shown below and try to sketch the graph of $y = \operatorname{cosec} x$ for
$-2\pi \leq x \leq 4\pi$ on the lower set of axes.

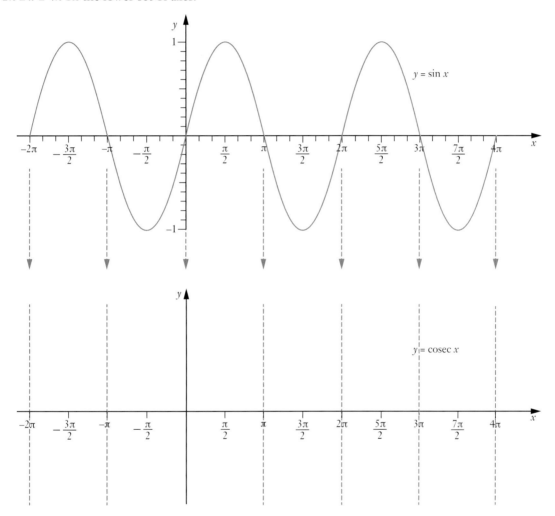

Then view the graph of $y = \operatorname{cosec} x$ on a graphic calculator or computer graphing package and see how your sketch compares.

Similarly produce sketches of $y = \sec x$ and $y = \cot x$ for $2\pi \leq x \leq 4\pi$ and then check the correctness of your sketches by comparing them to the display from a graphic calculator or computer.

How will the graphs of

$$y = a \operatorname{cosec} x, \qquad y = \operatorname{cosec} bx, \qquad y = \operatorname{cosec}(x - c), \qquad y = d + \operatorname{cosec} x$$

compare with that of $y = \operatorname{cosec} x$? Investigate.

Product to sum and sum to product

Given $\qquad\qquad\qquad\sin(A+B) = \sin A \cos B + \cos A \sin B$ [1]

and $\qquad\qquad\qquad\sin(A-B) = \sin A \cos B - \cos A \sin B$ [2]

[1] + [2] $\qquad\qquad\sin(A+B) + \sin(A-B) = 2\sin A \cos B$

writing $A + B = P$
and $A - B = Q$

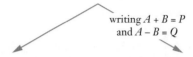

$$\sin A \cos B = \frac{1}{2}[\sin(A+B) + \sin(A-B)] \qquad\qquad \sin P + \sin Q = 2\sin\left(\frac{P+Q}{2}\right)\cos\left(\frac{P-Q}{2}\right)$$

The identity above right is sometimes remembered as

'sine + sine equals 2 sine semi-sum, cos semi-difference'.

[1] – [2] $\qquad\qquad\sin(A+B) - \sin(A-B) = 2\cos A \sin B$

writing $A + B = P$
and $A - B = Q$

$$\cos A \sin B = \frac{1}{2}[\sin(A+B) - \sin(A-B)] \qquad\qquad \sin P - \sin Q = 2\cos\left(\frac{P+Q}{2}\right)\sin\left(\frac{P-Q}{2}\right)$$

Given $\qquad\qquad\qquad\cos(A+B) = \cos A \cos B - \sin A \sin B$ [3]

and $\qquad\qquad\qquad\cos(A-B) = \cos A \cos B + \sin A \sin B$ [4]

[3] + [4] $\qquad\qquad\cos(A+B) + \cos(A-B) = 2\cos A \cos B$

writing $A + B = P$
and $A - B = Q$

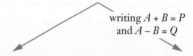

$$\cos A \cos B = \frac{1}{2}[\cos(A+B) + \cos(A-B)] \qquad\qquad \cos P + \cos Q = 2\cos\left(\frac{P+Q}{2}\right)\cos\left(\frac{P-Q}{2}\right)$$

[3] – [4] $\qquad\qquad\cos(A+B) - \cos(A-B) = -2\sin A \sin B$

writing $A + B = P$
and $A - B = Q$

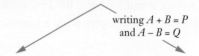

$$\sin A \sin B = \frac{1}{2}[\cos(A-B) - \cos(A+B)] \qquad\qquad \cos P - \cos Q = -2\sin\left(\frac{P+Q}{2}\right)\sin\left(\frac{P-Q}{2}\right)$$

Summary:

$$\sin A \cos B = \frac{1}{2}[\sin (A + B) + \sin (A - B)]$$

$$\sin P + \sin Q = 2 \sin \left(\frac{P+Q}{2}\right) \cos \left(\frac{P-Q}{2}\right)$$

$$\cos A \sin B = \frac{1}{2}[\sin (A + B) - \sin (A - B)]$$

$$\sin P - \sin Q = 2 \cos \left(\frac{P+Q}{2}\right) \sin \left(\frac{P-Q}{2}\right)$$

$$\cos A \cos B = \frac{1}{2}[\cos (A + B) + \cos (A - B)]$$

$$\cos P + \cos Q = 2 \cos \left(\frac{P+Q}{2}\right) \cos \left(\frac{P-Q}{2}\right)$$

$$\sin A \sin B = \frac{1}{2}[\cos (A - B) - \cos (A + B)]$$

$$\cos P - \cos Q = -2 \sin \left(\frac{P+Q}{2}\right) \sin \left(\frac{P-Q}{2}\right)$$

The identities on the left can be useful if we wish to express the products

‘cos cos’, ‘sin sin’, ‘sin cos’, ‘cos sin’

as sums or differences.

The identities on the right can be useful if we wish to express the sums or differences

‘cos ± cos’, ‘sin ± sin’

as products.

EXAMPLE 13

Solve $4 \sin 5x \sin 3x + 2 \cos 8x = 1$ for $0 \leq x \leq 360°$.

Solution

Given:

$$4 \sin 5x \sin 3x + 2 \cos 8x = 1$$

$\therefore$

$$4 \times \frac{1}{2} [\cos (5x - 3x) - \cos (5x + 3x)] + 2 \cos 8x = 1$$

$$2 \cos 2x - 2 \cos 8x + 2 \cos 8x = 1$$

$$2 \cos 2x = 1$$

$$\cos 2x = 0.5$$

$$2x = 60°, 300°, 420°, 660°$$

$$x = 30°, 150°, 210°, 330°$$

Solutions are $30°, 150°, 210°, 330°$.

EXAMPLE 14

Prove that $\dfrac{\sin 7x - \sin 3x}{\cos 6x + \cos 4x} = 2 \sin x$.

Solution

Left hand side $= \dfrac{\sin 7x - \sin 3x}{\cos 6x + \cos 4x}$

$= \dfrac{2 \cos\left(\dfrac{7x+3x}{2}\right) \sin\left(\dfrac{7x-3x}{2}\right)}{2 \cos\left(\dfrac{6x+4x}{2}\right) \cos\left(\dfrac{6x-4x}{2}\right)}$

$= \dfrac{2 \cos 5x \sin 2x}{2 \cos 5x \cos x}$

$= \dfrac{\sin 2x}{\cos x}$

$= \dfrac{2 \sin x \cos x}{\cos x}$

$= 2 \sin x$

$=$ Right hand side

Exercise 9F

Express each of the following as the sum or difference of two 'trig' functions.

1 $\cos 3x \cos 2x$ **2** $\sin 3x \sin x$ **3** $\sin 7x \cos x$ **4** $\cos 3x \sin x$

Express each of the following as the product of two 'trig' functions.

5 $\cos 5x + \cos x$ **6** $\cos 5x - \cos x$ **7** $\sin 6x + \sin 2x$ **8** $\sin 5x - \sin 3x$

9 Express the product $\sin 75° \cos 15°$ as an exact value.

10 Express $\sin 75° + \sin 15°$ as an exact value.

11 Solve: $4 \sin 7x \cos 2x = \sqrt{3} + 2 \sin 9x$ for $0 \le x \le 180°$.

12 Solve: $\sin 7x + \sin 3x = \sin 5x$ for $0 \le x \le \pi$.

13 Solve: $\sin 3x - \sin x = 0$ for $0 \le x \le 360°$

14 Solve: $\sin 5x \cos 3x = \sin 6x \cos 2x$ for $-\pi \le x \le \pi$.

15 Prove that: $\dfrac{\sin A + \sin B}{\cos A + \cos B} = \tan\left(\dfrac{A+B}{2}\right)$.

16 Prove that $\sqrt{2} \cos\left(2x - \dfrac{\pi}{4}\right) - \dfrac{\sin 7x + \sin 3x}{2 \sin 5x} = \sin 2x$.

17 Prove that: $\cos 8A \cos 2A - \cos 7A \cos 3A + \sin 5A \sin A = 0$.

18 Prove that: $4 \sin 3A \sin 2A \cos A = 1 + \cos 2A - \cos 4A - \cos 6A$.

General solutions of trigonometric equations

In all of the trigonometric equations we have been asked to solve so far in this text, the solutions we have been required to find have been restricted to some given interval. Suppose instead we were asked to find **all** of the solutions to a particular trigonometric equation?

When asked to solve the equation

$$\cos x = 0.5 \qquad \text{for } 0 \le x \le 360°$$

our knowledge of exact values and 'what's positive where' allows us to create the drawing shown on the right, and state the solutions in the required interval as

$$x = 60°, 300°.$$

Given a different interval, say $-180° \le x \le 180°$ our diagram again allows the solutions to be determined: $x = \pm60°$.

Asked to determine **all** solutions to the equation $\cos x = 0.5$ there will be an infinite number! However all of these solutions will be as shown in the diagram, but with some number of complete rotations added or subtracted.

Thus the general solution of the equation $\cos x = 0.5$ is

$$x = \begin{cases} 60° + n \times 360°, \\ -60° + n \times 360°, \end{cases} \text{for } n \in \{\ldots, -3, -2, -1, 0, 1, 2, 3, \ldots\}.$$

Writing $\mathbb{Z}$ for the set of integers $\{\ldots, -3, -2, -1, 0, 1, 2, 3, \ldots\}$ this general solution can be

written $\qquad x = \begin{cases} 60° + n \times 360°, \\ -60° + n \times 360°, \end{cases} \text{for } n \in \mathbb{Z},$

or as $\qquad x = n \times 360° \pm 60°$ for $n \in \mathbb{Z}$.

In this way the expression

$$x = n \times 360° \pm 60° \text{ for } n \in \mathbb{Z}, \text{ gives } \textbf{all} \text{ solutions to the equation } \cos x = 0.5.$$

$n = \ldots$

$n = -3$	gives	$x = -1140°, -1020°$
$n = -2$	gives	$x = -780°, -660°$
$n = -1$	gives	$x = -420°, -300°$
$n = 0$	gives	$x = -60°, 60°$
$n = 1$	gives	$x = 300°, 420°$
$n = 2$	gives	$x = 660°, 780°$
$n = 3$	gives	$x = 1020°, 1140°$

$n = \ldots$

$x = n \times 360° \pm 60°$ for $n \in \mathbb{Z}$ is referred to as the **general solution** of the equation.

Does your calculator give general solutions to trigonometric equations?
Investigate.

On the previous page, we used the solutions of $-60°$ and $+60°$ to obtain:

$$x = \begin{cases} 60° + n \times 360°, \\ -60° + n \times 360°, \end{cases} \text{for } n \in \mathbb{Z}.$$

Had we instead used $60°$ and $300°$ we would have written

$$x = \begin{cases} 60° + n \times 360°, \\ 300° + n \times 360°, \end{cases} \text{for } n \in \mathbb{Z},$$

Do these two general statements generate the same solutions? Investigate.

EXAMPLE 15

EXAMPLE 15

Find all solutions to the equation $\sqrt{3} + 2 \cos x = 0$, for x in radians.

Solution

If $\sqrt{3} + 2 \cos x = 0$

$$2 \cos x = -\sqrt{3}$$

$$\cos x = -\frac{\sqrt{3}}{2}$$

Thus $x = \begin{cases} \dfrac{5\pi}{6} + n \times 2\pi, \\ \dfrac{7\pi}{6} + n \times 2\pi, \end{cases} \text{for } n \in \mathbb{Z}.$

(Alternatively this could also be written as $x = 2n\pi \pm \dfrac{5\pi}{6}$.)

EXAMPLE 16

Find all solutions to the equation $\cos(3(x - 2)) = 0.4$, giving answers correct to 2 decimal places.

Solution

(With no indication to the contrary, we assume radians to be the unit of angle measure.)
Using a calculator to obtain the acute angle for which $\cos x = 0.4$, and an awareness of 'what's positive where' allows us to create the diagram on the right.

Thus $\quad 3(x - 2) = \begin{cases} 1.159 + n \times 2\pi, \\ -1.159 + n \times 2\pi, \end{cases} \text{for } n \in \mathbb{Z}.$

1.159 rads
1.159 rads

$$x - 2 = \begin{cases} 0.386 + \dfrac{2n\pi}{3}, \\ -0.386 + \dfrac{2n\pi}{3}, \end{cases} \text{for } n \in \mathbb{Z}.$$

$$x = \begin{cases} 2.39 + \dfrac{2n\pi}{3}, \\ 1.61 + \dfrac{2n\pi}{3}, \end{cases} \text{for } n \in \mathbb{Z}, \text{ correct to two decimal places.}$$

EXAMPLE 17

Find all solutions to the equation $\sin x = \dfrac{\sqrt{3}}{2}$, for x in degrees.

Solution

$$\sin x = \frac{\sqrt{3}}{2}$$

$$x = \begin{cases} 60° + n \times 360°, \\ 120° + n \times 360°, \end{cases} \text{for } n \in \mathbb{Z}.$$

Alternatively, if we consider the solutions to be generated by

adding $60°$ to even multiples of $180°$

and   subtracting $60°$ from odd multiples of $180°$,

the general solution could be written as

$$x = \begin{cases} 2n \times 180° + 60°, \\ (2n+1) \times 180° - 60°, \end{cases} \text{for } n \in \mathbb{Z}.$$

(which could then be simplified to give the earlier version.)

EXAMPLE 18

Find all solutions to the equation $\tan x = \sqrt{3}$.

Solution

(With no indication to the contrary, we assume radians to be the unit of angle measure.)

$$\tan x = \sqrt{3}$$

$$x = \begin{cases} \dfrac{\pi}{3} + 2n\pi, \\ \dfrac{4\pi}{3} + 2n\pi, \end{cases} \text{for } n \in \mathbb{Z}.$$

Alternatively, if we consider the solutions to be generated by adding $\dfrac{\pi}{3}$ to each multiple of π, the general solution could be written as

$$x = n\pi + \frac{\pi}{3} \text{ for } n \in \mathbb{Z}.$$

EXAMPLE 19

Find all solutions to the equation $\tan\left(\dfrac{\pi}{2}(2x-1)\right) = 0.6$, giving answers correct to 2 decimal places.

Solution

(With no indication to the contrary, we assume radians to be the unit of angle measure.)

Using a calculator to obtain the acute angle for which $\tan x = 0.6$, and an awareness of 'what's positive where' allows us to create the diagram on the right.

$$\frac{\pi}{2}(2x-1) = n\pi + 0.540 \text{ for } n \in \mathbb{Z}.$$

$$2x - 1 = \frac{2}{\pi}(n\pi + 0.54) \text{ for } n \in \mathbb{Z}.$$

$$= 2n + 0.344 \text{ for } n \in \mathbb{Z}.$$

$$2x = 2n + 1.344 \text{ for } n \in \mathbb{Z}.$$

$$x = n + 0.67 \text{ for } n \in \mathbb{Z}, \text{ correct to 2 decimal places.}$$

Exercise 9G

Find all solutions to each of the following equations for x in degrees.

(Give answers correct to 1 decimal place if rounding is necessary.)

1 $\sin x = 0.5$

2 $\cos x = 1$

3 $\tan x = -\dfrac{1}{\sqrt{3}}$

4 $\sin(2x + 30°) = 1$

5 $\cos(3(x - 20°)) = 0.7$

6 $\tan(2(x + 10°)) = 0.8$

Find all solutions to each of the following equations.

(Give answers correct to 2 decimal places if rounding is necessary and, with no indication to the contrary, assume radians to be the unit of angle measure.)

7 $4 \sin x \cos x = -1$

8 $\sin^3 x + \sin x \cos^2 x = \cos x$

9 $\cos^2 x - \sin^2 x = 1$

10 $\sin 2x \cos x + \cos 2x \sin x = 0.5$

11 $\cos(4(x - 1)) = 0.8$

12 $2 \sin 3x \sin x + \cos 4x = 0.5$

13 $\cos\left(3x - \dfrac{\pi}{4}\right) = 0$

14 $\sin\left(\dfrac{\pi}{4}(3x - 1)\right) = 0.25$

Obtaining the rule from the graph

The graph on the right looks like it could be the graph of $y = \sin x$ that has been

- moved right ten units

- stretched parallel to the y-axis, scale factor 2, and

- dilated parallel to x axis.

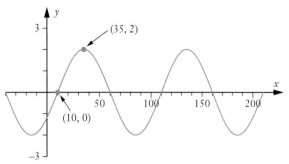

This would suggest that the given graph has an equation of the form $y = 2 \sin [b(x - 10)]$.

Now $y = \sin bx$ performs b cycles in 2π units of x-axis. (Using radians, not degrees).

I.e. $y = \sin bx$ has a period of $\dfrac{2\pi}{b}$.

The given graph has a period of 100, thus $100 = \dfrac{2\pi}{b}$

$$\text{giving } b = \frac{\pi}{50}$$

Hence the graph shown has equation $\qquad y = 2 \sin\left(\dfrac{\pi}{50}(x - 10)\right)$

Display such a function on a graphic calculator or computer graphing package to see if the rule does indeed give the graph shown above.

Of course, we could instead have considered the given graph to be that of $y = \cos x$ moved right 35 units, stretched parallel to the y-axis, and dilated horizontally. This would suggest the rule of the graph would be $y = 2 \cos\left(\dfrac{\pi}{50}(x - 35)\right)$.

However we already know that trigonometric expressions that appear different can sometimes be just different ways of expressing the same thing so the fact that there are two 'different' rules for the graph shown above should be no great surprise (and was also mentioned in the earlier *Preliminary work* section for this unit).

Use the fact that $\sin x = \cos\left(\dfrac{\pi}{2} - x\right)$ to show the equivalence of

$$y = 2 \sin\left(\frac{\pi}{50}(x - 10)\right) \qquad \text{and} \qquad y = 2 \cos\left(\frac{\pi}{50}(x - 35)\right)$$

Indeed, why stop at just two versions? Check that each of the following rules also give the graph shown at the top of this page:

$$y = 2 \sin\left(\frac{\pi}{50}(x + 90)\right) \qquad y = 2 \sin\left(\frac{\pi}{50}(x - 110)\right) \qquad y = -2 \sin\left(\frac{\pi}{50}(x + 40)\right)$$

Hence if asked to determine the rule for a given trigonometric graph do not be too quick to assume your answer is wrong even if it does not look quite the same as the answer obtained by someone else.

ISBN 9780170390477 **9.** Trigonometrical identities and equations ●●●●●●●●●○○○○

Sketching periodic functions—amplitude and period

Sketching periodic functions—phase and vertical shift

Sketching periodic functions

Modelling periodic motion

Some real-life situations follow periodical patterns in their variation that can be quite well modelled by an appropriate trigonometrical relationship.

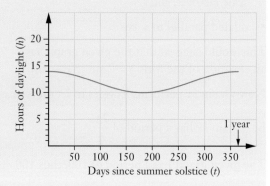

Shutterstock.com/vichie81

ISBN 9780170390477

Exercise 9H

1 Write the equation of each of the following in the form $y = a \sin x$.

a

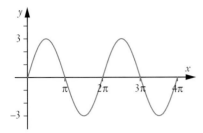

b

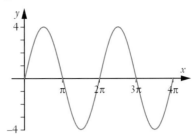

c

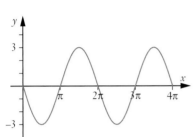

d

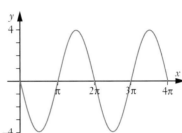

2 Write the equation of each of the following in the form $y = a \sin bx$.

a

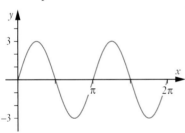

b

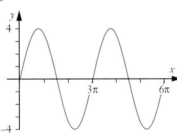

c

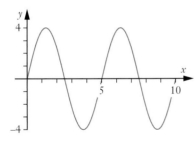

d

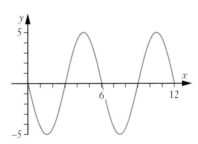

3 Write the equation of each of the following in the form $y = d + a \sin x$.

a

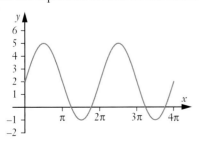

b

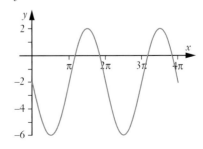

4 Write the equation of each of the following in the form $y = a \sin(x + c)$.

a
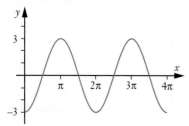

b
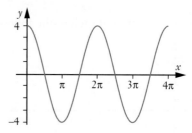

5 Write the equation of each of the following in the form $y = a \sin[b(x + c)]$.

a
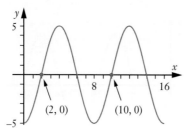

b
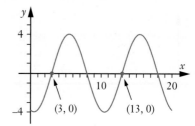

6 Write the equation of each of the following in the form $y = a \sin[b(x + c)] + d$.

a
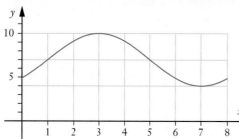

b
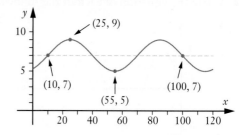

7 Let us suppose that in a particular part of the world the number of hours of daylight (h hours) can be reasonably well modelled by the graph shown on the right, with the graph commencing at the spring equinox (day and night of equal length) and showing one year (365 days).

Determine an equation for the graph in the form $h = a \sin(bt) + d$, given that the longest day had 17 hours of daylight and the shortest day had just 7 hours of daylight.

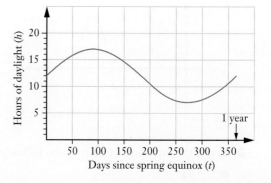

ISBN 9780170390477

8 An automatic device records the depth
of water (*d* metres) in a tidal harbour
from one low tide until two low tides later,
a time interval of 25 hours. It was found
that *d* plotted against *t*, the number
of hours since recording commenced,
fitted closely to the graph shown below.

a Determine an equation for the graph in the form $d = a \cos (bt) + e$.

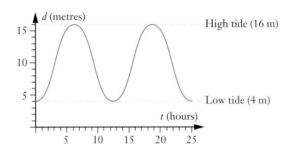

b Express the relationship between *d* and *t* as a sine function.

9 The graph below right shows the height above ground (*h* metres) of the tip of one of the blades
of a wind farm turbine at time *t* seconds.

a Express the relationship between *h* and *t* in the form $h = a \cos [b(t - c)] + d$.

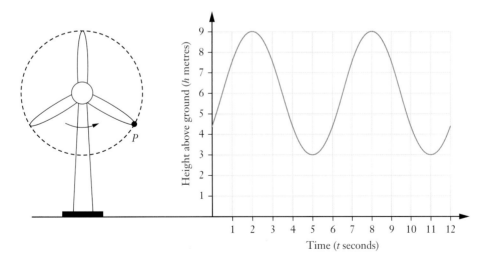

b Express the relationship between *h* and *t* as a sine function.

Miscellaneous exercise nine

This miscellaneous exercise may include questions involving the work of this chapter and the ideas mentioned in the Preliminary work section at the beginning of this unit.

1 Find all solutions to the equation $\sqrt{2}\sin 5x = 1$ lying in the interval $0 \le x \le \pi$.

2 Prove that $\cos 3\theta = 4\cos^3\theta - 3\cos\theta$

3 Solve $2\sin x \cos x = \sqrt{3} - 2\sqrt{3}\sin^2 x$ for $0 \le x \le 360°$.

4 Solve $\tan^2 x = 3(\sec x - 1)$ for $-\pi \le x \le \pi$.

5 Find all solutions to the equation $4\sin 3x \cos x = \sqrt{3} + 2\sin 2x$ for x in radians.

6 a Express $7\sin\theta - 10\cos\theta$ in the form $R\sin(\theta - \alpha)$ for α an acute angle in radians, correct to two decimal places and R exact.

b Hence determine the minimum value of $7\sin\theta - 10\cos\theta$ and the smallest positive value of θ for which it occurs. (Give θ in radians.)

7 The mean monthly daily temperatures for a particular location were as shown tabulated and graphed below:

Jan	Feb	Mar	Apr	May	Jun	Jul	Aug	Sept	Oct	Nov	Dec
27.2	26.8	24.6	22.0	19.5	17.5	17.0	17.7	19.8	22.0	24.6	26.5

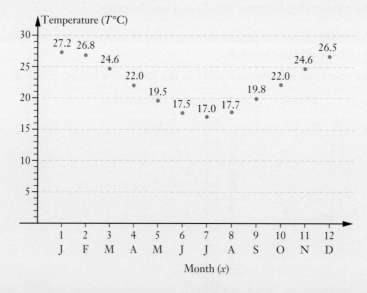

a *Without* using the ability of some calculators to determine a sinusoidal model to fit given data points, obtain an equation of the form $T = a\sin[b(x + c)] + d$ to model the above situation, explaining your reasoning.

b If you have a calculator or computer program that will find a sinusoidal model for given data points, use it to determine a suitable model and compare it to your answer for part **a**.

10.

Matrices

- Adding and subtracting matrices
- Multiplying a matrix by a number
- Equal matrices
- Multiplying matrices
- Zero matrices
- Multiplicative identity matrices
- The multiplicative inverse of a square matrix
- Using the inverse matrix to solve systems of equations
- Extension activity: Finding the determinant and inverse of a 3×3 matrix
- Miscellaneous exercise ten

Situation

A league soccer competition involves six teams:

Ajax,   Battlers,   Cloggers,

Devils,   Enzymes,   Flames.

Each team plays one game per week and, during the ten week competition, plays each other team twice, once in the first five weeks and once in the last five weeks. (All teams play at the same venue so no consideration needs to be made to balance home and away games.)

The results for the first five weeks gave rise to the following table:

	Played	Won	Drawn	Lost	Goals scored	
					For	Against
Ajax	5	2	1	2	10	5
Battlers	5	2	1	2	4	5
Cloggers	5	2	0	3	7	6
Devils	5	2	0	3	4	11
Enzymes	5	3	2	0	8	2
Flames	5	2	0	3	5	9

• Create a similar table for the last five weeks of the competition using the results stated below.

Week 6		
Ajax 3	1	Battlers
Cloggers 4	1	Devils
Enzymes 5	4	Flames

Week 7		
Ajax 1	2	Cloggers
Battlers 1	0	Enzymes
Devils 1	1	Flames

Week 8		
Ajax 2	2	Devils
Battlers 1	0	Flames
Cloggers 4	3	Enzymes

Week 9		
Ajax 0	1	Enzymes
Battlers 2	0	Devils
Cloggers 1	0	Flames

Week 10		
Ajax 1	3	Flames
Battlers 0	1	Cloggers
Devils 0	1	Enzymes

• Create a table like the one above for the complete ten week competition.

As part of the soccer league activity on the previous page we had to arrange information in a 'rows and columns' form of presentation. This rows and columns *rectangular array* presentation of numbers is called a **matrix**. (Plural: matrices).

If we remove the headings and indicate the start and end of the matrix with brackets, the table given on the previous page would be written as shown on the right.

$$\begin{bmatrix} 5 & 2 & 1 & 2 & 10 & 5 \\ 5 & 2 & 1 & 2 & 4 & 5 \\ 5 & 2 & 0 & 3 & 7 & 6 \\ 5 & 2 & 0 & 3 & 4 & 11 \\ 5 & 3 & 2 & 0 & 8 & 2 \\ 5 & 2 & 0 & 3 & 5 & 9 \end{bmatrix}$$

This matrix has 6 rows and 6 columns. We say it is a *six by six* matrix, (written 6×6). This gives the **size** or **dimensions** of the matrix.

Matrices do not have to have the same number of rows as they have columns, however those that do are called **square matrices**.

The matrix on the right is a 6×6 square matrix.

$$\begin{bmatrix} 1 & 0 & 4 \\ 3 & 2 & 0.5 \end{bmatrix}$$

2 rows
and
3 columns.

A 2 × 3 matrix.

$$\begin{bmatrix} 2 & 5 \\ 11 & -2 \end{bmatrix}$$

2 rows
and
2 columns.

A 2 × 2 matrix.
(A square matrix.)

$$\begin{bmatrix} 3 \\ 2 \\ -1 \end{bmatrix}$$

3 rows
and
1 column.

A 3 × 1 matrix.

$$\begin{bmatrix} 1 & 0 & 2 & 3 \\ 1 & 2 & -1 & 0 \\ 0 & 1 & 0 & 3 \end{bmatrix}$$

3 rows
and
4 columns.

A 3 × 4 matrix.

A matrix consisting of just one column, like the third matrix above, is called a **column matrix**.

Any matrix consisting of just one row is called a **row matrix**.

$$\begin{bmatrix} 5 & 0 & -2 & 1 \end{bmatrix}$$

A square matrix having zeros in all spaces that are not on the **leading diagonal** is called a **diagonal matrix**.

$$\begin{bmatrix} 5 & 0 & 0 & 0 \\ 0 & 0 & 0 & 0 \\ 0 & 0 & 2 & 0 \\ 0 & 0 & 0 & -4 \end{bmatrix}$$

We commonly use capital letters to label different matrices. The corresponding lower-case letters, with subscripted numbers, are then used to indicate the row and column a particular entry or **element** occupies.

For the matrix A shown on the right, the element occupying the 3rd row and 2nd column is the number 7.

$$A = \begin{bmatrix} 2 & 0 & -1 & 3 \\ 1 & 6 & -3 & 9 \\ 5 & 7 & 8 & 4 \end{bmatrix}$$

Thus $a_{32} = 7.$

Similarly $a_{11} = 2,$

$a_{12} = 0,$

$a_{13} = -1,$ etc.

ISBN 9780170390477

Adding and subtracting matrices

In the soccer competition activity earlier, you probably determined the matrix for the full ten weeks by adding the matrix for the first five weeks to the matrix for the last five weeks. To perform such addition it was natural to simply add elements occurring in corresponding locations. This is indeed how we add matrices. For example,

$$\text{If } A = \begin{bmatrix} 1 & 0 & 2 & 3 \\ 4 & -2 & 3 & 5 \\ 2 & 1 & -3 & 4 \end{bmatrix} \text{ and } B = \begin{bmatrix} 2 & 1 & -3 & 2 \\ 5 & 1 & 2 & 4 \\ 3 & 2 & 0 & -5 \end{bmatrix} \text{ then } A + B = \begin{bmatrix} 3 & 1 & -1 & 5 \\ 9 & -1 & 5 & 9 \\ 5 & 3 & -3 & -1 \end{bmatrix}$$

$$\text{and similarly} \qquad A - B = \begin{bmatrix} -1 & -1 & 5 & 1 \\ -1 & -3 & 1 & 1 \\ -1 & -1 & -3 & 9 \end{bmatrix}$$

Note: When adding or subtracting matrices there must be elements in corresponding spaces. Thus we can only add or subtract matrices that are the same size as each other.

Multiplying a matrix by a number

Suppose that the 3×2 matrix shown on the right shows the cost of three models of gas heater in two different shops. Now suppose that in a sale both shops offer 10% discount on all models.

	Shop One	Shop Two
Economy	$250	$280
Standard	$340	$330
Deluxe	$450	$450

The sale prices could be represented in a matrix formed by multiplying each element of the first matrix by 0.9. This is indeed how we multiply a matrix by a number: We multiply each element of the matrix by that number. (This is referred to as 'multiplication by a **scalar**'.)

	Shop One	Shop Two
Economy	$225	$252
Standard	$306	$297
Deluxe	$405	$405

Shutterstock.com/Csehak Szabolcs

Equal matrices

For two matrices to be equal, they must be of the same size and have all corresponding elements equal.

Thus if $\begin{bmatrix} a & b & c \\ d & e & f \end{bmatrix} = \begin{bmatrix} 2 & 3 & -5 \\ 1 & 0 & -2 \end{bmatrix}$ then $\begin{array}{ccc} a = 2 & b = 3 & c = -5 \\ d = 1 & e = 0 & f = -2 \end{array}$

EXAMPLE 1

If $A = \begin{bmatrix} 1 & 2 & 4 \\ 0 & -4 & 5 \end{bmatrix}$, $B = \begin{bmatrix} 3 & 5 & -2 \\ 1 & 0 & -2 \end{bmatrix}$ and $C = \begin{bmatrix} 2 & 3 \\ 1 & -5 \end{bmatrix}$ determine each of the following.

If any cannot be determined state this clearly and give the reason.

a $A + B$　　**b** $A + C$　　**c** $B - A$　　**d** $5C$　　**e** $3B - 2A$

Solution

a $A + B = \begin{bmatrix} 1 & 2 & 4 \\ 0 & -4 & 5 \end{bmatrix} + \begin{bmatrix} 3 & 5 & -2 \\ 1 & 0 & -2 \end{bmatrix}$

$= \begin{bmatrix} 4 & 7 & 2 \\ 1 & -4 & 3 \end{bmatrix}$

b A and C are not the same size. Thus A + C cannot be determined.

c $B - A = \begin{bmatrix} 3 & 5 & -2 \\ 1 & 0 & -2 \end{bmatrix} - \begin{bmatrix} 1 & 2 & 4 \\ 0 & -4 & 5 \end{bmatrix}$

$= \begin{bmatrix} 2 & 3 & -6 \\ 1 & 4 & -7 \end{bmatrix}$

d $5C = \begin{bmatrix} 10 & 15 \\ 5 & -25 \end{bmatrix}$

e $3B - 2A = \begin{bmatrix} 9 & 15 & -6 \\ 3 & 0 & -6 \end{bmatrix} - \begin{bmatrix} 2 & 4 & 8 \\ 0 & -8 & 10 \end{bmatrix}$

$= \begin{bmatrix} 7 & 11 & -14 \\ 3 & 8 & -16 \end{bmatrix}$

Many calculators will accept data in matrix form and can then manipulate these matrices in various ways.

Get to know the matrix capability of your calculator.

How does your calculator respond when you ask it to add together two matrices that are not of the same size?

$\begin{bmatrix} 1 & 2 & 4 \\ 0 & -4 & 5 \end{bmatrix} + \begin{bmatrix} 3 & 5 & -2 \\ 1 & 0 & -2 \end{bmatrix}$

$\begin{bmatrix} 4 & 7 & 2 \\ 1 & -4 & 3 \end{bmatrix}$

Exercise 10A

1 A matrix with m rows and n columns has size m × n. Write down the size of each of the following matrices.

$$A = \begin{bmatrix} 2 & -3 \\ 3 & 4 \\ 1 & 0 \\ 2 & -1 \end{bmatrix} \qquad B = \begin{bmatrix} 2 & 3 & -7 & 32 \\ 1 & 1 & 0 & 5 \end{bmatrix} \qquad C = \begin{bmatrix} -1 \\ 3 \\ 5 \\ 2 \end{bmatrix} \qquad D = \begin{bmatrix} 1 & 0 & 0 \\ 3 & 0 & 1 \\ 0 & 5 & 0 \\ 0 & 1 & 3 \end{bmatrix}$$

$$E = \begin{bmatrix} 2 & 5 \\ 0 & 1 \end{bmatrix} \qquad F = \begin{bmatrix} 1 & 11 & -2 \end{bmatrix} \qquad G = \begin{bmatrix} 12 & 3 \\ 0 & 5 \\ -5 & 2 \end{bmatrix} \qquad H = \begin{bmatrix} 1 & 0 & 0 & 0 \\ 0 & 1 & 0 & 0 \\ 0 & 0 & 1 & 0 \\ 0 & 0 & 0 & 1 \end{bmatrix}$$

2 If e_{mn} is the element situated in the mth row and nth column of matrix E determine

a e_{12} **b** e_{21} **c** f_{13} **d** g_{21} **e** g_{22} **f** g_{32}

where matrices E, F and G are as given below.

$$E = \begin{bmatrix} 5 & 4 & 13 \\ -4 & 2 & 0 \\ 1 & -8 & 12 \end{bmatrix} \qquad F = \begin{bmatrix} 1 & 5 & 7 & 2 \end{bmatrix} \qquad G = \begin{bmatrix} 1 & 2 \\ 7 & 3 \\ -2 & 0 \\ 4 & 11 \end{bmatrix}$$

3 If $A = \begin{bmatrix} 1 & 2 \\ 0 & -4 \end{bmatrix}$, $B = \begin{bmatrix} 3 & -1 \\ 2 & 4 \\ 0 & 3 \end{bmatrix}$, $C = \begin{bmatrix} 2 & -3 \\ 1 & -5 \end{bmatrix}$ and $D = \begin{bmatrix} 3 \\ 1 \\ -2 \end{bmatrix}$, determine each of the

following. (If any cannot be determined state this clearly.)

a A + B **b** A + C **c** C − A **d** 2D
e 3B **f** B + D **g** 2A **h** 2A − C

4 If $P = \begin{bmatrix} 3 & 2 & -1 \\ 1 & 4 & 3 \end{bmatrix}$, $Q = \begin{bmatrix} 2 & 1 & 0 \\ 0 & -1 & 0 \end{bmatrix}$ and $R = \begin{bmatrix} 1 & 2 & 1 \\ 2 & 1 & 2 \end{bmatrix}$, determine each of the following.

(If any cannot be determined state this clearly.)

a P + Q **b** Q − P **c** 3R **d** 3P − 2Q

5 If $A = \begin{bmatrix} 2 & 4 \\ 1 & 3 \end{bmatrix}$, $B = \begin{bmatrix} 2 & 1 & 3 \end{bmatrix}$, $C = \begin{bmatrix} 3 & 1 & 4 \end{bmatrix}$ and $D = \begin{bmatrix} 2 \\ 1 \\ 3 \end{bmatrix}$, determine each of the

following. (If any cannot be determined state this clearly.)

a A + B **b** 3A **c** B + 2C **d** C + D

6 If $A = \begin{bmatrix} 1 & 3 & 0 & 1 \\ 0 & 1 & 2 & 3 \\ 0 & 0 & 1 & 4 \end{bmatrix}$, $B = \begin{bmatrix} 3 & 1 & 4 \\ 2 & 1 & -3 \\ 0 & 1 & 2 \\ 1 & 0 & 0 \end{bmatrix}$ and $C = \begin{bmatrix} 5 & 1 & 3 & -1 \\ 2 & 1 & 4 & 3 \\ 1 & 5 & 2 & 0 \end{bmatrix}$, determine each of the

following. (If any cannot be determined state this clearly.)

a $A + B$ **b** $A + C$ **c** $2B$ **d** $5A - C$

7 $A = \begin{bmatrix} 1 & 2 \end{bmatrix}$, $B = \begin{bmatrix} 3 \\ 1 \end{bmatrix}$, $C = \begin{bmatrix} 5 & 2 \end{bmatrix}$ and $D = \begin{bmatrix} 1 & 7 \end{bmatrix}$.

For each of the following write 'Yes' if it can be determined and 'No' if it cannot be determined.

a $A + B$ **b** $B - A$ **c** $3C$ **d** $A + D$

e $A - 3D$ **f** $A + 3B$ **g** $B + B$ **h** $A + B + C$

8 Is matrix addition commutative? i.e. Does $A + B = B + A$ (assuming A and B are of the same size)?

9 Is matrix addition associative? i.e. Does $A + (B + C) = (A + B) + C$ (assuming A, B and C are all of the same size)?

10 If $A = \begin{bmatrix} 1 & -1 & 2 \\ 1 & 0 & 3 \end{bmatrix}$ and $B = \begin{bmatrix} 1 & -7 & 12 \\ 1 & 0 & 13 \end{bmatrix}$, determine matrix C given that the following equation is correct: $3A - 2C = B$.

11 For the first four games in a basketball season the points (P), assists (A), and blocks (B), that five members of one team carried out were as shown below.

	Game 1 P A B		Game 2 P A B		Game 3 P A B		Game 4 P A B
Alan	8 5 1	Alan	12 3 1	Alan	11 8 2	Alan	9 4 0
Bob	7 2 4	Bob	6 8 2	Bob	15 2 5	Bob	9 3 3
Dave	14 3 1	Dave	15 3 5	Dave	7 5 2	Dave	11 8 1
Mark	17 3 1	Mark	6 4 0	Mark	12 2 1	Mark	4 12 1
Roger	8 8 2	Roger	5 2 6	Roger	14 4 5	Roger	12 5 3

a Construct a single 5×3 matrix showing the total points, total assists and total blocks each of these five players achieved for the 4 game period.

b Construct a single 5×3 matrix showing the average points per game, average assists per game and average blocks per game for each of these five players for the 4 game period.

12 A company manufactures five types of lawn fertiliser:
- Basic (B)   • Feedit (F)   • Fertilawn (FL)   • Greenit (G)   • Growgrass (GG)

It sells these through its four garden centres (shops).

The number of bags of these fertilisers sold in these centres during the first and second halves of a year are given below:

1 January → 30 June

	B	F	FL	G	GG
Centre I	3100	550	1040	820	2250
Centre II	1640	420	720	480	1480
Centre III	2850	520	1320	640	1250
Centre IV	1240	300	800	360	960

1 July → 31 December

	B	F	FL	G	GG
Centre I	2500	1200	1280	950	2000
Centre II	1200	850	650	540	1240
Centre III	2200	950	1500	640	1450
Centre IV	950	640	720	480	820

The company predicts that at each shop the sales for the next year will increase by 10% due to a new sales campaign. Assuming this prediction is indeed correct produce a 4×5 matrix showing the number of bags of each fertiliser sold at each shop for the following 1st January → 31st December.

13 If a_{nm} is the element situated in the nth row and mth column of matrix A write down matrix A given that it is a 3×3 matrix with $a_{nm} = 2n + m$.

14 If a_{nm} is the element situated in the nth row and mth column of matrix A write down matrix A given that it is a 3×4 matrix with $a_{nm} = n^m$.

Multiplying matrices

An inter-school sports carnival involves five schools competing in seven sports. In each of these sports, medals, certificates and team points are awarded to teams finishing 1st, 2nd or 3rd.

	1st Place	2nd Place	3rd Place
School A	1	1	1
School B	3	1	0
School C	0	3	3
School D	1	2	0
School E	2	0	3

The 5×3 matrix on the right shows the number of first, second and third places gained by each of the five schools.

Suppose that points are awarded using the points system:

1st	3 points
2nd	2 points
3rd	1 point

The total points scored for each school are:

School A	School B	School C	School D	School E
$1 \times 3 +$	$3 \times 3 +$	$0 \times 3 +$	$1 \times 3 +$	$2 \times 3 +$
$1 \times 2 +$	$1 \times 2 +$	$3 \times 2 +$	$2 \times 2 +$	$0 \times 2 +$
1×1	0×1	3×1	0×1	3×1
6	11	9	7	9

Thus school B finished first with 11 points, followed by schools C and E equal second, school D was fourth and school A was fifth.

Note the way that each row of the 5×3 matrix is 'stood up' to align with the points matrix. This is indeed how we carry out matrix multiplication.

Shutterstock.com/Patchanee Samutarlai

Matrices can be multiplied together if the number of columns in the first matrix equals the number of rows in the second matrix.

If $A = \begin{bmatrix} 2 & 1 & 3 \\ 0 & -1 & 2 \end{bmatrix}$ and $B = \begin{bmatrix} 1 & 2 \\ 4 & -1 \\ 1 & -3 \end{bmatrix}$, the product AB is found as shown below.

Follow each step carefully to make sure you understand where each element in the final answer comes from.

First spin the 1st row of A to align with 1st column of B, multiply and then add:

Thus $\begin{bmatrix} 2 & 1 & 3 \\ 0 & -1 & 2 \end{bmatrix}\begin{bmatrix} 1 & 2 \\ 4 & -1 \\ 1 & -3 \end{bmatrix} = \begin{bmatrix} (2)(1)+(1)(4)+(3)(1) \end{bmatrix}$

Continue to use the first row of A, this time going *further across* to align with the 2nd column of B. We similarly go further across to place our answer:

$$\begin{bmatrix} 2 & 1 & 3 \\ 0 & -1 & 2 \end{bmatrix}\begin{bmatrix} 1 & 2 \\ 4 & -1 \\ 1 & -3 \end{bmatrix} = \begin{bmatrix} 9 & (2)(2)+(1)(-1)+(3)(-3) \end{bmatrix}$$

Having 'exhausted' the 1st row of A we now move *down* to use the 2nd row and similarly move *down* to place our answer:

$$\begin{bmatrix} 2 & 1 & 3 \\ 0 & -1 & 2 \end{bmatrix}\begin{bmatrix} 1 & 2 \\ 4 & -1 \\ 1 & -3 \end{bmatrix} = \begin{bmatrix} 9 & -6 \\ (0)(1)+(-1)(4)+(2)(1) & \end{bmatrix}$$

Continuing the process:

$$\begin{bmatrix} 2 & 1 & 3 \\ 0 & -1 & 2 \end{bmatrix}\begin{bmatrix} 1 & 2 \\ 4 & -1 \\ 1 & -3 \end{bmatrix} = \begin{bmatrix} 9 & -6 \\ -2 & (0)(2)+(-1)(-1)+(2)(-3) \end{bmatrix}$$

Thus $\begin{bmatrix} 2 & 1 & 3 \\ 0 & -1 & 2 \end{bmatrix}\begin{bmatrix} 1 & 2 \\ 4 & -1 \\ 1 & -3 \end{bmatrix} = \begin{bmatrix} 9 & -6 \\ -2 & -5 \end{bmatrix}$

Confirm this result using your calculator.

Using your calculator to determine the product of matrices can be useful but if the matrices are not too big you should be able to determine the answers mentally. You would not need to show each step of the process and, with practice, you should be able to write the answer directly, as shown at the top of the next page.

$$\begin{bmatrix} 2 & 1 & 3 \\ 0 & -1 & 2 \end{bmatrix} \times \begin{bmatrix} 1 & 2 \\ 4 & -1 \\ 1 & -3 \end{bmatrix}$$

$$\begin{bmatrix} 9 & -6 \\ -2 & -5 \end{bmatrix}$$

ISBN 9780170390477

$$\begin{bmatrix} 1 & 3 \end{bmatrix}\begin{bmatrix} 2 & 1 \\ -1 & 4 \end{bmatrix} = \begin{bmatrix} -1 & 13 \end{bmatrix}$$

$$\begin{bmatrix} 3 & 2 \\ 1 & 5 \end{bmatrix}\begin{bmatrix} 1 \\ 3 \end{bmatrix} = \begin{bmatrix} 9 \\ 16 \end{bmatrix}$$

$$\begin{bmatrix} 2 & 3 \\ -1 & 2 \end{bmatrix}\begin{bmatrix} 1 & 4 \\ 1 & 3 \end{bmatrix} = \begin{bmatrix} 5 & 17 \\ 1 & 2 \end{bmatrix}$$

$$\begin{bmatrix} 3 & 1 \end{bmatrix}\begin{bmatrix} 1 & 2 & 1 \\ -1 & 0 & 1 \end{bmatrix} = \begin{bmatrix} 2 & 6 & 4 \end{bmatrix}$$

As was mentioned earlier, this method of matrix multiplication means that:

Two matrices can be multiplied together provided the number of columns in the first matrix equals the number of rows in the second matrix.

Suppose matrix A has dimensions $m \times n$ and matrix B has dimensions $p \times q$.

• The product $A_{mn} B_{pq}$ can only be formed if $n = p$. In this case AB will have dimensions $m \times q$.
• The product $B_{pq} A_{mn}$ can only be formed if $q = m$. In this case BA will have dimensions $p \times n$.

Note: In the product AB we say that B is *premultiplied* by A
or that A is *postmultiplied* by B

EXAMPLE 2

If $A = \begin{bmatrix} 1 & 2 & 3 \\ 1 & 4 & 0 \end{bmatrix}$, $B = \begin{bmatrix} 2 & 1 \\ 3 & -1 \end{bmatrix}$ and $C = \begin{bmatrix} 1 \\ 4 \\ 1 \end{bmatrix}$ determine each of the following. If any cannot be

determined state this clearly and explain why.

a AB **b** BA **c** AC **d** CA **e** B^2

Solution

a $AB = \begin{bmatrix} 1 & 2 & 3 \\ 1 & 4 & 0 \end{bmatrix}\begin{bmatrix} 2 & 1 \\ 3 & -1 \end{bmatrix}$ which cannot be determined because the number of columns in

A ($2 \times \mathbf{3}$) ≠ the number of rows in B ($\mathbf{2} \times 2$).

b $BA = \begin{bmatrix} 2 & 1 \\ 3 & -1 \end{bmatrix}\begin{bmatrix} 1 & 2 & 3 \\ 1 & 4 & 0 \end{bmatrix} = \begin{bmatrix} 3 & 8 & 6 \\ 2 & 2 & 9 \end{bmatrix}$

c $AC = \begin{bmatrix} 1 & 2 & 3 \\ 1 & 4 & 0 \end{bmatrix}\begin{bmatrix} 1 \\ 4 \\ 1 \end{bmatrix} = \begin{bmatrix} 12 \\ 17 \end{bmatrix}$

d $CA = \begin{bmatrix} 1 \\ 4 \\ 1 \end{bmatrix}\begin{bmatrix} 1 & 2 & 3 \\ 1 & 4 & 0 \end{bmatrix}$ which cannot be determined because the number of columns in

C ($3 \times \mathbf{1}$) ≠ the number of rows in A ($\mathbf{2} \times 3$).

e $B^2 = \begin{bmatrix} 2 & 1 \\ 3 & -1 \end{bmatrix}\begin{bmatrix} 2 & 1 \\ 3 & -1 \end{bmatrix} = \begin{bmatrix} 7 & 1 \\ 3 & 4 \end{bmatrix}$

Confirm the above answers using your calculator.

EXAMPLE 3

A manufacturer makes three products A, B and C, each requiring a certain number of units of commodities P, Q, R, S and T. Matrix X below shows the number of units of each commodity required to make one of each product.

$$
\begin{array}{c}
 \\
\text{Product A} \\
\text{Product B} \\
\text{Product C}
\end{array}
\begin{array}{c}
\begin{array}{ccccc} P & Q & R & S & T \end{array} \\
\left[
\begin{array}{ccccc}
1 & 1 & 0 & 2 & 3 \\
1 & 1 & 2 & 1 & 2 \\
0 & 1 & 3 & 0 & 3
\end{array}
\right] = X
\end{array}
$$

a Each unit of P, Q, R, S and T costs the manufacturer $200, $100, $50, $400 and $300 respectively. Write this information as matrix Y which should be either a column matrix or a row matrix, whichever can form a product with X.

b Form the product referred to in **a** and explain what information it displays.

Solution

a As a column matrix, Y would have dimensions 5×1.

As a row matrix, Y would have dimensions 1×5.

Matrix $X_{3 \times 5}$ can form a product with $Y_{5 \times 1}$: $X_{3 \times 5}\, Y_{5 \times 1} = Z_{3 \times 1}$

$$
\text{Thus } Y =
\begin{bmatrix}
\$200 \\
\$100 \\
\$50 \\
\$400 \\
\$300
\end{bmatrix}
\begin{array}{l}
\leftarrow \text{Cost of 1 unit of P} \\
\leftarrow \text{Cost of 1 unit of Q} \\
\leftarrow \text{Cost of 1 unit of R} \\
\leftarrow \text{Cost of 1 unit of S} \\
\leftarrow \text{Cost of 1 unit of T}
\end{array}
$$

(The order P, Q, R, S, T being consistent with the order in X.)

b
$$
XY =
\begin{bmatrix}
1 & 1 & 0 & 2 & 3 \\
1 & 1 & 2 & 1 & 2 \\
0 & 1 & 3 & 0 & 3
\end{bmatrix}
\begin{bmatrix}
200 \\
100 \\
50 \\
400 \\
300
\end{bmatrix}
$$

$$
=
\begin{bmatrix}
2000 \\
1400 \\
1150
\end{bmatrix}
\begin{array}{l}
\leftarrow \text{total commodity cost (\$) for producing 1 unit of product A} \\
\leftarrow \text{total commodity cost (\$) for producing 1 unit of product B} \\
\leftarrow \text{total commodity cost (\$) for producing 1 unit of product C}
\end{array}
$$

Note: Whilst this chapter has considered

- adding and subtracting matrices,
- multiplying a matrix by a scalar
- multiplying matrices

the concept of dividing one matrix by another is undefined for matrices.

Exercise 10B

Determine each of the following products. If any are not possible state this clearly and explain why.

1 $\begin{bmatrix} 1 & 2 \end{bmatrix} \begin{bmatrix} 2 & 3 \\ 1 & 3 \end{bmatrix}$

2 $\begin{bmatrix} 2 & 3 \\ 1 & 3 \end{bmatrix} \begin{bmatrix} 1 & 2 \end{bmatrix}$

3 $\begin{bmatrix} 2 & -1 \\ 1 & 0 \end{bmatrix} \begin{bmatrix} 1 & 4 \\ 0 & -2 \end{bmatrix}$

4 $\begin{bmatrix} 3 & 1 \end{bmatrix} \begin{bmatrix} 1 \\ 4 \end{bmatrix}$

5 $\begin{bmatrix} 1 \\ 4 \end{bmatrix} \begin{bmatrix} 3 & 1 \end{bmatrix}$

6 $\begin{bmatrix} 2 & -3 \\ -1 & 4 \end{bmatrix} \begin{bmatrix} 2 & 1 \\ -3 & 2 \end{bmatrix}$

7 $\begin{bmatrix} 1 & 0 \\ 0 & 1 \end{bmatrix} \begin{bmatrix} 2 & 3 \\ 1 & -1 \end{bmatrix}$

8 $\begin{bmatrix} 1 & 4 \\ -1 & 3 \end{bmatrix} \begin{bmatrix} 1 & 0 \\ 0 & 1 \end{bmatrix}$

9 $\begin{bmatrix} 0 & 0 \\ 0 & 0 \end{bmatrix} \begin{bmatrix} 2 & 1 \\ 4 & 5 \end{bmatrix}$

10 $\begin{bmatrix} 3 & 1 \\ 5 & 2 \end{bmatrix} \begin{bmatrix} 2 & -1 \\ -5 & 3 \end{bmatrix}$

11 $\begin{bmatrix} 8 & -5 \\ -3 & 2 \end{bmatrix} \begin{bmatrix} 2 & 5 \\ 3 & 8 \end{bmatrix}$

12 $\begin{bmatrix} 3 & 1 \\ 1 & 1 \end{bmatrix} \begin{bmatrix} 0.5 & -0.5 \\ -0.5 & 1.5 \end{bmatrix}$

13 $\begin{bmatrix} 1 & 2 & 1 & 2 \end{bmatrix} \begin{bmatrix} 2 \\ 1 \\ 2 \\ 1 \end{bmatrix}$

14 $\begin{bmatrix} 1 & 0 & 1 & 0 \\ 0 & 1 & 0 & 1 \end{bmatrix} \begin{bmatrix} 1 & 0 & 1 \\ 3 & -1 & 0 \\ 2 & 2 & 2 \\ 1 & 4 & 1 \end{bmatrix}$

15 $\begin{bmatrix} 1 & 0 \\ 0 & 2 \\ 1 & 1 \end{bmatrix} \begin{bmatrix} 1 & 0 & 5 \\ 5 & 1 & -1 \end{bmatrix}$

16 $\begin{bmatrix} 1 & 3 & 1 \\ 3 & 0 & -2 \end{bmatrix} \begin{bmatrix} 1 & 2 \\ 4 & 1 \\ -3 & -2 \end{bmatrix}$

17 $\begin{bmatrix} 1 & 2 & 3 \\ 4 & 5 & 6 \end{bmatrix} \begin{bmatrix} 1 \\ 2 \\ 3 \end{bmatrix}$

18 $\begin{bmatrix} 2 & 1 & 0 \\ -1 & 3 & 2 \\ 0 & 2 & 4 \end{bmatrix} \begin{bmatrix} 1 & 1 & -1 \\ 0 & 2 & 3 \\ 3 & 1 & 4 \end{bmatrix}$

19 If $A = \begin{bmatrix} 1 & 0 & -1 \\ 2 & 0 & 1 \\ 0 & 1 & 1 \end{bmatrix}$ and $B = \begin{bmatrix} 0 & 1 & 2 \\ 2 & 1 & 0 \\ 0 & -1 & 1 \end{bmatrix}$, determine the following:

   a AB   **b** BA   **c** A^2 **d** B^2

20 Multiplication of numbers is commutative, i.e. if x and y represent numbers then xy is always equal to yx. Is matrix multiplication commutative for all pairs of matrices for which the necessary products can be formed? Justify your answer.

21 Provided the necessary products can be formed, matrix multiplication is associative, i.e. $(AB)C = A(BC)$. Verify this for

a $A = \begin{bmatrix} 1 & 2 \\ -1 & 0 \end{bmatrix}, B = \begin{bmatrix} 3 & 1 \\ 0 & -1 \end{bmatrix}, C = \begin{bmatrix} 1 & 2 \\ -1 & 1 \end{bmatrix}$.

b $A = \begin{bmatrix} 1 & 2 \end{bmatrix}, B = \begin{bmatrix} 1 & 0 & -1 \\ 2 & 1 & 1 \end{bmatrix}, C = \begin{bmatrix} 1 & 0 \\ -1 & 2 \\ 1 & 1 \end{bmatrix}$.

22 Provided the necessary sums and products can be formed, the distributive law:

$$A(B + C) = AB + AC$$

holds for matrices. Verify this for

a $A = \begin{bmatrix} 2 & 1 \\ 4 & 0 \end{bmatrix}, B = \begin{bmatrix} -1 & 1 \\ 0 & 1 \end{bmatrix}, C = \begin{bmatrix} 2 & 1 \\ -1 & 3 \end{bmatrix}$.

b $A = \begin{bmatrix} 2 & 0 \\ -3 & 1 \end{bmatrix}, B = \begin{bmatrix} 3 \\ 2 \end{bmatrix}, C = \begin{bmatrix} 1 \\ 4 \end{bmatrix}$.

23 If $A = \begin{bmatrix} a & b \\ c & d \end{bmatrix}$ and $B = \begin{bmatrix} e & f \\ g & h \end{bmatrix}$ and k is a number, prove that

$$(kA)B = A(kB) = k(AB)$$

24 A is a 3×2 matrix, B is a 3×2 matrix, C is a 2×3 matrix and D is a 1×3 matrix. State the dimensions of each of the following products. For any that cannot be formed state this clearly.

a AB	**b** BA	**c** BC	**d** CB
e AD	**f** DA	**g** BCA	**h** DAC

25 With $A = \begin{bmatrix} 2 & 1 \end{bmatrix}, B = \begin{bmatrix} 1 \\ 3 \end{bmatrix}, C = \begin{bmatrix} 1 & 4 \\ 2 & -1 \end{bmatrix}$ and $D = \begin{bmatrix} 1 \\ 3 \\ 1 \end{bmatrix}$, state whether each of the following

products can be formed or not.

a AB	**b** BA	**c** AC	**d** CA
e BD	**f** DB	**g** AD	**h** DA

26 If it is possible to form the matrix product AA what can we say about matrix A?

ISBN 9780170390477

27 BC is just one product that can be formed using two matrices selected from the three below. List all the other products that could be formed in this way. (The selection of the two matrices can involve the same matrix being selected twice.)

$$A = \begin{bmatrix} 1 & -1 \\ 2 & 1 \end{bmatrix} \qquad B = \begin{bmatrix} 1 & 3 \end{bmatrix} \qquad C = \begin{bmatrix} 4 \\ 1 \end{bmatrix}$$

28 a Premultiply $\begin{bmatrix} 2 & 0 \\ 3 & 2 \end{bmatrix}$ by $\begin{bmatrix} 1 & -1 \\ 2 & 0 \end{bmatrix}$.

b Postmultiply $\begin{bmatrix} 2 & 0 \\ 3 & 2 \end{bmatrix}$ by $\begin{bmatrix} 1 & -1 \\ 2 & 0 \end{bmatrix}$.

29 The 5×3 matrix shown on the right appeared on an earlier page. It shows the number of first, second and third places gained by each of five schools taking part in an inter-school sports carnival involving seven sports.

Determine the rank order for these schools using the points matrix

	1st Place	2nd Place	3rd Place
School A	1	1	1
School B	3	1	0
School C	0	3	3
School D	1	2	0
School E	2	0	3

a $\begin{array}{c} \text{1st} \\ \text{2nd} \\ \text{3rd} \end{array} \begin{bmatrix} \text{5 points} \\ \text{3 points} \\ \text{1 point} \end{bmatrix}$

b $\begin{array}{c} \text{1st} \\ \text{2nd} \\ \text{3rd} \end{array} \begin{bmatrix} \text{4 points} \\ \text{3 points} \\ \text{2 points} \end{bmatrix}$

30 A financial adviser sets up share portfolios for three clients. Each portfolio involves shares in 4 companies with the number of shares as shown below.

	Abel Co.	Big Co.	Con Co.	Down Co.
Client 1	1000	5000	400	270
Client 2	500	8000	500	250
Client 3	500	3000	500	500

Initially the value of each share is:

Abel Co.	$5
Big Co.	50 cents
Con Co.	$12
Down Co.	$10

Two years later the value of each share is:

Abel Co.	$4
Big Co.	60 cents
Con Co.	$20
Down Co.	$10

Use matrix multiplication to determine the value of each client's portfolio at each of these times.

31 A fast food outlet offers, amongst other things, Snack Packs and Family Packs. The contents of each of these are as in the contents matrix below:

	Drink (mL)	Number of Burgers
Each Snack Pack	375	1
Each Family Pack	1250	4

An order comes in for 15 Snack Packs and 10 Family Packs. Use matrix multiplication to determine a matrix that shows the total volume of drink and the total number of burgers this order requires.

32 Three hotels each have single rooms, double rooms and suites. The number of each of these in each hotel is as shown in matrix P below:

	Hotel A	Hotel B	Hotel C
Single	15	5	5
Double	25	25	14
Suite	2	1	3

= Matrix P

The three hotels are all owned by the same company and all operate the same pricing structure as shown in the tariff matrix, Q, shown below:

	Single	Double	Suite
Cost per night	$75	$125	$180

= Matrix Q

a Only one of PQ and QP can be formed. Which one?

b Determine the matrix product from part **a** and explain what information it is that this matrix displays.

c Suppose instead that the tariff matrix were written as a column matrix, R. The matrix product PR could be formed but would it give any useful data? Explain your answer.

33 A carpenter runs a business making three different models of cubby house for children. Each cubby house is made using four different sizes of treated pine timber. The number of metres of each size of timber required for each cubby house is shown below.

	Poles 120 mm diameter	Decking 90 mm × 22 mm	Framing 70 mm × 35 mm	Sheeting 120 mm × 12 mm
Cubby A	3	30	20	40
Cubby B	4	35	25	60
Cubby C	6	40	30	70

We will call this matrix P.

a The carpenter receives an order for 3 type As, 1 type B and 2 type Cs.

Write this information as matrix Q which should be either a row matrix or a column matrix, whichever can form a product with P.

b Determine the product referred to in part **a** and explain what this matrix represents.

c The poles cost $4 per metre, the decking $2 per metre, the framing $3 per metre and the sheeting $1.50 per metre. Write this information as matrix R which should be either a row matrix or a column matrix, dependent on which will form a product with P. What dimensions would this product matrix be and what information would it display?

Shutterstock.com/Olesia Bilkei

34 A manufacturer makes four different models of a particular product. Matrix D below gives the number of units of commodities A, B and C required to make one of each model type.

$$
\begin{array}{c}
\text{Commodity A} \\
\text{Commodity B} \\
\text{Commodity C}
\end{array}
\begin{array}{cccc}
\text{Model I} & \text{Model II} & \text{Model III} & \text{Model IV} \\
\end{array}
\left[
\begin{array}{cccc}
2 & 3 & 1 & 2 \\
20 & 30 & 50 & 40 \\
2 & 1 & 3 & 2
\end{array}
\right] = D
$$

a Each unit of the commodities A, B and C costs the manufacturer $800, $50 and $1000 respectively. Write this information as matrix E, either a column matrix or a row matrix, whichever can form a product with D.

b Form the product referred to in **a** and explain the information it displays.

35 A manufacturer makes three different models of a particular item. Matrix P below gives the number of minutes in the cutting area, the assembling area and the packing area required to make each model.

$$
\begin{array}{c}
\text{Each model A} \\
\text{Each model B} \\
\text{Each model C}
\end{array}
\begin{array}{ccc}
\text{Cutting} & \text{Assembling} & \text{Packing} \\
\end{array}
\left[
\begin{array}{ccc}
30 & 20 & 10 \\
20 & 30 & 10 \\
40 & 40 & 10
\end{array}
\right] = P
$$

The manufacturer receives orders for 50 As, 100 Bs and 80 Cs.

We could write this as a column matrix, Q:
$$
\left[
\begin{array}{c}
50 \\
100 \\
80
\end{array}
\right]
$$

or as a row matrix, R:
$$
\left[
\begin{array}{ccc}
50 & 100 & 80
\end{array}
\right]
$$

Both PQ and RP could be formed but only one of these will contain information likely to be useful.

a Which product is this?

b Form the product.

c Explain the information it gives.

Zero matrices

Any matrix which has every one of its entries as zero is called a **zero matrix**.

Thus the 2×2 zero matrix is $\begin{bmatrix} 0 & 0 \\ 0 & 0 \end{bmatrix}$. The 2×3 zero matrix is $\begin{bmatrix} 0 & 0 & 0 \\ 0 & 0 & 0 \end{bmatrix}$.

The letter O is used to indicate a zero matrix. If it is necessary to indicate that it is the zero matrix of a particular dimension, say 2×2 for example, then we write $O_{2 \times 2}$.

10. Matrices ●●●●●●●●●●○○○

There are obvious parallels between zero in the number system and a zero matrix in matrices.

With x representing a number:

$$x + 0 = x, \qquad 0 + x = x, \qquad x \times 0 = 0, \qquad 0 \times x = 0.$$

With A representing a matrix, and providing the necessary sums and products can be formed:

$$A + O = A, \qquad O + A = A, \qquad A \times O = O, \qquad O \times A = O.$$

Because a zero matrix leaves another matrix 'unchanged under addition', i.e. $A + O = A$ and also $O + A = A$, a zero matrix is sometimes referred to as an *additive identity matrix*.

In this text we will use the letter O rather than the number 0 (zero) for the zero matrix. The two symbols can easily be confused, especially when handwritten. However this should not cause a problem, and you don't need to take time making some distinction between the handwritten characters, because the context usually makes it obvious whether it is the number zero or a zero matrix that is being referred to.

Note: Care needs to be taken when working with matrices. We must guard against using rules and procedures that apply to numbers but that do not necessarily apply to matrices. For example, we have already seen that under matrix multiplication the matrix product AB is usually not the same as BA. Two more points to watch for are given below.

- With x and y representing numbers, a frequently used result in mathematics is that if $xy = 0$ then either $x = 0$ and/or $y = 0$.

  However, for matrices, if $AB = O$ it is not necessarily the case that A and/or $B = O$.

  For example, consider $\quad A = \begin{bmatrix} 2 & 2 \\ 1 & 1 \end{bmatrix} \quad$ and $\quad B = \begin{bmatrix} 1 & -2 \\ -1 & 2 \end{bmatrix}$.

  In this case $\quad AB = \begin{bmatrix} 2 & 2 \\ 1 & 1 \end{bmatrix}\begin{bmatrix} 1 & -2 \\ -1 & 2 \end{bmatrix}$

  $$= \begin{bmatrix} 0 & 0 \\ 0 & 0 \end{bmatrix}$$

  Thus $AB = O$ but neither A nor B equal O.

- With x and y representing numbers, if $xy = zy$, for $y \neq 0$, then $x = z$.

  However, for matrices, if $AB = CB$, $B \neq O$, matrix A is not necessarily equal to matrix C. i.e. we cannot simply cancel the Bs.

  For example consider $\quad A = \begin{bmatrix} 3 & -1 \\ 1 & 4 \end{bmatrix}, \quad B = \begin{bmatrix} 1 \\ 2 \end{bmatrix} \quad$ and $\quad C = \begin{bmatrix} -3 & 2 \\ 5 & 2 \end{bmatrix}$.

  In this case $\quad AB = \begin{bmatrix} 3 & -1 \\ 1 & 4 \end{bmatrix}\begin{bmatrix} 1 \\ 2 \end{bmatrix} = \begin{bmatrix} 1 \\ 9 \end{bmatrix}$

  and $\quad CB = \begin{bmatrix} -3 & 2 \\ 5 & 2 \end{bmatrix}\begin{bmatrix} 1 \\ 2 \end{bmatrix} = \begin{bmatrix} 1 \\ 9 \end{bmatrix}$.

  Thus $AB = CB$, $B \neq O$, but $A \neq C$.

ISBN 9780170390477

Multiplicative identity matrices

A multiplicative identity matrix leaves all other matrices unchanged under multiplication (provided the multiplication can be performed).

Thus if $I_{m \times n}$ is a multiplicative identity matrix then,

$$I_{m \times n} A_{n \times p} = A_{n \times p} \qquad \text{from which it follows that } m = n.$$

Also $\qquad B_{q \times m} I_{m \times n} = B_{q \times m} \qquad \text{from which it follows that } m = n.$

Thus multiplicative identity matrices are square matrices.

The letter I is used to indicate a multiplicative identity matrix. If clarification is needed as to the size of I we can write I_2 for the 2×2 multiplicative identity, I_3 for the 3×3 multiplicative identity etc.

A multiplicative identity matrix has every entry of its main or leading diagonal equal to one and every other entry equal to zero.

$$\begin{bmatrix} & & \\ & & \\ & & \end{bmatrix} \text{ Main or leading diagonal.}$$

The 2×2 multiplicative identity matrix is: $\begin{bmatrix} 1 & 0 \\ 0 & 1 \end{bmatrix}$

The 3×3 multiplicative identity matrix is: $\begin{bmatrix} 1 & 0 & 0 \\ 0 & 1 & 0 \\ 0 & 0 & 1 \end{bmatrix}$ etc.

There are obvious parallels between 1 in the number system and I in matrices.

With x representing a number: $\qquad 1 \times x = x, \qquad x \times 1 = x.$

With A representing a matrix: $\qquad I \times A = A, \qquad A \times I = A.$

Again be careful not to use rules and procedures that apply to numbers but that do not necessarily apply to matrices.

In numbers, if $\qquad xy = x, \quad$ then for $x \neq 0$, y must equal 1.

However, for matrices, if $\qquad AB = B, \quad B \neq O$, A does not necessarily equal I.

For example $\begin{bmatrix} 4 & -1 \\ -3 & 2 \end{bmatrix} \begin{bmatrix} 2 & 1 \\ 6 & 3 \end{bmatrix} = \begin{bmatrix} 2 & 1 \\ 6 & 3 \end{bmatrix}.$

Premultiplication by $\begin{bmatrix} 4 & -1 \\ -3 & 2 \end{bmatrix}$ has left $\begin{bmatrix} 2 & 1 \\ 6 & 3 \end{bmatrix}$ unchanged but $\begin{bmatrix} 4 & -1 \\ -3 & 2 \end{bmatrix} \neq I.$

Thus: Multiplication by I leaves a matrix unchanged, but a matrix being left unchanged under multiplication does not necessarily mean that it must have been I that we multiplied by.

I.e. For matrices A and B, even if we know that AB = B we cannot assume that matrix A = I, the identity matrix.

Thus whilst it is true that:

If matrix A is multiplied by the identity matrix I then A is left unchanged

the converse of this statement is not true.

What about the contrapositive statement?

The multiplicative inverse of a square matrix

The point made on an earlier page:

If $AB = CB$, $B \neq O$, then A is not necessarily equal to C

indicates that attempting to divide one matrix by another could present a problem. We might have expected that if $\qquad AB = CB$

then 'dividing by B' would give $\qquad A = C$.

However, we know that A is not necessarily equal to C.

The direct division of one matrix by another is undefined. However, we can 'undo' the effect of multiplying using the idea of a *multiplicative inverse*.

With numbers, multiplying by $\dfrac{1}{3}$ $(= 3^{-1})$ undoes the effect of multiplying by 3.

Similarly multiplying by a^{-1} undoes the effect of multiplying by a, $(a \neq 0)$.

We say that a^{-1} is the multiplicative inverse of a.

For $a \neq 0$ we have: $\qquad a \times a^{-1} = 1 = a^{-1} \times a$

With matrices, if for some square matrix A there exists a matrix B such that

$$AB = I = BA$$

then we say that B is the multiplicative inverse of A. This is usually simply referred to as the inverse of A, and is written A^{-1}.

Thus $\qquad \boxed{A A^{-1} = I = A^{-1} A}$

For example suppose that $A = \begin{bmatrix} 3 & 1 \\ 5 & 2 \end{bmatrix}$

If $A^{-1} = \begin{bmatrix} a & b \\ c & d \end{bmatrix}$ then $\begin{bmatrix} a & b \\ c & d \end{bmatrix}\begin{bmatrix} 3 & 1 \\ 5 & 2 \end{bmatrix} = \begin{bmatrix} 1 & 0 \\ 0 & 1 \end{bmatrix}$

from which $\qquad \left.\begin{array}{l} 3a + 5b = 1 \\ a + 2b = 0 \end{array}\right\}$ giving $a = 2$ and $b = -1$,

and $\qquad \left.\begin{array}{l} 3c + 5d = 0 \\ c + 2d = 1 \end{array}\right\}$ giving $c = -5$ and $d = 3$.

Thus for $A = \begin{bmatrix} 3 & 1 \\ 5 & 2 \end{bmatrix}$

$\qquad A^{-1} = \begin{bmatrix} 2 & -1 \\ -5 & 3 \end{bmatrix}$.

$\begin{bmatrix} 3 & 1 \\ 5 & 2 \end{bmatrix}^{-1}$

$\begin{bmatrix} 2 & -1 \\ -5 & 3 \end{bmatrix}$

Note
- Inverses are only defined for square matrices.

- For the 2×2 matrix $A = \begin{bmatrix} a & b \\ c & d \end{bmatrix}$, the inverse A^{-1} is $\begin{bmatrix} \dfrac{d}{ad-bc} & \dfrac{-b}{ad-bc} \\ \dfrac{-c}{ad-bc} & \dfrac{a}{ad-bc} \end{bmatrix}$.

The quantity $(ad - bc)$ is the **determinant** of A, written det A or $|A|$.

Thus the inverse of a 2×2 matrix can be found as follows:
1. Interchange the elements of the leading diagonal.
2. Change the signs of the elements of the other diagonal.
3. Divide by the determinant.

- The fact that this process involves dividing by the determinant should alert us to the fact that not all 2×2 matrices have inverses. Any matrix with a determinant of zero has no inverse. Such matrices are said to be **singular**.

- A matrix that has an inverse is said to be **invertible**.

EXAMPLE 4

If $A = \begin{bmatrix} 3 & -1 \\ 2 & 1 \end{bmatrix}$ and $B = \begin{bmatrix} 3 & 4 \\ 2 & 1 \end{bmatrix}$, determine

a $A + B$ **b** det A **c** A^{-1} **d** B^{-1}
e matrix C such that $AC = B$ **f** matrix D such that $DA = B$

Solution

a $A + B = \begin{bmatrix} 6 & 3 \\ 4 & 2 \end{bmatrix}$

b det $A = (3)(1) - (-1)(2)$
$\quad\quad = 5$

c $A^{-1} = \begin{bmatrix} \dfrac{1}{5} & \dfrac{1}{5} \\ -\dfrac{2}{5} & \dfrac{3}{5} \end{bmatrix}$ or $\dfrac{1}{5}\begin{bmatrix} 1 & 1 \\ -2 & 3 \end{bmatrix}$

d $B^{-1} = \begin{bmatrix} -\dfrac{1}{5} & \dfrac{4}{5} \\ \dfrac{2}{5} & -\dfrac{3}{5} \end{bmatrix}$ or $-\dfrac{1}{5}\begin{bmatrix} 1 & -4 \\ -2 & 3 \end{bmatrix}$

e $AC = B$
Premultiply both sides by A^{-1}:
$A^{-1}AC = A^{-1}B$
$\quad IC = A^{-1}B$
$\therefore \quad C = \dfrac{1}{5}\begin{bmatrix} 1 & 1 \\ -2 & 3 \end{bmatrix}\begin{bmatrix} 3 & 4 \\ 2 & 1 \end{bmatrix}$
$\quad\quad = \begin{bmatrix} 1 & 1 \\ 0 & -1 \end{bmatrix}$

f $DA = B$
Postmultiply both sides by A^{-1}:
$DAA^{-1} = BA^{-1}$
$\quad DI = BA^{-1}$
$\therefore \quad D = \begin{bmatrix} 3 & 4 \\ 2 & 1 \end{bmatrix}\dfrac{1}{5}\begin{bmatrix} 1 & 1 \\ -2 & 3 \end{bmatrix}$
$\quad\quad = \begin{bmatrix} -1 & 3 \\ 0 & 1 \end{bmatrix}$

Repeat this example using your calculator.

Note that parts **e** and **f** above make use of the fact that for matrices:
$\quad\quad (kP)Q = k(PQ) \quad\quad$ and $\quad\quad P(kQ) = k(PQ)$.

ISBN 9780170390477

EXAMPLE 5

Matrices P, Q and R are all 2×2 matrices and $R = P - QR$.

If we assume that $(I + Q)^{-1}$ exists, prove that:

$$R = (I + Q)^{-1} P$$

where I is the 2×2 identity matrix.

Hence determine R given that $P = \begin{bmatrix} 20 & 14 \\ 2 & 1 \end{bmatrix}$ and $Q = \begin{bmatrix} 1 & 4 \\ 1 & -1 \end{bmatrix}$.

Solution

Given that $\qquad\qquad R = P - QR$

then $\qquad\qquad R + QR = P$

$\qquad\qquad\qquad IR + QR = P$

$\qquad\qquad\qquad (I + Q)R = P$

Premultiply both sides by $(I + Q)^{-1}$

$\qquad\qquad (I + Q)^{-1}(I + Q)R = (I + Q)^{-1} P$

Therefore $\qquad\qquad R = (I + Q)^{-1} P$ as required.

Thus if $P = \begin{bmatrix} 20 & 14 \\ 2 & 1 \end{bmatrix}$ and $Q = \begin{bmatrix} 1 & 4 \\ 1 & -1 \end{bmatrix}$

$$R = \begin{bmatrix} 2 & 4 \\ 1 & 0 \end{bmatrix}^{-1} \begin{bmatrix} 20 & 14 \\ 2 & 1 \end{bmatrix}$$

$$= -\frac{1}{4} \begin{bmatrix} 0 & -4 \\ -1 & 2 \end{bmatrix} \begin{bmatrix} 20 & 14 \\ 2 & 1 \end{bmatrix}$$

$$= -\frac{1}{4} \begin{bmatrix} -8 & -4 \\ -16 & -12 \end{bmatrix}$$

$$= \begin{bmatrix} 2 & 1 \\ 4 & 3 \end{bmatrix}$$

Exercise 10C

Find the determinants of each of the following 2×2 matrices.

1 $\begin{bmatrix} 1 & 2 \\ 3 & 4 \end{bmatrix}$
　　　　2 $\begin{bmatrix} 4 & 3 \\ -2 & 1 \end{bmatrix}$
　　　　3 $\begin{bmatrix} -1 & -3 \\ 2 & -1 \end{bmatrix}$
　　　　4 $\begin{bmatrix} -1 & 3 \\ -2 & -1 \end{bmatrix}$

5 $\begin{bmatrix} 5 & 0 \\ 2 & 1 \end{bmatrix}$
　　　　6 $\begin{bmatrix} 1 & 1 \\ -1 & -1 \end{bmatrix}$
　　　　7 $\begin{bmatrix} x & 0 \\ y & -x \end{bmatrix}$
　　　　8 $\begin{bmatrix} x & y \\ y & x \end{bmatrix}$

Find the inverse of each of the following 2×2 matrices. If the matrix does not have an inverse, i.e. the matrix is singular, state this clearly.

9 $\begin{bmatrix} 2 & 1 \\ 1 & 1 \end{bmatrix}$
　　　10 $\begin{bmatrix} 3 & 2 \\ 4 & 3 \end{bmatrix}$
　　　11 $\begin{bmatrix} 2 & 1 \\ -1 & 1 \end{bmatrix}$
　　　12 $\begin{bmatrix} 4 & 3 \\ 1 & 2 \end{bmatrix}$

13 $\begin{bmatrix} 3 & -1 \\ 1 & 3 \end{bmatrix}$
　　　14 $\begin{bmatrix} -3 & 1 \\ -1 & -3 \end{bmatrix}$
　　　15 $\begin{bmatrix} 1 & 1 \\ 1 & 1 \end{bmatrix}$
　　　16 $\begin{bmatrix} 4 & -3 \\ -8 & 6 \end{bmatrix}$

17 $\begin{bmatrix} 0 & 0 \\ 1 & -1 \end{bmatrix}$
　　　18 $\begin{bmatrix} x & y \\ 0 & 1 \end{bmatrix}$
　　　19 $\begin{bmatrix} -1 & 0 \\ 0 & -1 \end{bmatrix}$
　　　20 $\begin{bmatrix} 1 & 0 \\ 0 & -1 \end{bmatrix}$

21 For each of the following, state whether the given statement is necessarily true for all matrices A, B and C for which the given operations can be determined.

a $AI = A$
　　　　　　b $IA = A$
　　　　　　c $AB = BA$

d $OA = O$
　　　　　　e $A^{-1} A = I$
　　　　　　f $A A^{-1} = I$

g $A(B + C) = AB + AC$
　　　h $(AB)C = A(BC)$

i If $AB = O$ then $A = O$ and/or $B = O$.

j If $AB = AC$ and $A \neq O$ then $B = C$.

For questions **22** to **25** determine matrices A, B, C and D.

22 $\begin{bmatrix} 5 & 3 \\ 3 & 2 \end{bmatrix} A = \begin{bmatrix} 7 \\ 5 \end{bmatrix}$
　　　　　　　　23 $\begin{bmatrix} 5 & 1 \\ 3 & 1 \end{bmatrix} B = \begin{bmatrix} 9 \\ 5 \end{bmatrix}$

24 $\begin{bmatrix} 4 & -3 \\ 2 & 1 \end{bmatrix} C = \begin{bmatrix} 2 \\ -4 \end{bmatrix}$
　　　　　　　　25 $\begin{bmatrix} 1 & 2 \\ 0 & -1 \end{bmatrix} D = \begin{bmatrix} 1 \\ 1 \end{bmatrix}$

26 Find the constant k given that $A = \begin{bmatrix} 3 & 4 \\ 1 & -1 \end{bmatrix}$ and $A^2 - 2A = kI$, where I is the 2×2 identity matrix.

27 Find k if $A = \begin{bmatrix} k & -2 \\ 5 & 0 \end{bmatrix}$ and $A + 10A^{-1} = 5I$, where I is the 2×2 identity matrix.

28 If $A = \begin{bmatrix} 3 & -1 \\ -2 & 1 \end{bmatrix}$ and $B = \begin{bmatrix} 16 & -5 \\ -14 & 5 \end{bmatrix}$ determine

    a $A - B$ **b** $\det B$ **c** A^{-1} **d** B^{-1}

    e   matrix C such that $AC = \begin{bmatrix} 9 \\ -5 \end{bmatrix}$ **f**   matrix D such that $DA = B$

29 If $P = \begin{bmatrix} -4 & -1 \\ 3 & 1 \end{bmatrix}$ and $Q = \begin{bmatrix} 6 & 2 \\ 0 & 1 \end{bmatrix}$ determine

    a $P + Q$ **b** P^{-1} **c** Q^{-1} **d** $(P + Q)^{-1}$
    e   matrix R such that $R(P + Q) = Q$

30 Find matrix A given that $AB = \begin{bmatrix} 3 & 1 \\ 22 & 7 \end{bmatrix}$ and $B^{-1} = \begin{bmatrix} 3 & -2 \\ -7 & 5 \end{bmatrix}$.

31 Find matrix D given that $CD = \begin{bmatrix} 7 \\ 5 \end{bmatrix}$ and $C^{-1} = \begin{bmatrix} 1 & -1 \\ -3 & 4 \end{bmatrix}$.

32 For each of the following determine the value(s) of x for which the matrix is singular.

    a $\begin{bmatrix} 3x & 4 \\ 6 & 1 \end{bmatrix}$ **b** $\begin{bmatrix} x & 8 \\ 2 & x \end{bmatrix}$ **c** $\begin{bmatrix} x & 2 \\ 10 & (x-1) \end{bmatrix}$

33 Find matrices F and G given that:

$$EF = \begin{bmatrix} -2 & 12 \\ 0 & 9 \end{bmatrix}, GE = \begin{bmatrix} -2 & -2 \\ 4 & -2 \end{bmatrix} \text{ and } E^{-1} = \frac{1}{2}\begin{bmatrix} -1 & 2 \\ -2 & 2 \end{bmatrix}.$$

34 Use your calculator to determine matrix C given that:

$$A = \begin{bmatrix} 4 & -2 & -1 \\ 2 & 1 & 0 \\ -1 & 3 & 1 \end{bmatrix}, B = \begin{bmatrix} -1 & -6 & -11 \\ 4 & 1 & 1 \\ 6 & 7 & 11 \end{bmatrix} \text{ and } AC = B.$$

35 Use your calculator to determine matrix C given that:

$$A = \begin{bmatrix} 2 & 3 & -3 \\ 1 & 2 & -1 \\ 0 & 1 & 2 \end{bmatrix}, B = \begin{bmatrix} 4 & 6 & -7 \\ 1 & 5 & 5 \\ 7 & 11 & -10 \end{bmatrix} \text{ and } CA = B.$$

36 The entry price matrix, A, for adults and concession card holders (i.e. pensioners and children) attending state and international cricket matches at a particular ground is as follows:

	State Match	International Match
Adult	$30	$70
Concession	$16	$36

To attract more people to matches one proposal under consideration is to reduce all adult non-concessional tickets by 20%, and leave concession tickets unchanged. Under this proposal the new entry price matrix for adults and concession card holders is matrix B, where B has the same row and column meanings as A.

a Write down matrix B. **b** Determine matrix C given that CA = B.

37 Matrices A, B and C are all 2×2 matrices and $C = A - CB$. Assuming that $(I + B)^{-1}$ exists, prove that $C = A(I + B)^{-1}$ where I is the 2×2 identity matrix.

Hence or otherwise determine C given that $A = \begin{bmatrix} -1 & 6 \\ 11 & 4 \end{bmatrix}$ and $B = \begin{bmatrix} 1 & 2 \\ -5 & 1 \end{bmatrix}$.

38 Matrices A, B and C are all 2×2 matrices and $A = BC - AC$.

Determine C given that $A = \begin{bmatrix} -3 & 5 \\ 1 & 6 \end{bmatrix}$ and $B = \begin{bmatrix} -1 & 0 \\ 2 & 4 \end{bmatrix}$.

39 Matrices P and Q are both 2×2 matrices and $P = Q + PQ + PQ^2$.

Determine P given that $Q = \begin{bmatrix} -1 & 0 \\ 5 & -2 \end{bmatrix}$.

40 A company makes two types of machine, the Oozenplatt and the Flibber. Each Oozenplatt requires six Flantles and five Creks whilst each Flibber requires eight Flantles and seven Creks.

a Copy and complete the following 2×2 matrix, matrix A, for this information.

	Flantles	Creks
Oozenplatt		
Flibber		

The company receives an order for x Oozenplatts and y Flibbers. Matrix B below:

Oozenplatt	Flibber
x	y

This order requires a total of 860 Flantles and 740 Creks.

b Write a matrix equation involving A, B and the information about the total numbers of Flantles and Creks required.

c Show how the multiplicative inverse of matrix A can be used to determine x and y (and determine these values in the process).

Using the inverse matrix to solve systems of equations

Consider the system of linear equations:

$$ax + by = c$$

$$dx + ey = f.$$

This could be written as the matrix equation

$$\begin{bmatrix} a & b \\ d & e \end{bmatrix} \begin{bmatrix} x \\ y \end{bmatrix} = \begin{bmatrix} c \\ f \end{bmatrix}, \qquad [1]$$

because when we multiply the matrices on the left hand side of this equation we obtain

$$\begin{bmatrix} ax + by \\ dx + ey \end{bmatrix} = \begin{bmatrix} c \\ f \end{bmatrix} \quad \text{i.e.} \quad \begin{matrix} ax + by = c \\ dx + ey = f \end{matrix} \quad \text{the original equations.}$$

The values of x and y can be determined from equation [1] by pre-multiplying both sides by the inverse of $\begin{bmatrix} a & b \\ d & e \end{bmatrix}$.

Note: For the system of equations to have a unique solution the matrix $\begin{bmatrix} a & b \\ d & e \end{bmatrix}$ must not be singular. Its determinant must not equal zero.

EXAMPLE 6

a   If $A = \begin{bmatrix} 1 & -1 \\ 2 & 3 \end{bmatrix}$, determine A^{-1}.

b   Use your answer from **a** to solve $\begin{cases} x - y = 7 \\ 2x + 3y = 4 \end{cases}$.

Solution

a $A^{-1} = \dfrac{1}{5} \begin{bmatrix} 3 & 1 \\ -2 & 1 \end{bmatrix}$

b   First write the equations in matrix form: $\begin{bmatrix} 1 & -1 \\ 2 & 3 \end{bmatrix} \begin{bmatrix} x \\ y \end{bmatrix} = \begin{bmatrix} 7 \\ 4 \end{bmatrix}.$

Thus

$$\begin{bmatrix} x \\ y \end{bmatrix} = \frac{1}{5} \begin{bmatrix} 3 & 1 \\ -2 & 1 \end{bmatrix} \begin{bmatrix} 7 \\ 4 \end{bmatrix}$$

$$= \frac{1}{5} \begin{bmatrix} 25 \\ -10 \end{bmatrix}$$

Hence $x = 5$ and $y = -2$.

ISBN 9780170390477

EXAMPLE 7

a If $A = \begin{bmatrix} 3 & 1 & 1 \\ 2 & -1 & -1 \\ 2 & 1 & 0 \end{bmatrix}$, use your calculator to determine A^{-1}.

b Use your answer from **a** to solve $\begin{cases} 3x + y + z = -1 \\ 2x - y - z = -4 \\ 2x + y = -5 \end{cases}$.

Solution

a $A^{-1} = \begin{bmatrix} 0.2 & 0.2 & 0 \\ -0.4 & -0.4 & 1 \\ 0.8 & -0.2 & -1 \end{bmatrix}$

or, if you prefer, $\frac{1}{5} \begin{bmatrix} 1 & 1 & 0 \\ -2 & -2 & 5 \\ 4 & -1 & -5 \end{bmatrix}$.

$$\begin{bmatrix} 3 & 1 & 1 \\ 2 & -1 & -1 \\ 2 & 1 & 0 \end{bmatrix}^{-1}$$

$$\begin{bmatrix} 0.2 & 0.2 & 0 \\ -0.4 & -0.4 & 1 \\ 0.8 & -0.2 & -1 \end{bmatrix}$$

b First express in matrix form: $\begin{bmatrix} 3 & 1 & 1 \\ 2 & -1 & -1 \\ 2 & 1 & 0 \end{bmatrix} \begin{bmatrix} x \\ y \\ z \end{bmatrix} = \begin{bmatrix} -1 \\ -4 \\ -5 \end{bmatrix}$.

Thus $\begin{bmatrix} x \\ y \\ z \end{bmatrix} = \frac{1}{5} \begin{bmatrix} 1 & 1 & 0 \\ -2 & -2 & 5 \\ 4 & -1 & -5 \end{bmatrix} \begin{bmatrix} -1 \\ -4 \\ -5 \end{bmatrix}$

$$= \frac{1}{5} \begin{bmatrix} -5 \\ -15 \\ 25 \end{bmatrix}$$

Hence $x = -1, y = -3$ and $z = 5$.

$$\begin{bmatrix} 3 & 1 & 1 \\ 2 & -1 & -1 \\ 2 & 1 & 0 \end{bmatrix}^{-1}$$

$$\begin{bmatrix} 0.2 & 0.2 & 0 \\ -0.4 & -0.4 & 1 \\ 0.8 & -0.2 & -1 \end{bmatrix}$$

$$\text{ans} \times \begin{bmatrix} -1 \\ -4 \\ -5 \end{bmatrix}$$

$$\begin{bmatrix} -1 \\ -3 \\ 5 \end{bmatrix}$$

Exercise 10D

Express each of the following in the form $\begin{bmatrix} * & * \\ * & * \end{bmatrix} \begin{bmatrix} x \\ y \end{bmatrix} = \begin{bmatrix} * \\ * \end{bmatrix}$.

1 $\begin{cases} 2x + 3y = 5 \\ x - 3y = 0 \end{cases}$ **2** $\begin{cases} -x + 2y = 6 \\ 6x - y = 4 \end{cases}$ **3** $\begin{cases} 3x + y = -2 \\ x - 3y = 1 \end{cases}$

Express each of the following in the form $\begin{bmatrix} * & * & * \\ * & * & * \\ * & * & * \end{bmatrix} \begin{bmatrix} x \\ y \\ z \end{bmatrix} = \begin{bmatrix} * \\ * \\ * \end{bmatrix}$.

4 $\begin{cases} x + y + z = 2 \\ 3x - 4y + 2z = 6 \\ x - y - z = 4 \end{cases}$ **5** $\begin{cases} x + 2y + 3z = 5 \\ 3x - 2y = 4 \\ 2x - 7z = 0 \end{cases}$ **6** $\begin{cases} 2x - 3y + z = 1 \\ x + y - 3z = 0 \\ -2y + 3z = 4 \end{cases}$

7 a If $A = \begin{bmatrix} 3 & -2 \\ -5 & 4 \end{bmatrix}$, determine A^{-1}.

b Use your answer from **a** to solve $\begin{cases} 3x - 2y = 4 \\ -5x + 4y = -9 \end{cases}$.

8 a If $A = \begin{bmatrix} -2 & 1 & -2 \\ 2 & -1 & 3 \\ 0 & 1 & 2 \end{bmatrix}$, use your calculator to determine A^{-1}.

b Use your answer from **a** to solve $\begin{cases} -2x + y - 2z = 3 \\ 2x - y + 3z = -1 \\ y + 2z = 9 \end{cases}$.

9 Using the idea of a multiplicative inverse of a matrix, and clearly showing that use, solve each of the following systems of equations.

a $\begin{cases} 3x + y = 2 \\ 5x + 2y = 1 \end{cases}$ **b** $\begin{cases} 3x + y = 8 \\ 7x + 3y = 13 \end{cases}$

10 a If $A = \begin{bmatrix} -2 & -1 & 2 \\ 1 & -1 & 3 \\ 3 & 2 & -2 \end{bmatrix}$ and $B = \begin{bmatrix} -4 & 2 & -1 \\ 11 & -2 & 8 \\ 5 & 1 & 3 \end{bmatrix}$, determine AB.

b Express A^{-1} in terms of B.

c Solve the system $\begin{cases} -2x - y + 2z = -3 \\ x - y + 3z = 7 \\ 3x + 2y - 2z = 5 \end{cases}$ clearly showing your use of A^{-1}.

ISBN 9780170390477

11 a Express the system of equations shown below left in the form AX = B, as shown below right.

$$v + w + x + y + z = 1$$
$$v - w + x + 2y - z = 13$$
$$2v - w + 3x - y + 2z = 2$$
$$3v + 2w - x - y - 2z = 4$$
$$2w + 3y - z = 8$$

$$\begin{bmatrix} * & * & * & * & * \\ * & * & * & * & * \\ * & * & * & * & * \\ * & * & * & * & * \\ * & * & * & * & * \end{bmatrix} \begin{bmatrix} v \\ w \\ x \\ y \\ z \end{bmatrix} = \begin{bmatrix} * \\ * \\ * \\ * \\ * \end{bmatrix}$$

b Use the ability of your calculator to determine A^{-1}, and to perform matrix multiplication, to solve the system.

Extension activity:
Finding the determinant and inverse of a 3 × 3 matrix

Fortunately the ready availability of calculators that can give the determinant and inverse of a square matrix, provided of course that the inverse does exist, means that we no longer have to determine these things 'by hand'.

This extension activity invites you to delve back into history and see how these things were determined prior to the ready availability of today's technology.

Writing the determinant of matrix A as $|A|$, the determinant of the 3×3 matrix

$$\begin{bmatrix} a & b & c \\ d & e & f \\ g & h & i \end{bmatrix}$$

is defined as

$$|A| = a \times \begin{vmatrix} e & f \\ h & i \end{vmatrix} - b \times \begin{vmatrix} d & f \\ g & i \end{vmatrix} + c \times \begin{vmatrix} d & e \\ g & h \end{vmatrix}$$

Check that this gives the same value for the determinants of

$$\begin{bmatrix} 4 & 1 & 1 \\ 3 & -1 & 1 \\ 1 & 1 & 0 \end{bmatrix} \text{ and } \begin{bmatrix} 3 & 1 & 2 \\ 5 & 0 & -1 \\ 2 & -3 & 4 \end{bmatrix}$$

as your calculator gives.

To determine the inverse of a 3×3 matrix, determinant $\neq 0$, *in the old days*,

one commonly used method involved   'row reduction'

and another used   'the adjoint matrix'.

Research these methods and write a brief report on each.

Miscellaneous exercise ten

This miscellaneous exercise may include questions involving the work of this chapter, the work of any previous chapters in this unit, and the ideas mentioned in the Preliminary work section at the beginning of this unit.

1 If $A = \begin{bmatrix} 2 & 0 \\ -4 & -3 \end{bmatrix}$ find

    a matrix B given that $A + B = O$, the 2×2 zero matrix.

    b matrix C given that $A + C = I$, the 2×2 identity matrix.

2 If $D = \begin{bmatrix} 5 & -1 \\ 2 & 0 \end{bmatrix}$, O is the 2×2 zero matrix and I is the 2×2 identity matrix, find

    a matrix E given that $DO = E$,

    b matrix F given that $D + O = F$,

    c matrix G given that $D + I = G$,

    d matrix H given that $DI = H$,

    e matrix K given that $ID = K$.

3 Solve $\sin\left(x + \dfrac{\pi}{4}\right) = \dfrac{\sqrt{3}}{2}$ for $0 \le x \le \pi$.

4 For $0 \le \beta \le \dfrac{\pi}{2}$ the equation $\tan\beta = k$ has the solution $\beta = p$ radians.

Find all of the solutions to the equation $k \sin 2\theta = 2 \sin^2 \theta$ that exist in the interval $0 \le \theta \le 2\pi$ giving answers in terms of p when it is suitable to do so.

5 Prove that $2 \sin^3 \theta \cos \theta + 2 \cos^3 \theta \sin \theta = \sin 2\theta$.

6 Prove that $\dfrac{\cos\theta - \sin\theta}{\cos\theta + \sin\theta} = \dfrac{1 - \sin 2\theta}{\cos 2\theta}$.

7 a Expand $(2y - 1)(y + 1)$.

    b Solve: $1 + \sin x = 2 \cos^2 x$ for $-2\pi \le x \le 2\pi$.

8 a Express $(2 \cos \theta + 5 \sin \theta)$ in the form $R \cos (\theta - \alpha)$ for α an acute angle in degrees, correct to one decimal place.

    b Hence determine the minimum value of $(2 \cos \theta + 5 \sin \theta)$ and the smallest positive value of θ (in degrees correct to one decimal place) for which it occurs.

    c Without the assistance of a calculator produce a sketch of the graph of

$$y = 2 \cos x + 5 \sin x$$

for x in degrees.

(Then use a graphic calculator to check the correctness of your sketch.)

9 If $A = \begin{bmatrix} 1 & 0 & 2 \\ 2 & -1 & 3 \end{bmatrix}$, $B = \begin{bmatrix} 1 & 2 \\ 0 & 1 \end{bmatrix}$, $C = \begin{bmatrix} 2 & 1 & 3 \\ -1 & 0 & 1 \end{bmatrix}$ and $D = \begin{bmatrix} 1 \\ 2 \end{bmatrix}$, determine each of the

following. If any cannot be determined state this clearly and explain why.

a $A + B$ **b** $2A - C$ **c** AB

d BA **e** AC **f** BD

10 A company makes three models, A, B and C, of a particular item. Each model requires a certain number of units of commodities P, Q and R. Matrix X below shows the number of units of each commodity required to make one of each model.

$$\begin{array}{c} \\ \text{Model A} \\ \text{Model B} \\ \text{Model C} \end{array} \begin{array}{ccc} P & Q & R \\ \begin{bmatrix} 2 & 2 & 1 \\ 3 & 1 & 1 \\ 1 & 3 & 1 \end{bmatrix} \end{array} = X$$

Each unit of P, Q and R costs the company $50, $60 and $200 respectively.

We could write this as a column matrix, Y: $\begin{bmatrix} 50 \\ 60 \\ 200 \end{bmatrix}$ or as a row matrix, Z: $\begin{bmatrix} 50 & 60 & 200 \end{bmatrix}$.

Both XY and ZX could be formed but only one of these will contain information likely to be useful.

a Which is the useful one?

b Form the product.

c Explain the information it displays.

11 Five soccer teams, A, B, C, D and E, take part in a league competition, playing each other once away and once at home. The number of wins, draws and losses of each team are given in matrix X shown on the right.

Points are awarded:

3 for each win, 1 for each draw and 0 for each loss.

$$X = \begin{array}{c} A \\ B \\ C \\ D \\ E \end{array} \begin{array}{ccc} \text{Won} & \text{Drawn} & \text{Lost} \\ \begin{bmatrix} 4 & 1 & 3 \\ 3 & 1 & 4 \\ 2 & 3 & 3 \\ 3 & 1 & 4 \\ 5 & 0 & 3 \end{bmatrix} \end{array}$$

This could be written as a row matrix, matrix Y shown below, or as a column matrix, matrix Z shown below.

$$Y = \begin{array}{c} \text{Win} \quad \text{Draw} \quad \text{Loss} \\ \begin{bmatrix} 3 & 1 & 0 \end{bmatrix} \end{array} \qquad Z = \begin{array}{c} \text{Win} \\ \text{Draw} \\ \text{Loss} \end{array} \begin{bmatrix} 3 \\ 1 \\ 0 \end{bmatrix}$$

For any of the following products that can be formed, determine the product and explain the information the product matrix displays.

$$XY, YX, XZ, ZX.$$

12 Write the equation of each of the following in the form $y = a \sin bx$.

a
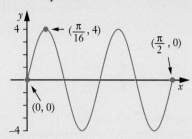

b
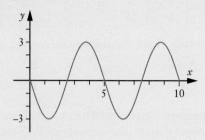

13 Write the equation of each of the following in the form $y = a \sin [b(x + c)]$.

a
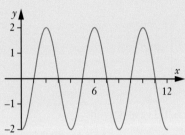

b
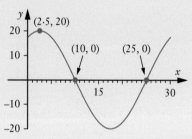

14 Write the equation of each of the following in the form $y = a \sin [b(x + c)] + d$.

a
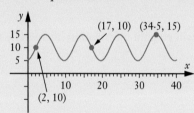

b
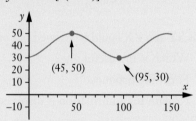

15 Given that $A = \begin{bmatrix} x & 2 \\ y & 1 \end{bmatrix}$, $B = \begin{bmatrix} 3 & 1 \\ -1 & 4 \end{bmatrix}$ and $AB = BA = \begin{bmatrix} p & q \\ r & s \end{bmatrix}$, determine x, y, p, q, r and s.

16 Find matrix P given that $A = \begin{bmatrix} 3 & -1 \\ 8 & 0 \end{bmatrix}$, $B = \begin{bmatrix} 4 & 3 \\ -1 & 1 \end{bmatrix}$, $Q = \begin{bmatrix} -2 & -2 \\ -1 & -3 \end{bmatrix}$ and $AP + BP + P = Q$.

17 Matrices A and B are both non singular square matrices.

If A and B commute for multiplication, i.e. $AB = BA$, prove that

a A and the multiplicative inverse of B commute for multiplication.

b B and the multiplicative inverse of A commute for multiplication.

ISBN 9780170390477

11.

Transformation matrices

- Transformations and matrices
- Determining the matrix for a particular transformation
- The determinant of a transformation matrix
- The inverse of a transformation matrix
- Combining transformations
- Further examples
- A general rotation about the origin
- A general reflection in a line that passes through the origin
- Miscellaneous exercise eleven

Transformations and matrices

Given some point in the cartesian plane with coordinates (x_1, y_1) we could transform this point to some point (x_2, y_2) using a 2×2 matrix, $\begin{bmatrix} a & b \\ c & d \end{bmatrix}$, as follows:

$$\begin{bmatrix} x_2 \\ y_2 \end{bmatrix} = \begin{bmatrix} a & b \\ c & d \end{bmatrix} \begin{bmatrix} x_1 \\ y_1 \end{bmatrix}$$

$$= \begin{bmatrix} ax_1 + by_1 \\ cx_1 + dy_1 \end{bmatrix}$$

In this way 2×2 matrices can be used to represent transformations.

Note that in this context we stand the coordinates of the point up as a column matrix and premultiply by the 2×2 **transformation matrix**.

- A transformation, T, is said to be **linear** if

$$T \begin{bmatrix} kx_1 \\ ky_1 \end{bmatrix} = kT \begin{bmatrix} x_1 \\ y_1 \end{bmatrix}$$

and

$$T \begin{bmatrix} x_1 + x_2 \\ y_1 + y_2 \end{bmatrix} = T \begin{bmatrix} x_1 \\ y_1 \end{bmatrix} + T \begin{bmatrix} x_2 \\ y_2 \end{bmatrix}.$$

The reader is left to confirm that this is the case for

$$T = \begin{bmatrix} a & b \\ c & d \end{bmatrix}.$$

- In the next example the images of points A, B and C under some transformation are written A′, B′ and C′, i.e. A *dash*, B *dash* and C *dash*. This dash notation commonly used for the image of a point under a transformation should not be confused with the use of a dash to indicate differentiation, a topic you will meet in the *Mathematics Methods* course, where we write $f'(x)$ for the derivative of $f(x)$ with respect to x. Which meaning is to be attributed to the dash will usually be obvious.

EXAMPLE 1

The points A(2, 1), B(3, –2) and C(0, 1) are transformed to A′, B′ and C′ by the matrix $\begin{bmatrix} 1 & 2 \\ 0 & 1 \end{bmatrix}$.

Find the coordinates of A′, B′ and C′.

Solution

$$\begin{bmatrix} 1 & 2 \\ 0 & 1 \end{bmatrix} \begin{bmatrix} 2 \\ 1 \end{bmatrix} = \begin{bmatrix} 4 \\ 1 \end{bmatrix}.$$ Thus A′ has coordinates (4, 1).

$$\begin{bmatrix} 1 & 2 \\ 0 & 1 \end{bmatrix} \begin{bmatrix} 3 \\ -2 \end{bmatrix} = \begin{bmatrix} -1 \\ -2 \end{bmatrix}.$$ Thus B′ has coordinates (–1, –2).

$$\begin{bmatrix} 1 & 2 \\ 0 & 1 \end{bmatrix} \begin{bmatrix} 0 \\ 1 \end{bmatrix} = \begin{bmatrix} 2 \\ 1 \end{bmatrix}.$$ Thus C′ has coordinates (2, 1).

Note: • All 2 × 2 matrices will transform the origin to itself:

$$\begin{bmatrix} a & b \\ c & d \end{bmatrix}\begin{bmatrix} 0 \\ 0 \end{bmatrix} = \begin{bmatrix} 0 \\ 0 \end{bmatrix}$$

The origin is an **invariant** point under these transformations.

It follows that translations cannot be represented by 2 × 2 matrices because, under a translation, the origin is moved. Indeed a translation of 'p units right and q units up' would be achieved by adding the column matrix $\begin{bmatrix} p \\ q \end{bmatrix}$, not by attempting to multiply by a 2 × 2 matrix.

• We will concentrate on matrices that perform reflections in lines passing through the origin, rotations about the origin, dilations and, going a little beyond the confines of the syllabus, shears. (The next dot point explains a *shear*.) In all of these transformations, straight lines are transformed to straight lines. i.e. Points that are collinear before the transformation will still be collinear after the transformation. However, we will also see some matrices that 'flatten' *all* points in the *x-y* plane onto a line, and some that 'collapse' all points lying on a line onto a single point.

• To understand a **shear**, consider four books stacked on a table as shown on the right. Suppose now that we move the top three books until the stack has the staggered arrangement shown in the lower diagram.

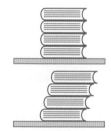

Notice that the higher up the pile a book is, the more it has been moved right. The bottom book remains in its original position and has not been moved at all.

In a shear transformation we have an invariant line which points move parallel to. The further a point is from the invariant line the more that point moves.

If the scale factor of the shear is k then a point that is p units from the invariant line moves a distance kp parallel to the invariant line.

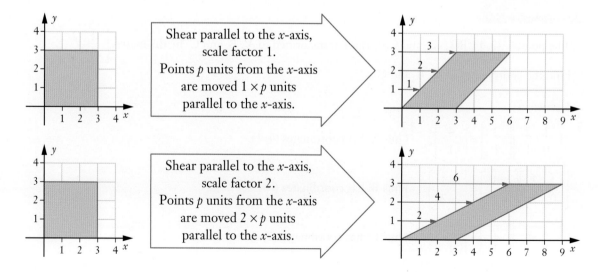

EXAMPLE 2

By considering the effect on the rectangle O(0, 0), A(2, 0), B(2, 1), C(0, 1), determine the

transformation represented by the matrix $\begin{bmatrix} 0 & 1 \\ -1 & 0 \end{bmatrix}$.

Solution

O': $\begin{bmatrix} 0 & 1 \\ -1 & 0 \end{bmatrix}\begin{bmatrix} 0 \\ 0 \end{bmatrix} = \begin{bmatrix} 0 \\ 0 \end{bmatrix}$.

A': $\begin{bmatrix} 0 & 1 \\ -1 & 0 \end{bmatrix}\begin{bmatrix} 2 \\ 0 \end{bmatrix} = \begin{bmatrix} 0 \\ -2 \end{bmatrix}$.

B': $\begin{bmatrix} 0 & 1 \\ -1 & 0 \end{bmatrix}\begin{bmatrix} 2 \\ 1 \end{bmatrix} = \begin{bmatrix} 1 \\ -2 \end{bmatrix}$.

C': $\begin{bmatrix} 0 & 1 \\ -1 & 0 \end{bmatrix}\begin{bmatrix} 0 \\ 1 \end{bmatrix} = \begin{bmatrix} 1 \\ 0 \end{bmatrix}$.

Plotting OABC and O'A'B'C' on a graph, see right, we see that

the matrix $\begin{bmatrix} 0 & 1 \\ -1 & 0 \end{bmatrix}$ represents a clockwise rotation about the

origin of 90°. (Or, if we use the convention of anticlockwise
rotations being positive, a rotation of –90°, or 270°, about
the origin.)

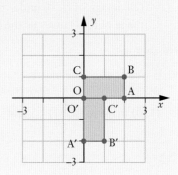

Exercise 11A

By considering the effect on the rectangle O(0, 0), A(2, 0), B(2, 1), C(0, 1) determine the transformation
represented by each of the following matrices.

1 $\begin{bmatrix} -1 & 0 \\ 0 & -1 \end{bmatrix}$

2 $\begin{bmatrix} 0 & -1 \\ 1 & 0 \end{bmatrix}$

3 $\begin{bmatrix} 1 & 0 \\ 0 & -1 \end{bmatrix}$

4 $\begin{bmatrix} -1 & 0 \\ 0 & 1 \end{bmatrix}$

5 $\begin{bmatrix} 0 & 1 \\ 1 & 0 \end{bmatrix}$

6 $\begin{bmatrix} 0 & -1 \\ -1 & 0 \end{bmatrix}$

7 $\begin{bmatrix} 2 & 0 \\ 0 & 1 \end{bmatrix}$

8 $\begin{bmatrix} 1 & 0 \\ 0 & 3 \end{bmatrix}$

9 $\begin{bmatrix} 2 & 0 \\ 0 & 3 \end{bmatrix}$

10 $\begin{bmatrix} 3 & 0 \\ 0 & 3 \end{bmatrix}$

11 $\begin{bmatrix} 1 & 2 \\ 0 & 1 \end{bmatrix}$

12 $\begin{bmatrix} 1 & 0 \\ 3 & 1 \end{bmatrix}$

13 For each of the matrices of questions 1 to 12 determine

    a   the absolute value of the determinant of the matrix.

    b   the value of $\dfrac{\text{Area O'A'B'C'}}{\text{Area OABC}}$ where O', A', B' and C' are the respective images of O, A, B and C

        under the transformation.

    c   What do you notice?

Determining the matrix for a particular transformation

With a bit of thought, it is easy to write down the matrix corresponding to each transformation you encountered in the last exercise. The method uses the fact that

$$\begin{bmatrix} a & b \\ c & d \end{bmatrix}\begin{bmatrix} 1 \\ 0 \end{bmatrix} = \begin{bmatrix} a \\ c \end{bmatrix}$$

and $\begin{bmatrix} a & b \\ c & d \end{bmatrix}\begin{bmatrix} 0 \\ 1 \end{bmatrix} = \begin{bmatrix} b \\ d \end{bmatrix}$.

i.e. The image of (1, 0) under the transformation gives us the first column of the matrix and the image of (0, 1) gives us the second column.

For example

- If we required the matrix that represents a 90° clockwise rotation about the origin, we note that under this transformation:

  (1, 0) maps to (0, –1)   and   (0, 1) maps to (1, 0).

  The required matrix is $\begin{bmatrix} 0 & 1 \\ -1 & 0 \end{bmatrix}$.

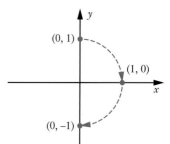

- If we required the matrix that represents a reflection in the line $y = x$, we note that under this transformation:

  (1, 0) maps to (0, 1)   and   (0, 1) maps to (1, 0).

  The required matrix is $\begin{bmatrix} 0 & 1 \\ 1 & 0 \end{bmatrix}$.

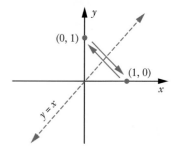

The determinant of a transformation matrix

Question 13 of Exercise 11A should have led you to conclude that when a shape with area A is transformed by a matrix with determinant k, the area of the image is $|k|$A.

The inverse of a transformation matrix

If matrix T transforms point A to its image A′ then T^{-1}, the multiplicative inverse of T, will transform A′ back to A (provided of course that T^{-1} exists).

ISBN 9780170390477

EXAMPLE 3

The triangle ABC is transformed to A'B'C' by the transformation matrix T where

$$T = \begin{bmatrix} 3 & 1 \\ 1 & 1 \end{bmatrix}.$$

If A', B' and C' have coordinates (22, 10), (–3, –5) and (–13, –5) respectively find the coordinates of A, B and C.

If triangle A'B'C' has an area of 75 units^2 determine the area of triangle ABC.

Solution

$$\left(\text{Note that if } T = \begin{bmatrix} 3 & 1 \\ 1 & 1 \end{bmatrix} \text{ then } T^{-1} = \frac{1}{2}\begin{bmatrix} 1 & -1 \\ -1 & 3 \end{bmatrix} \right)$$

If A has coordinates (a, b)

$$\begin{bmatrix} 3 & 1 \\ 1 & 1 \end{bmatrix}\begin{bmatrix} a \\ b \end{bmatrix} = \begin{bmatrix} 22 \\ 10 \end{bmatrix}$$

$$\begin{bmatrix} a \\ b \end{bmatrix} = \frac{1}{2}\begin{bmatrix} 1 & -1 \\ -1 & 3 \end{bmatrix}\begin{bmatrix} 22 \\ 10 \end{bmatrix}$$

$$= \frac{1}{2}\begin{bmatrix} 12 \\ 8 \end{bmatrix}$$

$$= \begin{bmatrix} 6 \\ 4 \end{bmatrix}$$

If B has coordinates (c, d)

$$\begin{bmatrix} 3 & 1 \\ 1 & 1 \end{bmatrix}\begin{bmatrix} c \\ d \end{bmatrix} = \begin{bmatrix} -3 \\ -5 \end{bmatrix}$$

$$\begin{bmatrix} c \\ d \end{bmatrix} = \frac{1}{2}\begin{bmatrix} 1 & -1 \\ -1 & 3 \end{bmatrix}\begin{bmatrix} -3 \\ -5 \end{bmatrix}$$

$$= \frac{1}{2}\begin{bmatrix} 2 \\ -12 \end{bmatrix}$$

$$= \begin{bmatrix} 1 \\ -6 \end{bmatrix}$$

If C has coordinates (e, f)

$$\begin{bmatrix} 3 & 1 \\ 1 & 1 \end{bmatrix}\begin{bmatrix} e \\ f \end{bmatrix} = \begin{bmatrix} -13 \\ -5 \end{bmatrix}$$

$$\begin{bmatrix} e \\ f \end{bmatrix} = \frac{1}{2}\begin{bmatrix} 1 & -1 \\ -1 & 3 \end{bmatrix}\begin{bmatrix} -13 \\ -5 \end{bmatrix}$$

$$= \frac{1}{2}\begin{bmatrix} -8 \\ -2 \end{bmatrix}$$

$$= \begin{bmatrix} -4 \\ -1 \end{bmatrix}$$

T has a determinant of $(3)(1) – (1)(1) = 2$

∴   Area △A'B'C' = 2 (Area △ABC)

Area △ABC = 37.5 units^2

The coordinates of A, B and C are (6, 4), (1, –6) and (–4, –1) respectively and triangle ABC has an area of 37.5 units^2.

Combining transformations

Suppose that (x, y) is transformed to (x', y') by matrix P and (x', y') is transformed to (x'', y'') by matrix Q.

$$\begin{bmatrix} x' \\ y' \end{bmatrix} = P \begin{bmatrix} x \\ y \end{bmatrix} \qquad \text{and} \qquad \begin{bmatrix} x'' \\ y'' \end{bmatrix} = Q \begin{bmatrix} x' \\ y' \end{bmatrix}$$

$$= QP \begin{bmatrix} x \\ y \end{bmatrix}$$

Thus the single matrix equivalent to applying 'P followed by Q' is QP.

This may initially seem the wrong way around but remember that in the expression $QP \begin{bmatrix} x \\ y \end{bmatrix}$ it is matrix P that operates on $\begin{bmatrix} x \\ y \end{bmatrix}$.

EXAMPLE 4

Triangle PQR is transformed to P'Q'R' by matrix $\begin{bmatrix} 1 & 3 \\ 0 & 1 \end{bmatrix}$.

Triangle P'Q'R' is transformed to P''Q''R'' by matrix $\begin{bmatrix} 2 & 0 \\ 1 & 2 \end{bmatrix}$.

Find the single matrix that will transform PQR directly to P''Q''R''.

Solution

$$\begin{bmatrix} x'' \\ y'' \end{bmatrix} = \begin{bmatrix} 2 & 0 \\ 1 & 2 \end{bmatrix} \begin{bmatrix} x' \\ y' \end{bmatrix}$$

$$= \begin{bmatrix} 2 & 0 \\ 1 & 2 \end{bmatrix} \begin{bmatrix} 1 & 3 \\ 0 & 1 \end{bmatrix} \begin{bmatrix} x \\ y \end{bmatrix}$$

$$= \begin{bmatrix} 2 & 6 \\ 1 & 5 \end{bmatrix} \begin{bmatrix} x \\ y \end{bmatrix}$$

The required matrix is $\begin{bmatrix} 2 & 6 \\ 1 & 5 \end{bmatrix}$.

ISBN 9780170390477

Further examples

EXAMPLE 5

Prove that $\begin{bmatrix} 3 & -1 \\ -6 & 2 \end{bmatrix}$ transforms **all** points in the x-y plane to the line $y = -2x$.

Solution

Consider some general point (a, b) transformed to (a', b').

$$\begin{bmatrix} a' \\ b' \end{bmatrix} = \begin{bmatrix} 3 & -1 \\ -6 & 2 \end{bmatrix}\begin{bmatrix} a \\ b \end{bmatrix} = \begin{bmatrix} 3a - b \\ -6a + 2b \end{bmatrix}$$

$$= \begin{bmatrix} 3a - b \\ -2(3a - b) \end{bmatrix}$$

Thus for all image points: y-coordinate = -2 (x-coordinate).

Therefore all image points lie on the line $y = -2x$ as required.

EXAMPLE 6

All points on the line $y = 2x - 3$ are transformed by the matrix $\begin{bmatrix} 1 & 2 \\ 1 & 0 \end{bmatrix}$. Find the equation of the image line.

Solution

Consider some general point $(k, 2k - 3)$ on $y = 2x - 3$.

$$\begin{bmatrix} 1 & 2 \\ 1 & 0 \end{bmatrix}\begin{bmatrix} k \\ 2k - 3 \end{bmatrix} = \begin{bmatrix} 5k - 6 \\ k \end{bmatrix}$$

Thus points (x, y) on the image line are of the form $x = 5k - 6$ and $y = k$.

Eliminating k gives $x = 5y - 6$, i.e. $y = 0.2x + 1.2$, the equation of the image line.

(Alternatively we could take any two points lying on $y = 2x - 3$, determine their images and find the equation of the straight line through these two images.)

iStock.com/olegkalina

Exercise 11B

1 Matrix A represents a rotation of $-90°$ (i.e. $90°$ clockwise) about the origin, matrix B represents a rotation of $180°$ about the origin and matrix C represents a rotation of $90°$ (i.e. $90°$ anticlockwise) about the origin.

   a Determine A, B and C.

Show that:

   b $A^2 = B$ **c** $C^2 = B$ **d** $A^3 = C$

   e $B^2 = I$ where I is the 2×2 identity matrix

   f $A^{-1} = C$ **g** $B^{-1} = B$

2 Determine the 2×2 transformation matrix representing:

   a a reflection in the x-axis,

   b a reflection in the y-axis,

   c a $180°$ rotation about the origin.

Hence show that a reflection in the x-axis followed by a reflection in the y-axis is the same as:

   d a reflection in the y-axis followed by a reflection in the x-axis,

   e a $180°$ rotation about the origin.

3 Matrix P represents a reflection in the line $y = -x$.

Determine P and show that matrix P is its own inverse.

4 A stretch parallel to the x-axis, scale factor 3, should multiply the area of an original shape by 3. Write down the 2×2 matrix representing this transformation and confirm that the determinant is of magnitude 3.

5 a Determine the matrix product $\begin{bmatrix} 1 & 0 \\ -2 & 1 \end{bmatrix}\begin{bmatrix} 0 & 1 & 3 & 2 \\ 1 & 0 & 1 & -1 \end{bmatrix}$.

   b Hence write the coordinates of A′, B′, C′ and D′ the images of A(0, 1), B(1, 0), C(3, 1) and D(2, −1) under the transformation $\begin{bmatrix} 1 & 0 \\ -2 & 1 \end{bmatrix}$.

6 The triangle ABC is transformed to triangle A′B′C′ by the transformation matrix T where

$$T = \begin{bmatrix} 1 & 2 \\ 0 & 1 \end{bmatrix}.$$

If A′, B′ and C′ have coordinates (7, 3), (3, 1) and (−2, −3) respectively, find the coordinates of A, B and C.

7 The triangle ABC is transformed to triangle A′B′C′ by the transformation matrix T where

$$T = \begin{bmatrix} 2 & 0 \\ -3 & 1 \end{bmatrix}.$$

If A′, B′ and C′ have coordinates (2, 0), (−2, 5) and (0, 2) respectively, find the coordinates of A, B and C.

8 Triangle PQR is transformed to triangle P′Q′R′ by matrix $\begin{bmatrix} 0 & 1 \\ 1 & 0 \end{bmatrix}$.

Triangle P′Q′R′ is transformed to triangle P″Q″R″ by matrix $\begin{bmatrix} 1 & 4 \\ 0 & 1 \end{bmatrix}$.

Find the single matrix that will transform PQR directly to P″Q″R″.

9 Triangle PQR is transformed to triangle P′Q′R′ by matrix $\begin{bmatrix} 1 & 0 \\ 2 & 1 \end{bmatrix}$.

Triangle P′Q′R′ is transformed to triangle P″Q″R″ by matrix $\begin{bmatrix} 1 & 0 \\ 0 & -1 \end{bmatrix}$.

a Find the single matrix that will transform PQR directly to P″Q″R″.
b Find the single matrix that will transform P″Q″R″ directly to PQR.

10 Find the single 2 × 2 matrix representing the combination of a shear parallel to the y-axis, scale factor 3, followed by a clockwise rotation of 90° about the origin.

11 Find the single matrix representing the combination of a clockwise rotation of 90° about the origin, followed by a shear parallel to the y-axis, scale factor 3.

12 If the matrix $\begin{bmatrix} a & b \\ c & d \end{bmatrix}$ maps (1, 2) to (12, 7) and (–3, 1) to (–1, 0) then

$$\begin{bmatrix} a & b \\ c & d \end{bmatrix}\begin{bmatrix} 1 & -3 \\ 2 & 1 \end{bmatrix} = \begin{bmatrix} 12 & -1 \\ 7 & 0 \end{bmatrix}$$

Determine a, b, c and d.

13 Quadrilateral ABCD is transformed to $A_1B_1C_1D_1$ by the matrix $\begin{bmatrix} 1 & 4 \\ 0 & 1 \end{bmatrix}$.

Quadrilateral $A_1B_1C_1D_1$ is transformed to $A_2B_2C_2D_2$ by the matrix $\begin{bmatrix} 1 & 0 \\ 2 & 1 \end{bmatrix}$.

Quadrilateral $A_2B_2C_2D_2$ is transformed to $A_3B_3C_3D_3$ by the matrix $\begin{bmatrix} 0 & 1 \\ -1 & 0 \end{bmatrix}$.

a Find the single matrix that will transform ABCD to $A_2B_2C_2D_2$.
b Find the single matrix that will transform ABCD to $A_3B_3C_3D_3$.
c Find the single matrix that will transform $A_2B_2C_2D_2$ to $A_1B_1C_1D_1$.
d Find the single matrix that will transform $A_3B_3C_3D_3$ to $A_1B_1C_1D_1$.

14 Use transformation matrices to prove that if a shape is
- reflected in the x-axis, and then
- reflected in the line $y = x$, and then
- rotated 90° clockwise about the origin,

it ends up in its original position.

15 The transformation matrix $T = \begin{bmatrix} 4 & -2 \\ 1 & 1 \end{bmatrix}$ transforms the rectangle OABC to the parallelogram O'A'B'C'.

If O, A, B and C have coordinates $(0, 0)$, $(3, 0)$, $(3, 2)$ and $(0, 2)$ respectively, the area of OABC is 6 square units.

    a   Use the determinant of T to determine the area of O'A'B'C'.

    b   Determine the coordinates of O', A', B' and C'.

    c   Draw OABC and O'A'B'C' on square grid paper.

    d   Hence confirm your answer to part **a**, the area of O'A'B'C'.

16 The transformation matrix $M = \begin{bmatrix} 1 & 2 \\ -1 & 3 \end{bmatrix}$ transforms the square ABCD to the parallelogram A'B'C'D'.

A, B, C and D have coordinates $(-2, 0)$, $(0, -2)$, $(2, 0)$ and $(0, 2)$ respectively.

    a   With x and y-axes each from -6 to 6 show ABCD on grid paper.

    b   Determine the area of ABCD.

    c   Use the determinant of M to determine the area of A'B'C'D'.

    d   Show A'B'C'D' on the grid and confirm that its area agrees with your answer from **b**.

17 Prove that the matrix $\begin{bmatrix} 2 & -1 \\ -2 & 1 \end{bmatrix}$ transforms all points on the line $y = 2x + 3$ to the single point $(-3, 3)$.

18 Determine the equation of the image line formed when all points on the line $y = x - 1$ are transformed by the matrix $\begin{bmatrix} 1 & 0 \\ 2 & 1 \end{bmatrix}$.

19 Prove that the matrix $\begin{bmatrix} 1 & 3 \\ 3 & 9 \end{bmatrix}$ transforms **all** points to the line $y = 3x$.

20 Show that the matrix $\begin{bmatrix} 6 & 2 \\ 3 & 1 \end{bmatrix}$ transforms all points

    a   on the line $y = 5 - 3x$ to a single point, and find the coordinates of that point,

    b   in the x-y plane to one line, and find the equation of that line.

21 Prove that the transformation matrix $A = \begin{bmatrix} 3 & 0 \\ 2 & 1 \end{bmatrix}$ transforms the straight line $y = m_1 x + p$

to the straight line $y = m_2 x + p$ and find m_2 in terms of m_1.

A pair of straight lines are perpendicular to each other both before and after transformation by matrix A. Find the gradients of the two lines before the transformation.

A general rotation about the origin

The diagram below shows square OABC, of side one unit, rotated anticlockwise about the origin through an angle θ.

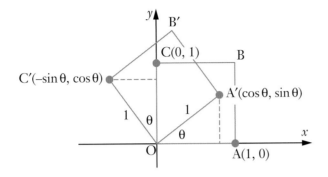

$(1, 0)$ maps to $(\cos\theta, \sin\theta)$ and $(0, 1)$ maps to $(-\sin\theta, \cos\theta)$

The required matrix is $\begin{bmatrix} \cos\theta & -\sin\theta \\ \sin\theta & \cos\theta \end{bmatrix}$.

A general reflection in a line that passes through the origin

The diagram below shows square OABC, of side one unit, reflected in the straight line $y = mx$ where $m = \tan\theta$.

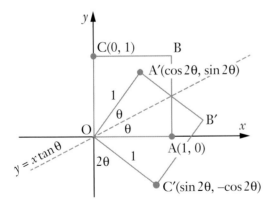

$(1, 0)$ maps to $(\cos 2\theta, \sin 2\theta)$ and $(0, 1)$ maps to $(\sin 2\theta, -\cos 2\theta)$

The required matrix is $\begin{bmatrix} \cos 2\theta & \sin 2\theta \\ \sin 2\theta & -\cos 2\theta \end{bmatrix}$.

Exercise 11C

1 Find the 2×2 transformation matrix representing a rotation of

   a   30° anticlockwise about the origin,

   b   45° anticlockwise about the origin,

   c   60° anticlockwise about the origin,

   d   90° anticlockwise about the origin.

Use your answers to show that

   e   a 30° anticlockwise rotation about the origin followed by another 30° anticlockwise rotation about the origin is equivalent to a 60° anticlockwise rotation about the origin.

   f   a 30° anticlockwise rotation about the origin followed by a 60° anticlockwise rotation about the origin is equivalent to a 90° anticlockwise rotation about the origin.

   g   a 45° rotation about the origin followed by a 45° rotation about the origin is equivalent to a 90° rotation about the origin.

2 Find the 2×2 transformation matrix representing

   a   a reflection in the line $y = x \tan 30°$,

   b   a reflection in the line $y = x \tan 60°$.

Show that for each of parts **a** and **b** the square of the matrix is equal to the identity matrix and explain why this should be so.

3 Write a 2×2 matrix representing a clockwise rotation of θ about the origin.

4 Use transformation matrices to show that a reflection in the line $y = x \tan 45°$ followed by a reflection in the line $y = x \tan 60°$ is equivalent to an anticlockwise rotation about the origin of 30°.

5 By considering a rotation of angle A followed by a rotation of angle B, use the fact that the transformation matrix

$$\begin{bmatrix} \cos\theta & -\sin\theta \\ \sin\theta & \cos\theta \end{bmatrix}$$

represents an anticlockwise rotation of θ about the origin to prove that

$$\sin(A + B) = \sin A \cos B + \cos A \sin B,$$

and $$\cos(A + B) = \cos A \cos B - \sin A \sin B.$$

6 Prove that a reflection in the line $y = m_1 x$, with $m_1 = \tan\theta$, followed by a reflection in the line $y = m_2 x$, with $m_2 = \tan\phi$, is equivalent to an anticlockwise rotation about the origin of angle α, and find α in terms of θ and ϕ.

ISBN 9780170390477

7 a The diagram on the right shows the unit square OABC and its image O′A′B′C′ after rotation of 180° about the point (3, 2).

By considering the transformation as a rotation about the origin followed by a translation, write the 180° rotation about (3, 2) in the form:

$$\begin{bmatrix} x' \\ y' \end{bmatrix} = \begin{bmatrix} ? & ? \\ ? & ? \end{bmatrix} \begin{bmatrix} x \\ y \end{bmatrix} + \begin{bmatrix} ? \\ ? \end{bmatrix}$$

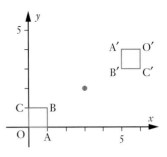

b What 2 × 2 matrix will rotate O′A′B′C′ clockwise about the origin such that point O′ lies on the *x*-axis?

c Find the coordinates of O″, A″, B″ and C″, the images of O′, A′, B′ and C′ under the transformation described in part **b**.

Miscellaneous exercise eleven

This miscellaneous exercise may include questions involving the work of this chapter, the work of any previous chapters in this unit, and the ideas mentioned in the Preliminary work section at the beginning of this unit.

1 Prove that $\cos^4\theta - \sin^4\theta = \cos 2\theta$.

2 Solve $2\cos^2 x + \sin x = 2\cos 2x$ for $0 \le x \le 2\pi$.

3 If $\cos 3\theta = a\cos^3\theta + b\cos^2\theta + c\cos\theta + d$ determine a, b, c and d.

4 When a shape is transformed under a reflection, or a rotation or a shear parallel to a coordinate axis, the area of the shape does not alter. Write the matrix representing each of the following transformations and confirm that for each matrix the absolute value of the determinant is equal to 1.

a Rotate 90° anticlockwise about the origin.

b Rotate 180° about the origin.

c Reflect in the *x*-axis.

d Reflect in the line $y = x$.

e Shear parallel to the *x*-axis, scale factor 4.

f Shear parallel to the *y*-axis, scale factor 3.

5 The matrices A, B and C shown below can be multiplied together to form a single matrix if A, B and C are placed in an appropriate order. What is the order and what is the single matrix this order produces?

$$A = \begin{bmatrix} 2 & 1 & 3 \\ 0 & -1 & 2 \end{bmatrix}, B = \begin{bmatrix} 1 & -1 \end{bmatrix}, C = \begin{bmatrix} 1 & 1 & 0 & -1 \\ 0 & 1 & -1 & 3 \\ 3 & 1 & 4 & 0 \end{bmatrix}.$$

6 If $A = \begin{bmatrix} 1 & 3 & 0 \\ 1 & 0 & -1 \end{bmatrix}$, $B = \begin{bmatrix} 2 & 1 \\ 1 & 3 \end{bmatrix}$ and $C = \begin{bmatrix} 1 & 0 \\ -1 & 2 \end{bmatrix}$, determine each of the following.

If any cannot be determined state this clearly and explain why.

a $A + B$ **b** $B + C$ **c** AC **d** CA

e BC **f** B^2 **g** $BA + C$

7 Explain why $\begin{bmatrix} 2x & -1 \\ 4 & x \end{bmatrix}$ cannot be a singular matrix for real x.

8 If $A = \begin{bmatrix} k & 4 \\ -3 & -1 \end{bmatrix}$, $A^2 + A = \begin{bmatrix} 0 & p \\ q & -12 \end{bmatrix}$ and $p > 0$, find k, p and q.

9 If $A = \begin{bmatrix} 1 & -2 & 2 \end{bmatrix}$ and $B = \begin{bmatrix} 2 \\ 0 \\ -1 \end{bmatrix}$ determine   **a** AB **b** BA

10 Find x and y given that $\begin{bmatrix} 45 & y^2 \\ y^2 & 6x \end{bmatrix} - \begin{bmatrix} x^2 & y \\ -5y & 5 \end{bmatrix} = \begin{bmatrix} 4x & 4-y \\ -6 & x^2 \end{bmatrix}$.

11 With $A = \begin{bmatrix} 3 & 0 \\ 0 & 1 \end{bmatrix}$ and $B = \begin{bmatrix} x & y \\ 0 & z \end{bmatrix}$, what restrictions must be put on the values that x, y and z can take if we require $AB = BA$?

12 If $M = \begin{bmatrix} 0 & -1 \\ 2 & a \end{bmatrix}$ and $M^{-1}M^{-1} = \begin{bmatrix} b & 1 \\ c & d \end{bmatrix}$, find a, b, c and d.

13 a Determine the 2×2 transformation matrix that will transform the point $(1, 0)$ to $(3, -1)$ and the point $(0, 1)$ to $(-2, 1)$.

     b Determine the 2×2 transformation matrix that will transform the point $(2, 1)$ to $(5, 3)$ and the point $(1, -1)$ to $(4, 0)$.

14 Triangle PQR is transformed to P′Q′R′ by matrix $\begin{bmatrix} 2 & -1 \\ 0 & 1 \end{bmatrix}$.

Triangle PQR is transformed to P″Q″R″ by matrix $\begin{bmatrix} 2 & 0 \\ 0 & 3 \end{bmatrix}$.

Find the single matrix that will transform P′Q′R′ to P″Q″R″.

15 Find all solutions to the equation

$$\tan [2(x - 1.5)] = 2.3,$$

rounding answers to two decimal places when rounding is appropriate.

12.

Proof

- Proof by exhaustion
- Proof by induction
- Extension activity: Investigating some conjectures
- Miscellaneous exercise twelve

In Unit One of this *Mathematics Specialist* course we met the idea of *proof*.

In particular • we deduced a number of geometrical truths by reasoning from other accepted truths, i.e. we used *deductive proof*,

• we used our understanding of *vectors* to prove a number of geometrical truths,

• we used *proof by contradiction*, in which the technique is to assume that the opposite of what we are trying to prove is true and then follow correct logical argument only to arrive at a contradiction, thus showing that our initial assumption must be wrong.

In this chapter, we will continue our consideration of proof but now our emphasis is not so much on proving geometrical truths but instead we concentrate more on proving various truths involving real numbers, $\mathbb{R}$. The methods of *proof by exhaustion* and *proof by induction* are then particularly useful.

Real numbers are either rational (can be expressed as a fraction) or irrational (cannot be expressed as a fraction). To define rational and irrational numbers more formally we would say that rational numbers can be expressed in the form $\frac{a}{b}$ where a and b are integers with $b \neq 0$, whilst irrational numbers cannot be expressed in this form. Every real number has a decimal equivalent. The decimal equivalents of rational numbers are either terminating decimals or recurring decimals.

EXAMPLE 1

Decimal representation

Find the following recurring decimals as fractions: a   0.222 222 222 ...

b   0.212 121 212 ...

Solution

a   Let $\quad\quad\quad\quad A = 0.222\,222\,222\,\ldots \quad\quad\quad$ [1]

then $\quad\quad\quad 10A = 2.222\,222\,222\,\ldots \quad\quad$ [2]

[2] − [1] $\quad\quad 9A = 2$

Hence $\quad\quad\quad A = \dfrac{2}{9}$

b   Let $\quad\quad\quad\quad B = 0.212\,121\,212\,\ldots \quad\quad\quad$ [3]

then $\quad\quad 100B = 21.212\,121\,212\,\ldots \quad\quad$ [4]

[4] − [3] $\quad\quad 99A = 21$

Hence $\quad\quad\quad A = \dfrac{21}{99}$

$\quad\quad\quad\quad\quad\quad = \dfrac{7}{33}.$

You should also be familiar with the idea that one *counterexample* can show a general *conjecture* to be false.

Consider, for example, the claim:

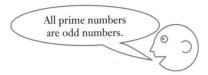

All prime numbers are odd numbers.

Checking some prime numbers: 13 – an odd number
11 – an odd number
7 – an odd number
23 – an odd number

might lead us to believe the statement to be true but with just one counterexample, the number 2, a prime number but not an odd number, we show the general statement to be false.

We might then adjust the statement in the light of the counter example:

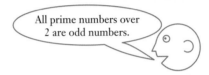

All prime numbers over 2 are odd numbers.

In some cases we may be able to prove a general statement to be true.

Consider, for example, the claim:

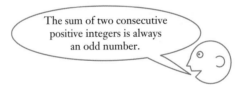

The sum of two consecutive positive integers is always an odd number.

Considering some specific cases:

For the consecutive positive integers 5 and 6: $5 + 6 = 11$,   an odd number.
For the consecutive positive integers 12 and 13: $12 + 13 = 25$,   an odd number.
For the consecutive positive integers 21 and 22: $21 + 22 = 43$,   an odd number.

To prove the statement true we could proceed as follows:

If x is a positive integer then we can represent two consecutive positive integers as x and $x + 1$.
The sum of these two integers is then $x + x + 1 = 2x + 1$.
Now with x an integer $2x$ must be even.
Hence $2x + 1$ must be odd and the statement is proved to be true.

Exercise 12A

For questions 1 to 10 state whether you think the given conjecture is true or false.

If you think it is false, give one example of when it is false.

If you think it is true, give three examples of when it is true, and try to prove it to be true.

1 If we square any even counting number greater than 2 and then subtract 1 we get a multiple of 5.

2 The cube of any even integer is always a multiple of 8.

3 All multiples of 5 are also multiples of 10.

4 All right triangles are isosceles.

5 If we add together an integer squared, six times the integer and 9 we get a square number.

6 The sum of three consecutive positive integers will always be a multiple of 3.

7 The product of two even numbers is always even.

8 The square of an odd number is always an odd number.

9 The product of two consecutive even whole numbers is always a multiple of 8.

10 Multiplying any odd counting number by itself and then adding 7 always gives a multiple of 8.

11 Express each recurring decimal as a fraction.

    a $0.555\,555\,555\,\ldots$

    b $0.\overline{75}$

    c $0.636\,363\,636\,\ldots$

    d $2.\overline{231}$

    e $0.231\,444\,444\,\ldots$

12 By assuming that $\sqrt{2} = \dfrac{a}{b}$, a fraction expressed with a and b having no common factors (i.e. fully cancelled) and with a and b as integers, $b \neq 0$, use the method of proof by contradiction to prove that $\sqrt{2}$ is in fact irrational.

Proof by exhaustion

In this sense the word exhaustion is not used to mean that the proof tires us out and makes us exhausted! Instead the use of the word exhaustion means that the proof 'exhausts all possibilities', it 'considers completely all possible options'. For example consider the following claim:

> The square of any integer is always either   a multiple of 5
>   or   1 more than, or 4 more than, a multiple of 5.

Now the integer to be squared could be

a multiple of 5 itself.	Which we could represent as $5x$ for integer x.
1 more than a multiple of 5.	Represented by $5x + 1$, for integer x.
2 more than a multiple of 5.	Represented by $5x + 2$, for integer x.
3 more than a multiple of 5.	Represented by $5x + 3$, for integer x.
or 4 more than a multiple of 5.	Represented by $5x + 4$, for integer x.

These possibilities together exhaust all options. Hence if we can prove the statement true for all these options we will have proved the statement true for all integers. Completing this proof is one of the questions of the next exercise.

Exercise 12B

Use proof by exhaustion for each of the following.

1 Prove that:

The square of any integer always has the same parity as the integer.

(The parity of a number refers to it being even or odd.)

2 Prove that:

The square of any integer is always either   a multiple of 5
                                      or   1 or 4 more than a multiple of 5.

(Hint: See earlier on this page.)

3 By considering integers as   multiples of 3

or   1 more than a multiple of 3

or   … ,

prove that:

The cube of any integer is always either   a multiple of 9

or   1 more or 1 less than a multiple of 9.

4 A family of sequences is defined by the rule

$$T_{n+1} = 3T_n + 2, \text{ where } T_n \text{ is the } n\text{th term.}$$

For example,

with $T_1 = 3$, $T_2 = 3(3) + 2$ = 11   with $T_1 = 4$, $T_2 = 3(4) + 2$ = 14

$T_3 = 3(11) + 2$ = 35 $T_3 = 3(14) + 2$ = 44

$T_4 = 3(35) + 2$ = 107 $T_4 = 3(44) + 2$ = 134

Prove that for sequences in this family, whatever the parity of a particular term is then the next term will have the same parity. (The parity of a number refers to it being even or odd.)

5 Prove that:

For integer x, $x > 1$, $x^5 - x$ is always a multiple of 5.

(See the factorisation on the right for a clue.)

Is it always a multiple of 10?

Is it always a multiple of 20? Justify your answers.

> factor($x^5 - x$)
>
> $x \cdot (x - 1) \cdot (x + 1) \cdot (x^2 + 1)$

6 Prove that:

For integer x, $x > 1$, $x^7 - x$ is always a multiple of 7.

(See the factorisation on the right for a clue.)

> factor($x^7 - x$)
>
> $x \cdot (x - 1) \cdot (x + 1) \cdot (x^2 + x + 1) \cdot (x^2 - x + 1)$

7 Noticing that $3^3 - 3$ =   24

$4^3 - 4$ =   60

$5^3 - 5$ = 120

John conjectured (suggested) that

For x any integer greater than 2,
the expression $x^3 - x$ is always divisible by 12.

Is John's conjecture correct?

If yes, prove it. If no, make a similar conjecture of your own involving the divisibility of $x^3 - x$ and prove your conjecture true.

iStock.com/UroshPetrovic

Proof by induction

Proof by induction

Consider the following sums of square numbers:

$$1^2 \qquad\qquad\qquad\qquad = \quad 1 \qquad\qquad\qquad\qquad = \quad 1$$

$$1^2 + 2^2 \qquad\qquad\qquad = \quad 1 + 4 \qquad\qquad\qquad = \quad 5$$

$$1^2 + 2^2 + 3^2 \qquad\qquad = \quad 1 + 4 + 9 \qquad\qquad = \quad 14$$

$$1^2 + 2^2 + 3^2 + 4^2 \qquad = \quad 1 + 4 + 9 + 16 \qquad = \quad 30$$

$$1^2 + 2^2 + 3^2 + 4^2 + 5^2 \quad = \quad 1 + 4 + 9 + 16 + 25 \qquad = \quad 55$$

$$1^2 + 2^2 + 3^2 + 4^2 + 5^2 + 6^2 \quad = \quad 1 + 4 + 9 + 16 + 25 + 36 \quad = \quad 91$$

Verify that for each of the above the following formula is true:

$$1^2 + 2^2 + 3^2 + \ldots + n^2 = \frac{n}{6}(n+1)(2n+1)$$

Consider the following:

$$1 \times 2 \qquad\qquad\qquad\qquad = \quad 2 \qquad\qquad\qquad\qquad = \quad 2$$

$$1 \times 2 + 2 \times 3 \qquad\qquad\quad = \quad 2 + 6 \qquad\qquad\qquad = \quad 8$$

$$1 \times 2 + 2 \times 3 + 3 \times 4 \qquad = \quad 2 + 6 + 12 \qquad\qquad = \quad 20$$

$$1 \times 2 + 2 \times 3 + 3 \times 4 + 4 \times 5 \qquad = \quad 2 + 6 + 12 + 20 \qquad = \quad 40$$

$$1 \times 2 + 2 \times 3 + 3 \times 4 + 4 \times 5 + 5 \times 6 \quad = \quad 2 + 6 + 12 + 20 + 30 \quad = \quad 70$$

Verify that for each of the above the following formula is true:

$$1 \times 2 + 2 \times 3 + 3 \times 4 + \ldots + n(n+1) = \frac{n}{3}(n+1)(n+2)$$

FORMULA CAR

Shutterstock.com/Wikrom Kitsamritchai

248 MATHEMATICS SPECIALIST Units 1 & 2

ISBN 9780170390477

The previous page involved two rules,

$$1^2 + 2^2 + 3^2 + \ldots + n^2 \quad = \frac{n}{6}(n+1)(2n+1)$$

and

$$1 \times 2 + 2 \times 3 + 3 \times 4 + \ldots + n(n+1) \quad = \frac{n}{3}(n+1)(n+2).$$

We could verify the rules to be true for various positive values of n but how would we **prove** the above formulae true for **all** positive integer values of n?

One suitable method of proof for these situations is **proof by induction**.

In proof by induction, we follow two steps:

(1) Prove that **if** the statement is true for some general value of n, say $n = k$, then it must also be true for the next value of n, i.e. $n = k + 1$.

(2) Prove that there is a value of n, usually $n = 1$, for which the statement is true.

Question: Why do these two steps form a proof?

Answer: Step (2) proves that the rule is true for $n = 1$ but then, by step (1), it must therefore be true for $n = 2$.

But if it is true for $n = 2$, step (1) means that it must be true for $n = 3$.

But if it is true for $n = 3$, step (1) means that it must be true for $n = 4$.

But if … etc, etc.

Hence the statement must be true for all positive integer n.

Proof by induction is like 'an infinite ladder'.

If we can prove that • if any rung exists then the next rung must also exist,

and that • at least one rung does exist,

then the infinite ladder must exist.

EXAMPLE 2

Use the method of proof by induction to prove that

$$1^2 + 2^2 + 3^2 + \ldots + n^2 = \frac{n}{6}(n + 1)(2n + 1)$$

for all integer $n \geq 1$.

Solution

Let us assume that the rule applies for $n = k$, i.e.

$$1^2 + 2^2 + 3^2 + \ldots + k^2 = \frac{k}{6}(k + 1)(2k + 1).$$

Now consider the situation for $n = k + 1$, i.e. consider

$$1^2 + 2^2 + 3^2 + \ldots + k^2 + (k + 1)^2$$

It follows that

$$
\begin{aligned}
1^2 + 2^2 + 3^2 + \ldots + k^2 + (k + 1)^2 &= \frac{k}{6}(k + 1)(2k + 1) + (k + 1)^2 \\
&= \frac{k+1}{6}[k(2k + 1) + 6(k + 1)] \\
&= \frac{k+1}{6}(2k^2 + 7k + 6) \\
&= \frac{k+1}{6}(k + 2)(2k + 3)
\end{aligned}
$$

Thus $\qquad 1^2 + 2^2 + 3^2 + \ldots + (k + 1)^2 = \dfrac{k+1}{6}(k + 1 + 1)[2(k + 1) + 1]$

i.e. the initial rule applied for $n = k + 1$.

Hence, if the initial rule is true for $n = k$, it is also true for $n = k + 1$.

If $n = 1$, the rule claims that $\qquad 1^2 = \dfrac{1}{6}(2)(3)$

$$= 1 \text{ which is true.}$$

Thus: If the initial rule is true for $n = k$, it is also true for $n = k + 1$.
And: The rule is true for $n = 1$.

Hence, by induction, $\qquad 1^2 + 2^2 + 3^2 + \ldots + n^2 = \dfrac{n}{6}(n + 1)(2n + 1)$ for all integer $n \geq 1$.

Exercise 12C

1 Use proof by induction to prove that

$$1 + 2 + 3 + 4 \ldots n = \frac{1}{2}n(n + 1)$$

for all integer $n \geq 1$.

2 Prove, by induction, that

$$1 \times 2 + 2 \times 3 + 3 \times 4 + 4 \times 5 + \ldots + n(n + 1) = \frac{n}{3}(n + 1)(n + 2)$$

for all integer $n \geq 1$.

3 Prove, by induction, that

$$2 + 4 + 8 + 16 + 32 + \ldots + 2^n = 2^{n+1} - 2$$

for all integer $n \geq 1$.

4 Use proof by induction to prove that

$$1^3 + 2^3 + 3^3 + 4^3 + 5^3 + \ldots + n^3 = \frac{n^2}{4}(n + 1)^2$$

for all integer $n \geq 1$.

5 a Verify that the statements

$$
\begin{array}{lcl}
1 + 3 & = & 4 \\
1 + 3 + 5 & = & 9 \\
1 + 3 + 5 + 7 & = & 16 \\
1 + 3 + 5 + 7 + 9 & = & 25
\end{array}
$$

are consistent with the rule

$$1 + 3 + 5 + 7 + \ldots + (2n - 1) = n^2.$$

b Use the method of proof by induction to prove the above rule to be true for all integer $n \geq 1$.

6 Use proof by induction to prove that

$$\frac{1}{2} + \frac{1}{2^2} + \frac{1}{2^3} + \frac{1}{2^4} + \frac{1}{2^5} + \ldots + \frac{1}{2^n} = \frac{2^n - 1}{2^n}$$

for all integer $n \geq 1$.

7 Use proof by induction to prove that

$$\frac{1}{1 \times 2} + \frac{1}{2 \times 3} + \frac{1}{3 \times 4} + \frac{1}{4 \times 5} + \frac{1}{5 \times 6} + \ldots + \frac{1}{n(n + 1)} = \frac{n}{n + 1}$$

for all integer $n \geq 1$.

8 Prove, by induction, that
$$1 \times 3 \times 5 + 2 \times 4 \times 6 + \ldots + n(n+2)(n+4) = \frac{n}{4}(n+1)(n+4)(n+5)$$
for all integer $n \geq 1$.

9 Use proof by induction to prove that
$$(x-1) \text{ is a factor of } x^n - 1$$
for all positive integer values of n.

10 Use proof by induction to prove that
$$1 \times 2 \times 3 \times 4 \times 5 \times 6 \ldots \times n \geq 3^n$$
for all integer values of $n > 6$.

11 Use the method of proof by induction to prove that
$$7^n + 2 \times 13^n \text{ is a multiple of three}$$
for all $n \geq 0$.

12 Prove, by induction, that
$$2 - 4 + 8 - 16 + 32 \ldots (-1)^{n+1} 2^n = \frac{2}{3}[1 + (-1)^{n+1} 2^n]$$
for all integer $n \geq 1$.

Note

Many questions in the previous exercise involved expressions like

$$1 + 2 + 3 + 4 + 5 + \ldots$$
$$1 + 3 + 5 + 7 + 9 + \ldots$$
$$1^3 + 2^3 + 3^3 + 4^3 + 5^3 + \ldots$$

A shorthand way of writing $1 + 2 + 3 + 4 + 5 + 6 + 7$ is $\displaystyle\sum_{i=1}^{7} i$

This is read as 'sum all the i values starting from $i = 1$ and finishing at $i = 7$', (where i takes integer values).

Using this, **summation notation**, question 4, for example, could be written:

Prove, by induction, that $\displaystyle\sum_{i=1}^{n} i^3 = \frac{n^2}{4}(n+1)^2$

RESEARCH

Extension activity: Investigating some conjectures

Do you understand the difference between a **conjecture** and a **theorem**?

If, based on our opinion or perhaps some observations or maybe some research, we think something to be true we might make a conjecture suggesting it as a truth. A conjecture could be our 'best guess' at what seems to be the case. It may be based on incomplete information and has not been proven. Such a conjecture may later be proved to be true, in which case it would then become a theorem. On the other hand, perhaps someone, or some event, may prove the conjecture to be false.

You may convince others into believing your conjecture is true even though no proof is forthcoming. Just because a conjecture has not been proven true it may also not have been proven false and may be considered by all to be a truth, even though unproven. A theorem on the other hand is a statement that has been proved to be true, often by reasoning from other known truths.

Investigate each of the following famous conjectures. What does each conjecture claim? Give some examples of what it is claiming to be the case. What is the history of the conjecture? Who made the conjecture? When? Where? Has it since been proven to be true, or perhaps false? Etc.

Write a report about each conjecture.

> # Goldbach's conjecture

> # The twin prime conjecture

> # Fermat's conjecture

> # The four-colour conjecture

Shutterstock.com/Rose Carson

Miscellaneous exercise twelve

This miscellaneous exercise may include questions involving the work of this chapter, the work of any previous chapters, and the ideas mentioned in the Preliminary work section at the beginning of the book.

1 If $A = \begin{bmatrix} 3 \\ 1 \end{bmatrix}$, $B = \begin{bmatrix} -1 & 2 \\ 1 & 4 \end{bmatrix}$, $C = \begin{bmatrix} 1 & -1 & 1 \\ -1 & 1 & -1 \end{bmatrix}$, and $D = \begin{bmatrix} 2 & 1 & 0 \\ 1 & 1 & 1 \end{bmatrix}$, determine each of the following. If any cannot be determined, state this clearly.

    a AB   **b** BA   **c** BC   **d** CD   **e** BD

2 If $A = \begin{bmatrix} 2 & 3 \\ -1 & 1 \end{bmatrix}$, determine matrices B, C, D and E given that

    $AB = \begin{bmatrix} 13 \\ -4 \end{bmatrix}$, $AC = \begin{bmatrix} 13 \\ 6 \end{bmatrix}$, $DA = \begin{bmatrix} 6 & 19 \end{bmatrix}$ and $EA = \begin{bmatrix} 5 & 0 \end{bmatrix}$.

3 If $A = \begin{bmatrix} 2 & 3 \\ -1 & 4 \end{bmatrix}$, $B = \begin{bmatrix} 4 & 21 \\ 9 & 17 \end{bmatrix}$ and $AC = B$, find C.

4 In the first copy of a new magazine for 'would-be stamp collectors', an invitation is made to each purchaser of the magazine to complete a six-month subscription order and receive a bonus 'free starter pack'. Two types of pack are available with the contents of each as shown below.

	Number of Australian stamps	Number of Rest of the world stamps
Each *Mainly Australian* starter pack:	75	25
Each *Rest of the World* starter pack:	20	80

We will call this matrix X.

The offer prompts 210 requests for the *Mainly Australian* starter pack and 120 requests for the *Rest of the World* starter pack.

We could write this as a column matrix, Y: $\begin{bmatrix} 210 \\ 120 \end{bmatrix}$

or as a row matrix, Z: $\begin{bmatrix} 210 & 120 \end{bmatrix}$

    a Which of the following matrix products could be formed:
        XY, YX, XZ, ZX?

    b Of those matrix products in **a** that can be formed, which will contain information that is likely to be of use?

    c Determine the useful products from **b** and explain the information displayed.

5 Given that $A = \begin{bmatrix} x & 1 \\ 0 & 3 \end{bmatrix}$ and $A^2 + A = \begin{bmatrix} 6 & x^2 - 8 \\ p & q \end{bmatrix}$ determine p, q and x.

6 Prove that $\sin 2\theta = \dfrac{2\tan\theta}{\tan^2\theta + 1}$.

7 Prove that $\sin 5x \cos 3x - \cos 6x \sin 2x = \sin 3x \cos x$.

8 a Express $(5\cos\theta - 3\sin\theta)$ in the form $R\cos(\theta + \alpha)$ for α an acute angle in radians and correct to two decimal places.

b Hence determine the minimum value of $(5\cos\theta - 3\sin\theta)$ and the smallest positive value of θ (in radians and correct to two decimal places) for which it occurs.

9 The matrices A, B and C shown below can be multiplied together to form a single matrix if A, B and C are placed in an appropriate order. What is the order and what is the single matrix this order produces?

$$A = \begin{bmatrix} 3 \\ 1 \\ 4 \end{bmatrix}, \qquad B = \begin{bmatrix} 2 & 0 & 1 \\ -1 & 3 & 2 \end{bmatrix}, \qquad C = \begin{bmatrix} 1 & 0 & 1 & 1 \end{bmatrix}.$$

10 If $A = \begin{bmatrix} 2x & x \\ 4 & y \end{bmatrix}$ and $A^2 = \begin{bmatrix} 24 & p \\ 0 & q \end{bmatrix}$, find all possible values of x, y, p and q.

11 If $AB = AC$, $A \neq O$, then matrix B does not necessarily equal matrix C, as the following examples show:

Example 1: $\qquad A = \begin{bmatrix} 1 & 3 \end{bmatrix}, \qquad B = \begin{bmatrix} 1 & 2 \\ 2 & -2 \end{bmatrix}, \qquad C = \begin{bmatrix} 4 & -4 \\ 1 & 0 \end{bmatrix}.$

$$AB = \begin{bmatrix} 1 & 3 \end{bmatrix}\begin{bmatrix} 1 & 2 \\ 2 & -2 \end{bmatrix} = \begin{bmatrix} 7 & -4 \end{bmatrix} \qquad AC = \begin{bmatrix} 1 & 3 \end{bmatrix}\begin{bmatrix} 4 & -4 \\ 1 & 0 \end{bmatrix} = \begin{bmatrix} 7 & -4 \end{bmatrix}$$

Thus $AB = AC$, $A \neq O$, but $B \neq C$.

Example 2: $\qquad A = \begin{bmatrix} 4 & 6 \\ 2 & 3 \end{bmatrix}, \qquad B = \begin{bmatrix} 2 & -1 \\ 1 & 2 \end{bmatrix}, \qquad C = \begin{bmatrix} -1 & 2 \\ 3 & 0 \end{bmatrix}.$

$$AB = \begin{bmatrix} 4 & 6 \\ 2 & 3 \end{bmatrix}\begin{bmatrix} 2 & -1 \\ 1 & 2 \end{bmatrix} = \begin{bmatrix} 14 & 8 \\ 7 & 4 \end{bmatrix} \qquad AC = \begin{bmatrix} 4 & 6 \\ 2 & 3 \end{bmatrix}\begin{bmatrix} -1 & 2 \\ 3 & 0 \end{bmatrix} = \begin{bmatrix} 14 & 8 \\ 7 & 4 \end{bmatrix}$$

Thus $AB = AC$, $A \neq O$, but $B \neq C$.

Do the examples above conflict with the following proof that if $AB = AC$ then $B = C$?

If $\qquad AB = AC$

then $\qquad A^{-1}AB = A^{-1}AC$

$\qquad\qquad IB = IC$

and so $\qquad B = C$

12 BC is just one product that can be formed using two matrices selected from the four below. List all the other products that could be formed in this way. (The selection of the two matrices can involve the same matrix being selected twice.)

$$A = \begin{bmatrix} 1 & 2 & 1 \\ 2 & 2 & 0 \end{bmatrix}, \qquad B = \begin{bmatrix} 2 & -3 & 1 \end{bmatrix}, \qquad C = \begin{bmatrix} 1 \\ 0 \\ 1 \end{bmatrix}, \qquad D = \begin{bmatrix} 0 & 1 & 0 \\ 3 & 0 & 1 \\ 2 & 1 & -1 \end{bmatrix}.$$

13 Triangle ABC has vertices A(2, 0), B(2, 3) and C(4, 3). Find the coordinates of the vertices of triangle A′B′C′, the image of ABC when transformed using the transformation matrix $\begin{bmatrix} 1 & 3 \\ 0 & 1 \end{bmatrix}$.

Show both △ABC and △A′B′C′ on grid paper.

What is the transformation this matrix represents?

14 Prove that

$$\sec x \cosec x \cot x = 1 + \cot^2 x.$$

15 Find all solutions to the equation

$$7 \sin x + \cos x = 5$$

rounding answers to two decimal places when rounding is appropriate.

16 Prove, by induction, that

$$12 + 19 + 31 + 53 + \ldots + [5(1 + 2^{n-1}) + 2n] = n(n + 6) + 5(2^n - 1)$$

for all integer $n \geq 1$.

17 Prove by induction that

$$3^{2n+4} - 2^{2n} \text{ is divisible by 5}$$

for all positive integer n.

18 Prove that

$$5^n + 7 \times 13^n \text{ is a multiple of 8}$$

for all integer $n \geq 1$.

19 Prove, by induction, that for $r \neq 1$ and all integer $n \geq 1$,

$$r + r^2 + r^3 + r^4 + \ldots + r^n = \frac{r(r^n - 1)}{r - 1}.$$

ISBN 9780170390477

13.

Complex numbers

- Complex numbers
- Complex number arithmetic
- The conjugate of a complex number
- Equal complex numbers
- Linear factors of quadratic polynomials
- Argand diagrams
- Miscellaneous exercise thirteen

Consider the three quadratic equations

$$(x - 3)(2x + 1) = 0,$$
$$x^2 - 2x - 8 = 0,$$
and
$$x^2 + 6x + 3 = 0.$$

Asked to solve these equations you might opt to solve each using the ability of your calculator but hopefully you would realise that:

- the first, being in factorised form, is readily solved without the assistance of a calculator (to give: $x = 3$, $x = -0.5$)

- the second is readily factorised and hence can also be solved without the assistance of a calculator (to give: $x = 4$, $x = -2$)

The third equation, not being in factorised form and not being readily factorised, could be solved using the ability of some calculators to solve such equations, as the display on the right suggests, or by completing the square (as shown below left), or by use of the quadratic formula (as shown below right).

$\boxed{\begin{array}{l} \text{solve}(x^2 + 6 \cdot x + 3 = 0, x) \\ \quad \{x = -5.449489743, x = -0.5505102572\} \\ \text{solve}(x^2 + 6 \cdot x + 3 = 0, x) \\ \quad \{x = -\sqrt{6} - 3, x = \sqrt{6} - 3\} \end{array}}$

Completing the square.

$$x^2 + 6x + 3 = 0$$

Create a 'gap'

$$x^2 + 6x \quad = -3$$

Insert the square of half the coefficient of x

$$x^2 + 6x + 3^2 = -3 + 3^2$$
$$x^2 + 6x + 9 = 6$$
$$(x + 3)^2 = 6$$
$$x + 3 = \pm\sqrt{6}$$
$$x = -3 \pm \sqrt{6}$$

Use of the quadratic formula.

Comparing $x^2 + 6x + 3 = 0$

with $ax^2 + bx + c = 0$

gives $a = 1, b = 6$ and $c = 3$.

Using $x = \dfrac{-b \pm \sqrt{b^2 - 4ac}}{2a}$

gives $x = \dfrac{-6 \pm \sqrt{6^2 - 4(1)(3)}}{2(1)}$

$= \dfrac{-6 \pm \sqrt{24}}{2}$

$= \dfrac{-6 \pm 2\sqrt{6}}{2}$

$\therefore \qquad x = -3 \pm \sqrt{6}$

Now consider $\qquad x^2 + 6x + 13 = 0$.

If we approach solving this quadratic equation by using the quadratic formula, we encounter the square root of a negative number and correctly conclude that there are 'no real solutions':

Comparing $\quad x^2 + 6x + 13 = 0$
with $\qquad ax^2 + bx + c = 0$ gives $a = 1$, $b = 6$ and $c = 13$.

Thus $\qquad\qquad\qquad x = \dfrac{-6 \pm \sqrt{6^2 - 4(1)(13)}}{2(1)}$

$$= \dfrac{-6 \pm \sqrt{-16}}{2}$$

$\therefore$ $\qquad\qquad\qquad\qquad\qquad$ No real solutions.

However if we attempt to solve this same equation using a calculator we may be given a message indicating there are no solutions, as in the display below left, or we may be given solutions that involve 'i', as in the display below right. Which of these types of response we get depends on whether our calculator is set to solve for real solutions only or is set to include 'complex' solutions.

solve(x^2 + 6·x + 13 = 0, x) No Solution	solve(x^2 + 6·x + 13 = 0, x) {x = −3 − 2·i, x = −3 + 2·i}

Let us now consider what this 'i' means in the 'complex' solutions.

There are indeed no *real* solutions to the equation

$$x^2 + 6x + 13 = 0$$

but a calculator display showing complex solutions uses the concept of $\sqrt{-1}$ being an *imaginary*, non-real, number. Using i to represent $\sqrt{-1}$ gives us a way of representing the non-real solutions of an equation symbolically as shown below.

With $a = 1$, $b = 6$ and $c = 13$ $\qquad x = \dfrac{-b \pm \sqrt{b^2 - 4ac}}{2a}$

$$x = \dfrac{-6 \pm \sqrt{-16}}{2}$$

$$= \dfrac{-6 \pm \sqrt{(16)(-1)}}{2}$$

$$= \dfrac{-6 \pm 4\sqrt{-1}}{2}$$

$$= \dfrac{-6 \pm 4i}{2}$$

$$= -3 \pm 2i$$

$$= -3 + 2i \ \text{ or } \ -3 - 2i \qquad \text{as displayed earlier.}$$

ISBN 9780170390477

As you would be familiar with from your study of *Mathematics Methods*, the quantity $(b^2 - 4ac)$ is called the **discriminant** of the quadratic equation. Its value allows us to *discriminate* between the various types of solution the equation may have.

If $b^2 - 4ac > 0$, we have two real solutions.

If $b^2 - 4ac = 0$, there is just one solution, and it will be real.

If $b^2 - 4ac < 0$, we have two *complex* solutions.

Note: The solutions of an equation $f(x) = 0$ are also referred to as the *roots* of the equation.

The 'just one solution' situation is sometimes referred to as a 'repeated root' because the 'two' solutions to the equation are the same, i.e. 'repeated'.

Complex numbers

Numbers like $-3 + 2i$ and $-3 - 2i$ have the general form $a + bi$ (sometimes written $x + iy$) where a and b are real and $i = \sqrt{-1}$. Such numbers are called **complex** numbers. They consist of a real part, a, and an imaginary part, b.

If $\quad z = a + bi \quad$ we say that the real part of z is a: $\quad \text{Re}(z) = a$
and the imaginary part of z is b: $\quad \text{Im}(z) = b$.

Thus if $\quad z = 4 + 5i \quad$ then $\quad \text{Re}(z) = 4$
and $\quad \text{Im}(z) = 5$.

Introducing an i to represent $\sqrt{-1}$ allows us to give more informative solutions to quadratics in which $(b^2 - 4ac)$ is negative, than simply saying 'no real solutions'. However, if that was the only benefit gained from expanding our number system to include the concept of a complex number, it would hardly be worth the effort. We do make the effort though because, as you will find if you continue your mathematical studies, complex numbers do prove to be very useful in some branches of mathematics.

The idea of expanding our number system to include the concept of a complex number should not be seen as anything particularly strange. When you first started counting, the number system as far as you were concerned would have consisted only of the counting numbers:

$$1, 2, 3, 4, 5, \ldots$$

As the concepts in which you used number became more involved your number system needed to develop to include ways of representing fractions, zero and negatives. Irrational numbers like $\sqrt{2}$ and π then became necessary and now we need to expand the number system beyond $\mathbb{R}$ so that when we are asked to solve equations like

$$x^2 = -4,$$
$$x^2 + 8 = 0,$$
$$x^2 + 6x + 13 = 0, \quad \text{etc.}$$

we can be more informative than simply saying *no real solutions*.

EXAMPLE 1

Use the fact that if $ax^2 + bx + c = 0$

then
$$x = \frac{-b \pm \sqrt{b^2 - 4ac}}{2a}$$

to determine the exact solutions of the following quadratic equations giving your answers in the form $d + ei$ where d and e are real numbers and $i = \sqrt{-1}$.

a $x^2 + 4x + 5 = 0$

b $2x^2 - 3x + 2 = 0$

Solution

a Comparing $\quad x^2 + 4x + 5 = 0$
with $\quad\quad\quad ax^2 + bx + c = 0$
gives $\quad\quad a = 1, b = 4$ and $c = 5$.

Thus $x = \dfrac{-4 \pm \sqrt{4^2 - 4(1)(5)}}{2(1)}$

$\quad\quad = \dfrac{-4 \pm \sqrt{-4}}{2}$

$\quad\quad = \dfrac{-4 \pm \sqrt{(4)(-1)}}{2}$

$\quad\quad = \dfrac{-4 \pm 2i}{2}$

$\quad\quad = -2 + i \text{ or } -2 - i$

b Comparing $\quad 2x^2 - 3x + 2 = 0$
with $\quad\quad\quad ax^2 + bx + c = 0$
gives $\quad\quad a = 2, b = -3$ and $c = 2$.

Thus $x = \dfrac{3 \pm \sqrt{(-3)^2 - 4(2)(2)}}{2(2)}$

$\quad\quad = \dfrac{3 \pm \sqrt{-7}}{4}$

$\quad\quad = \dfrac{3 \pm \sqrt{(7)(-1)}}{4}$

$\quad\quad = \dfrac{3 \pm \sqrt{7}i}{4}$

$\quad\quad = \dfrac{3}{4} + \dfrac{\sqrt{7}}{4}i \text{ or } \dfrac{3}{4} - \dfrac{\sqrt{7}}{4}i$

Shutterstock.com/bluecrayola

EXAMPLE 2

For the complex number $z = 2 + 3i$ state

a   Re(z)   **b**   Im(z).

Solution

If $z = 2 + 3i$ then

a   Re(z) = 2,   **b**   Im(z) = 3.

Note: Re(z) and Im(z) are both real numbers. Im(z) = 3, not $3i$.

```
re(2 + 3·i)
                          2
im(2 + 3·i)
                          3
```

Exercise 13A

Write each of the following in the form ai where a is real and $i = \sqrt{-1}$.

1 $\sqrt{-25}$ **2** $\sqrt{-144}$ **3** $\sqrt{-9}$ **4** $\sqrt{-49}$

5 $\sqrt{-400}$ **6** $\sqrt{-5}$ **7** $\sqrt{-8}$ **8** $\sqrt{-45}$

9 For the complex number $z = 3 + 5i$ state

   a   Re(z)   **b**   Im(z)

10 For the complex number $z = -2 + 7i$ state

   a   Re(z)   **b**   Im(z)

11 For the complex number $z = 3 - i$ state

   a   Re(z)   **b**   Im(z)

Use the fact that if $ax^2 + bx + c = 0$ then $x = \dfrac{-b \pm \sqrt{b^2 - 4ac}}{2a}$ to determine the *exact* solutions of the following quadratic equations giving your answers in the form $d + ei$ where d and e are real numbers and $i = \sqrt{-1}$.

12 $x^2 + 2x + 5 = 0$ **13** $x^2 + 2x + 3 = 0$

14 $x^2 + 4x + 6 = 0$ **15** $x^2 + 2x + 10 = 0$

16 $x^2 - 4x + 6 = 0$ **17** $2x^2 - x + 1 = 0$

18 $2x^2 + x + 1 = 0$ **19** $2x^2 + 6x + 5 = 0$

20 $2x^2 - 2x + 25 = 0$ **21** $5x^2 - 2x + 13 = 0$

22 $x^2 - x + 1 = 0$ **23** $5x^2 - 3x + 1 = 0$

Complex number arithmetic

The following example demonstrates complex number arithmetic. Note that in each case the answer is given in the form $a + bi$ and especially note the following technique used to achieve this when one complex number is divided by another:

If the denominator is $(a + bi)$, we multiply by $\dfrac{(a - bi)}{(a - bi)}$.

(I.e. we multiply by 1, but the 1 is written in a very specific and helpful form.)

EXAMPLE 3

If $w = 2 + 3i$ and $z = 5 - 4i$ determine:

a $w + z$ **b** $w - z$ **c** $3w - 2z$

d wz **e** z^2 **f** $\dfrac{w}{z}$

Solution

a
$$\begin{aligned} w + z &= (2 + 3i) + (5 - 4i) \\ &= 7 - i \end{aligned}$$

b
$$\begin{aligned} w - z &= (2 + 3i) - (5 - 4i) \\ &= 2 + 3i - 5 + 4i \\ &= -3 + 7i \end{aligned}$$

c
$$\begin{aligned} 3w - 2z &= 3(2 + 3i) - 2(5 - 4i) \\ &= 6 + 9i - 10 + 8i \\ &= -4 + 17i \end{aligned}$$

d
$$\begin{aligned} wz &= (2 + 3i)(5 - 4i) \\ &= 10 - 8i + 15i - 12i^2 \\ &= 10 - 8i + 15i + 12 \\ &= 22 + 7i \end{aligned}$$

e
$$\begin{aligned} z^2 &= (5 - 4i)(5 - 4i) \\ &= 25 - 20i - 20i + 16i^2 \\ &= 25 - 20i - 20i - 16 \\ &= 9 - 40i \end{aligned}$$

f
$$\frac{w}{z} = \frac{(2 + 3i)}{(5 - 4i)}$$
$$\therefore \frac{w}{z} = \frac{(2 + 3i)}{(5 - 4i)} \frac{(5 + 4i)}{(5 + 4i)}$$
$$= \frac{10 + 8i + 15i + 12i^2}{25 + 20i - 20i - 16i^2}$$
$$= \frac{-2 + 23i}{41}$$
$$= -\frac{2}{41} + \frac{23}{41}i$$

Alternatively, and as the reader should verify, these same answers can be obtained from a calculator.

ISBN 9780170390477

The conjugate of a complex number

Complex conjugates

If $z = a + bi$, we say that $a - bi$ is the **conjugate** of z. We use the symbol $\bar{z}$ for the conjugate of z.

Thus
if $z = 2 + 3i$	then	$\bar{z} = 2 - 3i$,
if $z = 5 - 7i$	then	$\bar{z} = 5 + 7i$,
if $z = -2 + 8i$	then	$\bar{z} = -2 - 8i$,
if $z = -3 - 4i$	then	$\bar{z} = -3 + 4i$, etc.

Note that for any complex number z $(= a + bi)$, both the sum $z + \bar{z}$ and the product $z\bar{z}$ are real.

Proof: If $z = a + bi$ then $\bar{z} = a - bi$.

Hence $z + \bar{z} = (a + bi) + (a - bi)$
$= 2a$ a real number.

and $z\bar{z} = (a + bi)(a - bi)$
$= a^2 - abi + abi - b^2i^2$
$= a^2 + b^2$ a real number.

- The fact that the product $z\bar{z}$ is real is used when dividing complex numbers, as in **example 3** part **f** shown earlier and in part **d** of the next example.

EXAMPLE 4

If $z = 12 - 5i$ determine

a $\bar{z}$ b $z\bar{z}$ c $z + \bar{z}$ d $\dfrac{z}{\bar{z}}$

Solution

a $\bar{z} = 12 + 5i$

b $z\bar{z} = (12 - 5i)(12 + 5i)$
$= 144 + 60i - 60i - 25i^2$
$= 169$

c $z + \bar{z} = (12 - 5i) + (12 + 5i)$
$= 24$

d $\dfrac{z}{\bar{z}} = \dfrac{(12 - 5i)}{(12 + 5i)}$

$= \dfrac{(12 - 5i)}{(12 + 5i)} \dfrac{(12 - 5i)}{(12 - 5i)}$

$= \dfrac{119}{169} - \dfrac{120}{169}i$

Equal complex numbers

Complex number operations

When we say that two complex numbers, w and z, are equal then we mean that

$$\text{Re}(w) = \text{Re}(z) \quad \textbf{and} \quad \text{Im}(w) = \text{Im}(z).$$

Thus if $a - 3i = 5 + bi$
then $a = 5$ and $b = -3$.

Complex numbers

Linear factors of quadratic polynomials

A linear factor is of the form $ax + b$, i.e. a linear expression. Hence writing $x^2 + 6x - 16$ in the form $(x + 8)(x - 2)$ is expressing $x^2 + 6x - 16$ in terms of linear factors.

Now that we have been introduced to complex numbers, these linear factors could involve complex numbers, as the next example shows.

EXAMPLE 5

Express $x^2 + 2x + 5$ as the product of two linear factors.

Solution

Using the quadratic formula.

If $\quad x^2 + 2x + 5 = 0$

Then $\quad x = \dfrac{-2 \pm \sqrt{-16}}{2}$

$\qquad\qquad = -1 \pm 2i$

Thus $\quad x^2 + 2x + 5 = [x - (-1 + 2i)][x - (-1 - 2i)]$

$\qquad\qquad\qquad\quad = (x + 1 - 2i)(x + 1 + 2i)$

Using completing the square.

$x^2 + 2x + 5 = x^2 + 2x \qquad + 5$

$\qquad\qquad\quad = x^2 + 2x + 1 + 5 - 1$

$\qquad\qquad\quad = (x + 1)^2 + 4$

(Write as difference of 2 squares)

$\qquad\qquad\quad = (x + 1)^2 - (2i)^2$

$\qquad\qquad\quad = (x + 1 - 2i)(x + 1 + 2i)$

Can your calculator factorise $x^2 + 2x + 5$ in this way?

Exercise 13B

Simplify

1 $(2 + 3i) + (5 - i)$

2 $(5 - 6i) - (2 + 4i)$

3 $(2 + 3i) - (5 - i)$

4 $(5 - 6i) + (2 + 4i)$

5 $2 + 3i - 5 - i$

6 $5 - 6i + 2 + 4i$

7 $(3 + i) + (4 - 2i) + (6 + 5i)$

8 $2(3 + 2i) + 3(2 + i)$

9 $5(2 + i) + 3(1 - i)$

10 $5(2 + i) - 3(1 - i)$

11 $3(1 - 5i) + 7i$

12 $3(1 - 5i) + 7$

13 $\text{Re}(2 + 3i) + \text{Re}(5 - 2i)$

14 $\text{Im}(-1 + 4i) + \text{Im}(3 + i)$

15 $(3 + 2i)(2 + 5i)$

16 $(1 + 3i)(3 + 2i)$

17 $(2 + i)(1 - i)$

18 $(-2 + 3i)(5 + i)$

Express each of the following in the form $a + bi$ where a and b are real numbers.

19 $\dfrac{(3 + 2i)}{(1 + 5i)}$

20 $\dfrac{(3 + i)}{(1 - 2i)}$

21 $\dfrac{4}{(1 + 3i)}$

22 $\dfrac{2i}{(1 + 4i)}$

23 $\dfrac{(-3 + 2i)}{(2 + 3i)}$

24 $\dfrac{(5 + i)}{(2i + 3)}$

25 If $w = 5 - 2i$ and $z = 4 + 3i$ determine exactly

   a $w + z$ **b** $w - z$ **c** $3w - 2z$ **d** wz **e** z^2 **f** $\dfrac{w}{z}$

26 If $Z_1 = 3 + 5i$ and $Z_2 = 1 - 5i$ determine exactly

   a $Z_1 + Z_2$ **b** $Z_2 - Z_1$ **c** $Z_1 + 3Z_2$ **d** $Z_1 Z_2$ **e** Z_1^2 **f** $\dfrac{Z_1}{Z_2}$

27 If $z = 24 - 7i$ determine

   a $\overline{z}$ **b** $z + \overline{z}$ **c** $z\overline{z}$ **d** $\dfrac{z}{\overline{z}}$

28 If $z = 4 + 9i$ determine

   a $\overline{z}$ **b** $z - \overline{z}$ **c** $2z + 3\overline{z}$ **d** $2z - 3\overline{z}$ **e** $z\overline{z}$ **f** $\dfrac{z}{\overline{z}}$

29 Given that $z = 2 + ci$, $w = d + 3i$, c and d are real numbers and $z = w$ determine c and d.

30 If $a + bi = (2 - 3i)^2$, where a and b are real numbers, determine a and b.

31 If $z = 5 - (c + 3)i$ (c real), $w = d + 1 + 7i$ (d real) and $z = w$ determine c and d.

32 If $(a + 3i)(5 - i) = p$ where a and p are real numbers, determine a and p.

33 State whether each of the following are correct for all complex z and w? (If your answer is 'no' then give an example to support your claim.)

   a   If $w = \overline{z}$, the conjugate of z, then $\text{Im}(w) = -\text{Im}(z)$.

   b   If $\text{Im}(z) = -\text{Im}(w)$ then $w = \overline{z}$.

34 Express each of the following as the product of two linear factors. (Not all will involve complex numbers.)

   a $x^2 - 4x + 13$

   b $x^2 - 2x + 10$

   c $x^2 - 6x + 1$

   d $x^2 + 10x + 26$

   e $x^2 + 14x + 53$

   f $x^2 + 4x - 3$

35 a Use the quadratic formula to prove that if a quadratic equation has any non-real roots. then it must have two and they must be conjugates of each other.

b One root of $x^2 + bx + c = 0$ is $x = 3 + 2i$. Find b and c.

c One root of $x^2 + dx + e = 0$ is $x = 5 - 3i$. Find d and e.

36 Simplify

a $\dfrac{c + di}{-c - di}$

b $\dfrac{c + di}{d - ci}$

c $\dfrac{c - di}{-d - ci}$

37 Find all possible real number pairs p, q such that $\dfrac{3 + 5i}{1 + pi} = q + 4i$.

38 The complex numbers z and w are such that for the real variable x, $(x - z)(x - w) = ax^2 + bx + c$ for real a, b and c.

a Determine the value of a.

b Prove that $(z + w)$ and (zw) must both be real.

c By letting $z = p + qi$ and $w = r + si$, prove that z and w must be the conjugates of each other.

39 Given that $z = a + bi$, $w = c + di$, $\bar{z}$ is the conjugate of z and $\bar{w}$ is the conjugate of w, prove that each of the following are true.

a $\overline{z}\,\overline{w} = \overline{zw}$

b $\overline{\left(\dfrac{z}{w}\right)} = \dfrac{\bar{z}}{\bar{w}}$

40 The complex number $a + bi$ can be expressed as the 'ordered pair' (a, b).

Express each of the following in this form.

a $2 + 3i$ **b** $-5 + 6i$ **c** $7i$ **d** 3

In the following, each ordered pair represents a complex number.

Simplify the following sums, giving answers in ordered pair form.

e $(3, 8) + (-2, 1)$ **f** $(3, -5) + (3, 5)$

Simplify the following differences, giving answers in ordered pair form.

g $(5, 3) - (2, 0)$ **h** $(2, 7) - (2, -7)$

Simplify the following products, giving answers in ordered pair form.

i $(0, 2) \times (3, 5)$ **j** $(-3, 1) \times (-3, -1)$

Simplify the following quotients, giving answers in ordered pair form.

k $(3, 0) \div (2, -4)$ **l** $(3, -8) \div (3, 8)$

41 Showing full algebraic reasoning, determine exactly, and in the form $a + bi$, the complex number z for which:

$$\frac{1}{z} = \frac{2 + 7i}{1 - i}$$

Argand diagrams

In this chapter we had written complex numbers in the form $a + bi$ until question 40 of the previous exercise, when we saw that they could also be written as an ordered pair (a, b). Clearly this ordered pair form of writing the complex number $a + bi$ is similar to the way we label a point on a graph by stating its coordinates. If instead of x and y axes we have real and imaginary axes we can use this similarity to provide us with a diagrammatic way of representing a complex number.

• Such a graphical representation is called an **Argand diagram**.

• The plane containing the real and imaginary axes is referred to as the complex plane.

• The complex number $a + bi$ can be thought of as the point (a, b) on the Argand diagram or as the vector from the origin to the point (a, b).

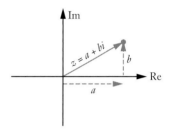

The diagram on the right uses these ideas to show the complex numbers

$$Z_1 = 6 + 4i,$$
$$Z_2 = -7 + 6i,$$
$$Z_3 = -6 - 3i,$$
and $$Z_4 = 4 - 3i.$$

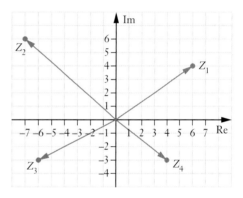

Notice that we can then add complex numbers in the complex plane using the 'nose to tail' method of vector addition.

$$Z_2 + Z_3 = (-7 + 6i) + (-6 - 3i) \qquad\qquad Z_1 + Z_4 = (6 + 4i) + (4 - 3i)$$
$$= -13 + 3i \qquad\qquad\qquad\qquad\qquad = 10 + i$$

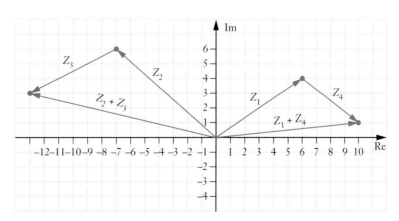

Exercise 13C

1 Express each of the complex numbers Z_1 to Z_8, shown on the Argand diagram below, in the form $a + bi$.

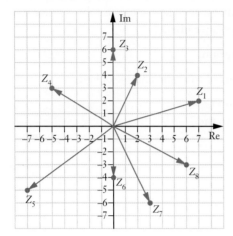

2 Express each of the complex numbers Z_1 to Z_8, shown on the Argand diagram below, in the ordered pair form (a, b).

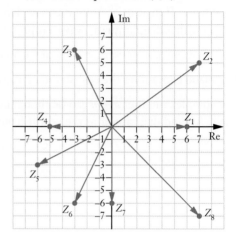

3 Show the following numbers as vectors on a single Argand diagram.

$Z_1 = 4 + 3i$ $\qquad$ $Z_2 = 4i$ $\qquad$ $Z_3 = -3 + 5i$ $\qquad$ $Z_4 = -5 + 2i$

$Z_5 = -3i$ $\qquad$ $Z_6 = 2 - 4i$ $\qquad$ $Z_7 = \overline{Z_1}$ $\qquad$ $Z_8 = \overline{Z_3}$

4 The Argand diagram on the right shows four complex numbers Z_1, Z_2, Z_3 and Z_4.

Given that $\quad \dfrac{\text{Re}(Z_1)}{\text{Im}(Z_1)} > 0$

$$\dfrac{\text{Re}(Z_2)}{\text{Im}(Z_2)} > 1$$

and $\qquad Z_3 = \overline{Z_2}$

determine Z_1, Z_2, Z_3 and Z_4.

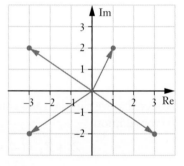

5 With $z = 3 + 5i$, display z, iz, i^2z and i^3z as vectors on a single Argand diagram.

6 Display Z_1, Z_2, ... Z_8 as vectors on a single Argand diagram where:

$Z_1 = 2 + i$ $\qquad$ $Z_2 = (2 + i)(1 + i)$ $\qquad$ $Z_3 = (2 + i)(1 + i)^2$ $\qquad$ $Z_4 = (2 + i)(1 + i)^3$

$Z_5 = (2 + i)(1 + i)^4$ $\qquad$ $Z_6 = (2 + i)(1 + i)^5$ $\qquad$ $Z_7 = (2 + i)(1 + i)^6$ $\qquad$ $Z_8 = (2 + i)(1 + i)^7$

Miscellaneous exercise thirteen

This miscellaneous exercise may include questions involving the work of this chapter, the work of any previous chapters in this unit, and the ideas mentioned in the Preliminary work section at the beginning of the unit.

1 Simplify each of the following.

    a $(2 + 5i)(2 - 5i)$ **b** $(3 + i)(3 - i)$ **c** $(6 + 2i)(6 - 2i)$

    d $(3 + 4i)^2$ **e** $\dfrac{2 - 3i}{3 + i}$ **f** $\dfrac{3 + i}{2 - 3i}$

2 Given that $z = 2 - 3i$ and $w = -3 + 5i$ determine

    a $z + w$ **b** zw **c** $\bar{z}$, the conjugate of z

    d $\overline{zw}$ **e** z^2 **f** $(zw)^2$

    g   the complex number p such that $\text{Re}(p) = \text{Re}(\bar{z})$
                                    and $\text{Im}(p) = \text{Im}(\bar{w})$.

3 a $2x^3 - 5x^2 + 8x - 3 = (px - q)(x^2 + rx + 3)$ for real integers p, q and r. Determine p, q and r.

    b   Without the assistance of a calculator, but using the fact that the quadratic equation $ax^2 + bx + c = 0$ has solutions given by

$$x = \frac{-b \pm \sqrt{b^2 - 4ac}}{2a}$$

       find *exactly*, all values of x, real and complex, for which: $2x^3 - 5x^2 + 8x - 3 = 0$.

4 Find

    a $(\text{Re}(2 + 3i))(\text{Re}(5 - 4i))$ **b** $\text{Re}((2 + 3i)(5 - 4i))$

5 Given that $z = 5\sqrt{2}i$ determine each of the following exactly.

    a $\bar{z}$ **b** z^2 **c** $(1 + z)^2$

6 Find the real numbers a and b given that $(a + bi)^2 = 5 - 12i$.

7 The triangle ABC is transformed to A′B′C′ by the transformation matrix T where

$$T = \begin{bmatrix} 1 & 3 \\ 1 & 0 \end{bmatrix}.$$

    a   If A′, B′ and C′ have coordinates $(-1, 2)$, $(10, -2)$ and $(-4, -4)$ respectively, find the coordinates of A, B and C.

    b   Draw triangles ABC and A′B′C′ on grid paper and confirm that
                      Area $\triangle$A′B′C′ $= |\det T|$ Area $\triangle$ABC.

8 The transformation matrix $\begin{bmatrix} a & b \\ c & d \end{bmatrix}$ maps the point $(1, -1)$ to $(4, 1)$ and maps the point $(2, -3)$ to $(9, 1)$. Determine a, b, c and d.

9 Find all real or complex solutions to the equation $(z - 2 + 7i)^2 = -25$

10 A particular 2×2 matrix transforms all points in the x-y plane to a straight line.
   a   What does this suggest about the determinant of the matrix?
   b   Why must the straight line pass through the origin?

11 Given that $z = a + bi$, determine a and b given that $z + 2\bar{z} = 9 + 5i$, where $\bar{z}$ is the complex conjugate of z.

12 Using your calculator if you wish
   a   express $(2 + 3i)^4$ in the form $a + bi$,
   b   determine $\text{Im}((1 - 3i)^5)$.

13 Determine the quadratic equation $x^2 + bx + c = 0$ for which one of the solutions is $x = 2 + 3i$.

14 Determine the complex number z given that $3z + 2\bar{z} = 5 + 5i$, where $\bar{z}$ is the complex conjugate of z.

15 Showing full algebraic reasoning and giving your answer in the form $a + bi$, determine *exactly* the complex number z for which $z(2 - 3i) = 5 + i$.

16 The diagram on the right shows three complex numbers, w, $\bar{w}$ and z, with $\bar{w}$ representing the complex conjugate of w. Write z in the form $a + bi$ and determine z^2.

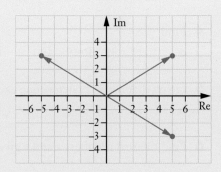

17 Prove that $\dfrac{\sin\theta}{\cos\left(\dfrac{1}{2}\theta\right)} = 2\sin\left(\dfrac{1}{2}\theta\right)$.

18 Solve $\tan 2x + \tan x = 0$ for $0 \le x \le 360°$.

19 Prove that matrices of the form $\begin{bmatrix} a & b \\ ka & kb \end{bmatrix}$ transform **all** points to the straight line $y = kx$.

20 a Premultiply $\begin{bmatrix} -3 & 5 \end{bmatrix}$ by $\begin{bmatrix} 4 \\ 1 \end{bmatrix}$.

b Postmultiply $\begin{bmatrix} -3 & 5 \end{bmatrix}$ by $\begin{bmatrix} 4 \\ 1 \end{bmatrix}$.

21 With the complex number w as defined by the Argand diagram on the right, state which of the diagrams below show a complex number z for which:

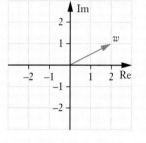

a $z = \bar{w}$

b $z + w$ is real

c zw is real

d $\text{Im}(w) = \text{Im}(z)$

e $\text{Im}(w) = |\text{Im}(z)|$

f $|\text{Im}(w)| = \text{Im}(z)$

g $z = iw$

h $\dfrac{\bar{w}}{z}$ is real

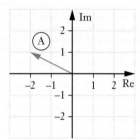
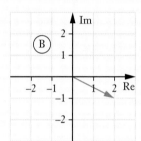
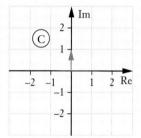

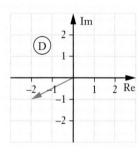
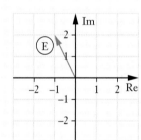
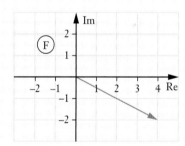

22 Matrices A, B and C are all 2×2 matrices, matrix B is not a singular matrix and $A = BCB^{-1}$. Determine simplified expressions for

a A^2 **b** A^3 **c** A^n.

23 Use the method of proof by induction to prove that

$$1 \times 2 \times 3 + 2 \times 3 \times 4 + 3 \times 4 \times 5 + \ldots + n(n+1)(n+2) = \frac{n}{4}(n+1)(n+2)(n+3)$$

for all integer n, $n \geq 1$.

24 Prove, by induction, that $2^{n-1} + 3^{2n+1}$ is a multiple of seven for all integer n, $n \geq 1$.

25 Prove that $5^n + 3 \times 9^n$ is a multiple of 4 for all $n \geq 0$.

26 Prove that:

$$\sin\theta\,(\sin\theta + \sin 2\theta) = 1 + 2\cos\theta - \cos^2\theta - 2\cos^3\theta.$$

27 Find all solutions to the equation

$$4\sin x \cos^2 x - \cos x = 0,$$

giving answers as exact values.

28 Write down:

a the 2×2 transformation matrix that will perform a 30° anticlockwise rotation about the origin,

b the 2×2 transformation matrix that will perform a 60° anticlockwise rotation about the origin,

c the 2×2 transformation matrix that will reflect a shape in the line $y = \dfrac{\sqrt{3}}{3}x$.

d The diagram on the right shows square 1 reflected in

the line $y = \dfrac{\sqrt{3}}{3}x$ to give square 2, which is then rotated

60° anticlockwise about the origin to give square 3.

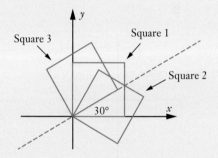

It would appear that square 3 could be obtained directly from square 1 by an anticlockwise rotation of 30° about the origin.

Use your answers from the earlier parts of this question to show that a reflection in the line

$y = \dfrac{\sqrt{3}}{3}x$ followed by an anticlockwise rotation of 60° about the origin is not equivalent to an anticlockwise rotation of 30° about the origin.

Explain the apparent equivalence suggested by the diagram.

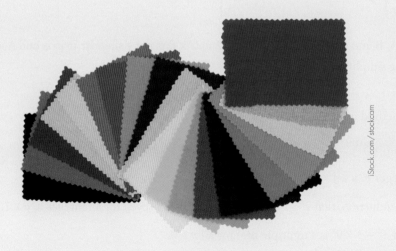

ISBN 9780170390477

ANSWERS UNIT ONE

Exercise 1A   PAGE 9

1 At least one of the seven questions will be done by two or more of the eight students.

2 One of the year three classes will have at least two of the Singh triplets.

3 At least one of the variations of the genetic marker is possessed by more than one human.

4 At least two of the socks will be of the same colour.

5 There are people in Australia who have the same number of hairs on their head as do other people in Australia.

6 Some people who have existed occupy more than one space on my ancestral tree. I.e. some great great great … grandfather on my mother's side was also a great great great … grandfather on my father's side.

7 a 14

  b 0

  c No. If some person A shook hands with all 14 others then none of the other 14 could have shaken hands with nobody because they all at least shook hands with person A.

  Similarly if some person shook hands with none of the others then no one could have shaken hands with everyone.

  Hence there are 15 people to either assign to the numbers 0, 1, 2, 3, 4, 5, 6, 7, 8, 9, 10, 11, 12, 13

  or to assign to the numbers 1, 2, 3, 4, 5, 6, 7, 8, 9, 10, 11, 12, 13, 14.

  Either way we have 15 people to assign to 14 integers so at least two people will have shaken hands with the same number of people.

8 If a polygon is a triangle then the polygon has exactly three sides. True. $P \Leftrightarrow Q$.

9 If Jenny's mouth is open then she is talking. False. $P \not\Leftrightarrow Q$.

10 If the animal is a mammal then it is a platypus. False. $P \not\Leftrightarrow Q$.

11 If the car will not start it is out of fuel. False. $P \not\Leftrightarrow Q$.

12 If points are collinear then they lie on the same straight line. True. $P \Leftrightarrow Q$.

13 If tomorrow is not Friday then today is not Thursday. True

14 If a number is not a multiple of two then it is not even. True.

15 If a triangle does not have three different length sides then it is not scalene. True.

16 If my lawn is not wet then my sprinklers are not on. True.

17 If Armand does get up before 8 am then it is a school day. True

18 a True

  b If a polygon is not a triangle then its angles do not add up to 180°.

  c True

19 a True

  b If a positive integer does not have exactly 2 factors then it is not a prime number.

  c True

20 a True

  b If the car battery is not flat then the car will start.

  c False

21 a True

  b If there are no letters in my mail box the post person has not been to our road.

  c False

22 a False

b If a number is not even then it is not a multiple of 4.

c True

23 Converse: If a polygon is a pentagon then the polygon is five sided. True.

Inverse: If a polygon is not five sided then the polygon is not a pentagon. True.

Contrapostive: If a polygon is not a pentagon then the polygon is not five sided. True.

24 Converse: If the four angles of a quadrilateral are all right angles then the quadrilateral is a square. False.

Inverse: If a quadrilateral is not a square then the four angles of the quadrilateral are not all right angles. False.

Contrapostive: If the four angles of a quadrilateral are not all right angles then the quadrilateral is not a square. True.

Miscellaneous exercise one   PAGE 11

1 The ladder will make an angle of 71° with the ground (to nearest degree).

Exercise 2A   PAGE 19

1 a 6   **b** 8   **c** 120

   d 11   **e** 110   **f** 15

   g 20   **h** 210   **i** 56

2 6   **3** 16   **4** 719   **5** 243

6 a 5040   **b** 823 543

7 a 2520   **b** 16807

8 a 3375   **b** 2730

9 a 57 600   **b** 12 441 600   **c** 311 040 000

10 132

11 5040. PIN: **P**ersonal **I**dentification **N**umber

12 665 280   **13** 1 048 576

14 2730   **15** 336, 40 320

16 1024

Exercise 2B   PAGE 24

1 360   **2** 840

3 420   **4** 239 500 800

5 34 650, 3150   **6** 75 600, 7560, 68 040

7 80   **8** 18 252

9 16 250

10 a 36   **b** 12

11 a 150   **b** 80

12 1465   **13** 90   **14** 468

15 a 33 280 (8^5 long key base + 8^3 short key base)

b 7056 (6720 long key base + 336 short key base)

16 a 59 778 (9^5 long key base + 9^3 short key base)

b 15 624 (15 120 long key base + 504 short key base)

17 36   **18** 1200   **19** 79

20 30   **21** 78

22 a 199   **b** 142   **c** 313

23 $n(A \cup B \cup C) = 40$

Venn diagram below confirms this answer of 40.

24 74   **25** 413

26 $|A \cup B \cup C \cup D| = |A| + |B| + |C| + |D|$
$- |A \cap B| - |A \cap C| - |A \cap D| - |B \cap C|$
$- |B \cap D| - |C \cap D| + |A \cap B \cap C|$
$+ |A \cap B \cap D| + |A \cap C \cap D| + |B \cap C \cap D|$
$- |A \cap B \cap C \cap D|$

Exercise 2C   PAGE 29

1 a 24   **b** 625

2 a 360   **b** 72

3 a 720   **b** 240   **c** 24   **d** 144

4 120   **a** 72   **b** 48

5 120   **a** 24   **b** 24   **c** 6

6 a 5040   **b** 2160   **c** 360

7 a 24   **b** 6   **c** 3

8 a 486 720   **b** 650 000   **c** 421 200   **d** 117 000

9 a 144   **b** 24   **c** 72

10 a 1 757 600   **b** 1 404 000   **c** 1 134 000

   d 6500   **e** 2400   **f** 216

11 a 3 628 800   **b** 40 320   **c** 241 920

   d 5040

Exercise 2D   PAGE 35

1 a 750   **b** 180   **c** 108

2 a 7992   **b** 840   **c** 700

3 a 2160   **b** 600

4 120   **a** 24   **b** 24

   c 6   **d** 42

5 a 40 320   **b** 10 080   **c** 30 240

   d 1440   **e** 9360

6 a 210 **b** 30 **c** 30
d 5 **e** 55 **f** 120
g 30 **h** 90 **i** 40
7 a 6 **b** 6 **c** 2
d 10 **e** 4 **f** 12
8 a 70 560 **b** 25 200
9 29 030 400
10 a 130 **b** 26 **c** 5 **d** 1
e 30 **f** 5 **g** 125

Exercise 2E   PAGE 43

1 In a combination lock the order of the numbers is important. Thus to be more correct it should really be called a permutation lock. Hence a combination lock is not correctly named.

2 10 **3** 4845 **4** 4200 **5** 700
6 36 **7** 495, 240 **8** 128 **9** 510
10 163 800 **a** 13 650 **b** 65 520 **c** 73 710
11 a 210 **b** 28 **c** 98 **d** 182
12 a 1400 **b** 8 **c** 0 **d** 176
e 1016
13 a 752 538 150 **b** 115 775 100 **c** 171 028 000
d 73 629 072 **e** 80 672 868
14 715, 360 **15** 70, 22
16 399 **17** 5472

Miscellaneous exercise two   PAGE 47

1 39.7 **2** 8.4
3 Converse: If $x^2 = 64$ then $x = 8$. False.
Contrapositive: If $x^2 \neq 64$ then $x \neq 8$. True.
4 There are 24 different permutations $(4 \times 3 \times 2 \times 1)$ and twenty five responses from the students. Hence, by the pigeon hole principle, at least two pieces of paper will feature the same permutation.
5 There are 44 352 different bets.
6 15 120, 7200 **7** 15 504, 3072

Exercise 3A   PAGE 53

1 a 11.5 km, 071° **b** 251°
2 a 5.5 km, 027° **b** 207°
3 a 47 km, 304° **b** 124°
4 a 1150 m, 089° **b** 269°
5 a 87 m, 046° **b** 226°
6 a 66 km, 235° **b** 055°
7 Approximately 41 m. **8** 8.9 km, 145°

9 3.2 km, 095° **10** 071°, 349°
11 286 m in direction 060°.

Exercise 3B   PAGE 57

1 8.3 N at 27° to the vertical.
2 16.3 N at 22° to the vertical.
3 28.3 N at 0° to the vertical.
4 24.4 N at 25° to the vertical.
5 $5\sqrt{3}$ N, 090° **6** $2\sqrt{31}$ N, 159°
7 12.1 N, 018° **8** 11.7 N, 158°
9 47 N at 66° to the slope.
10 90 N at 78° to the slope.
11 38 N at 67° to the slope.
12 ~9.2 N at ~42° to the larger force.
13 ~23.2 N at ~27° to the smaller force.

Exercise 3C   PAGE 59

1 4.5 m/s at 63° to the bank.
2 3.6 m/s at 85° to the bank.
3 5.5 m/s at 34° to the bank
4 353°. Approximately 15.3 km
5 170° at 72 km/h, 194°
6 a 180 m **b** $\sqrt{10}$ m/s (≈ 3.2 m/s)
c ~72°
7 a Upstream at 73° to the bank, 30 seconds.
b Upstream at 66° to the bank, 31 seconds.
c Upstream at 53° to the bank, 36 seconds.
8 356° **9** 005°
10 a 048° **b** 1 h 34 min **c** 1 h 20 min
11 46 secs (19.1 + 14.0 + 13.3)

Exercise 3D   PAGE 66

1 a d and e **b** c and d or c and e
c a and b or a and f **d** b and f
2 a $b + c = a$ **b** $a + b = c$ **c** $a + c = b$
3 $b = a$, $c = \frac{1}{2}a$, $d = -a$, $e = -\frac{1}{2}a$,
$f = -\frac{1}{4}a$, $g = \frac{3}{2}a$, $h = \frac{3}{4}a$.
4 $p = 2m$, $q = -n$, $r = 2n$, $s = -m$,
$t = \frac{1}{2}n$, $u = m + n$, $v = m + 2n$.
5 $c = a + b$, $d = a - b$, $e = b - a$,
$f = 2b + a$, $g = b + 2a$.

6 $u = s + t$, $\quad v = 2s + t$, $\quad w = s - t$, $\quad x = -s + t$,

$y = 2t + 2s$, $\quad z = 3s + \dfrac{3}{2}t$.

7 (Reduced scale)

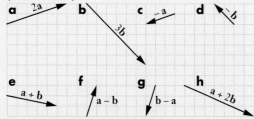

8 a 5.8 units in direction 028°

b 6.9 units in direction 105°

9 a 65 units in direction 151°

b 91 units in direction 100°

10 1.9 m/s² in direction 235°

11 4.6 m/s² in direction 238°

12 a $\lambda = 0, \mu = 0$ **b** $\lambda = 0, \mu = 0$

    c $\lambda = 3, \mu = -4$ **d** $\lambda = 2, \mu = 5$

    e $\lambda = 5, \mu = -2$ **f** $\lambda = 1, \mu = 3$

    g $\lambda = 2, \mu = -1$ **h** $\lambda = 3, \mu = -2$

    i $\lambda = 1, \mu = -3$ **j** $\lambda = 4, \mu = -2$

13 a a   **b** −a   **c** c

    d −c   **e** $\dfrac{1}{2}c$ **f** $c + \dfrac{1}{2}a$

    g $a + \dfrac{1}{2}c$ **h** $\dfrac{1}{2}c - \dfrac{1}{2}a$

14 a $b - a$ **b** $\dfrac{3}{4}(b - a)$ **c** $\dfrac{1}{4}(b - a)$

    d $\dfrac{1}{4}a + \dfrac{3}{4}b$

15 a $a + b$ **b** $\dfrac{1}{3}b$ **c** $\dfrac{1}{2}a$

    d $a + \dfrac{1}{3}b$ **e** $b + \dfrac{1}{2}a$ **f** $b - \dfrac{1}{2}a$

    g $a - \dfrac{2}{3}b$ **h** $\dfrac{2}{3}b - \dfrac{1}{2}a$

16 a $a + b$ **b** $2b$ **c** $b - a$

    d $\dfrac{1}{2}(b - a)$ **e** $\dfrac{1}{2}a + \dfrac{3}{2}b$

17 a $\dfrac{1}{2}a$ **b** $b - a$ **c** $\dfrac{2}{3}(b - a)$

    d $\dfrac{2}{3}b - \dfrac{1}{6}a$ **e** $h = 3, k = 2$

18 $h = \dfrac{3}{2}, k = \dfrac{5}{4}$

Miscellaneous exercise three   **PAGE 70**

1 There are 64 different settings for the system.

2 337°, 3.4 km   **3** 495   **4** 25

5 Converse: If you attend XYZ high school then you are in my Specialist Mathematics class.   False.

    Contrapositive: If you do not attend XYZ high school then you are not in my Specialist Mathematics class.   True.

6 259 459 200

7 a 18.1 units in direction 121°

    b 12.7 units in direction 222°

    c 29.0 units in direction 109°

8 a $h = 0, k = 0$ **b** $h = 0, k = 1$ **c** $h = 3, k = -1$

    d $h = -5, k = 0$ **e** $h = 1, k = -2$ **f** $h = 4, k = -1$

9 a 1800   **b** 252   **c** 3312   **d** 1056

Exercise 4A   **PAGE 78**

Note: In this and future vector exercises the choice as to whether answers are presented as $a\mathbf{i} + b\mathbf{j}$,

    $<a, b>$ or $\begin{pmatrix} a \\ b \end{pmatrix}$ is determined by the notation

    used in the question.

1 14.3 N, 334°   **2** 13.2 m/s, 074°

3 10.5 units, 142°   **4** 15.7 N, 318°

5 a $= 3\mathbf{i} + 2\mathbf{j}$ **b** $= 3\mathbf{i} + \mathbf{j}$ **c** $= 2\mathbf{i} + 2\mathbf{j}$

    d $= -\mathbf{i} + 3\mathbf{j}$ **e** $= 2\mathbf{j}$ **f** $= -\mathbf{i} + 2\mathbf{j}$

    g $= \mathbf{i} - 2\mathbf{j}$ **h** $= 4\mathbf{i}$ **k** $= 2\mathbf{i} - 4\mathbf{j}$

    l $= 4\mathbf{i} - \mathbf{j}$ **m** $= -4\mathbf{i} - \mathbf{j}$ **n** $= 9\mathbf{i} + 2\mathbf{j}$

6 $|\mathbf{a}| = \sqrt{13}$ units $|\mathbf{b}| = \sqrt{10}$ units

    $|\mathbf{c}| = 2\sqrt{2}$ units $|\mathbf{d}| = \sqrt{10}$ units

    $|\mathbf{e}| = 2$ units $|\mathbf{f}| = \sqrt{5}$ units

    $|\mathbf{g}| = \sqrt{5}$ units $|\mathbf{h}| = 4$ units

    $|\mathbf{k}| = 2\sqrt{5}$ units $|\mathbf{l}| = \sqrt{17}$ units

    $|\mathbf{m}| = \sqrt{17}$ units $|\mathbf{n}| = \sqrt{85}$ units

7 25 Newtons

8 a $(4.3\mathbf{i} + 2.5\mathbf{j})$ units   **b** $(3.5\mathbf{i} + 6.1\mathbf{j})$ units

    c $(9.1\mathbf{i} + 4.2\mathbf{j})$ units   **d** $(5.4\mathbf{i} + 4.5\mathbf{j})$ N

    e $(-4\mathbf{i} + 6.9\mathbf{j})$ m/s   **f** $(9.4\mathbf{i} - 3.4\mathbf{j})$ N

    g $(-2.6\mathbf{i} + 3.1\mathbf{j})$ units   **h** $(7.3\mathbf{i} - 3.3\mathbf{j})$ units

    i $(-4.6\mathbf{i} - 3.9\mathbf{j})$ units   **j** $(-6.4\mathbf{i} + 7.7\mathbf{j})$ m/s

    k $(-7.3\mathbf{i} - 3.4\mathbf{j})$ N   **l** $(4.1\mathbf{i} + 2.9\mathbf{j})$ m/s

---

9 a 5 units, 53.1°   **b** $\sqrt{29}$ units, 21.8°

c $\sqrt{13}$ units, 123.7°   **d** 5 units, 53.1°

e $\sqrt{41}$ units, 38.7°   **f** $4\sqrt{2}$ units, 45°

10 a −330 km/h (To nearest 10 km/h.)

b 120 km/h (To nearest 10 km/h.)

11 $\sqrt{89}$ units in direction 328°

12 a $3i + 7j$ **b** $i - j$

c $-i + j$ **d** $4i + 6j$

e $3i + 12j$ **f** $7i + 18j$

g $i - 6j$ **h** $-i + 6j$

i $\sqrt{13}$ units   **j** $\sqrt{17}$ units

k $\sqrt{13} + \sqrt{17}$ units   **l** $\sqrt{58}$ units

13 a $4i - j$ **b** $-i - 2j$

c $i + 2j$ **d** $5i - 5j$

e $7i - 4j$ **f** $9i - 3j$

g $12i + 3j$ **h** $-3j$

i 3 units   **j** $\sqrt{2} + \sqrt{5}(\approx 3.65)$ units

k 3 units   **l** $\sqrt{5}$ units

14 a <7, 1>   **b** <3, 7>   **c** <10, 8>

d <17, 9>   **e** <−1, −10>   **f** $\sqrt{41}$ units

g $5\sqrt{2}$ units   **h** $\sqrt{41} + \sqrt{13}$ units

15 a $\begin{pmatrix} 2 \\ 4 \end{pmatrix}$ **b** $\begin{pmatrix} 4 \\ 4 \end{pmatrix}$ **c** $\begin{pmatrix} -4 \\ -4 \end{pmatrix}$

d $\begin{pmatrix} 5 \\ 8 \end{pmatrix}$ **e** $\begin{pmatrix} 1 \\ 4 \end{pmatrix}$ **f** $\begin{pmatrix} 5 \\ 4 \end{pmatrix}$

g $\sqrt{41}$ units   **h** $\sqrt{41}$ units

16 a $\sqrt{53}$ units   **b** $\sqrt{13}$ units   **c** $2\sqrt{53}$ units

d 10 units   **e** $4\sqrt{2}$ units

17 a ~3760 N   **b** ~1370 N

18 $(17.7i + 9.2j)$ N   **19** $(2.3i + 9.2j)$ N

20 $(9.2i + 8.6j)$ N   **21** $(5.9i + 3.5j)$ m/s

22 $(16.2i + 3.9j)$ N   **23** $(10.3i + 1.1j)$ N

24 $2\sqrt{17}$ N   **25** $a = 2i - 3j, b = i + 4j$

26 $c = -2i + 11j, d = 3i - 16j$

Exercise 4B   PAGE 85

1 For vector **a**   **i** $4i + 3j$ **ii** $8i + 6j$

iii $\frac{4}{5}i + \frac{3}{5}j$ **iv** $\frac{8}{5}i + \frac{6}{5}j$

For vector **b**   **i** $4i - 3j$ **ii** $8i - 6j$

iii $\frac{4}{5}i - \frac{3}{5}j$ **iv** $\frac{8}{5}i - \frac{6}{5}j$

For vector **c**   **i** $2i + 2j$ **ii** $4i + 4j$

iii $\frac{1}{\sqrt{2}}i + \frac{1}{\sqrt{2}}j$ **iv** $\sqrt{2}i + \sqrt{2}j$

For vector **d**   **i** $3i - 2j$ **ii** $6i - 4j$

iii $\frac{3}{\sqrt{13}}i - \frac{2}{\sqrt{13}}j$ **iv** $\frac{6}{\sqrt{13}}i - \frac{4}{\sqrt{13}}j$

2 a $\frac{2}{\sqrt{5}}i + \frac{1}{\sqrt{5}}j$ **b** $2\sqrt{5}i + \sqrt{5}j$

c $-\frac{3\sqrt{13}}{5}i + \frac{4\sqrt{13}}{5}j$ **d** $\frac{10}{\sqrt{13}}i + \frac{15}{\sqrt{13}}j$

3 a **a** and **d**   **b** $12i - 14j$

c $2\sqrt{85}$ units   **d** 139°

4 $w = -4, x = 0.75, y = \pm\frac{\sqrt{3}}{2}, z = -9$ or 15.

5 $a = 0.8, b = 3, c = -4, d = 5, e = -12, f = \frac{25}{13}, g = -\frac{60}{13}$.

6 21.3 units, $-9.9i - 18.9j$

7 7.8, 50°   **8** 11.9, 31°   **9** 9.5, 36°

10 $T_1 = T_2 = \frac{100}{\sqrt{3}}$ N   **11** $T_1 = T_2 = 100$ N

12 $T_1 = 50\sqrt{3}$ N, $T_2 = 50$ N

13 Particle A is moving the fastest.

14 The body will move approximately 323 metres.

15 a $75j$ m/s. Appoximately 67 minutes.

b $(-21i + 72j)$ m/s. Approximately 61 minutes.

16 $(-21i - 72j)$ m/s. Approximately 81 minutes.

17 a $= -p$ **b** $= 2p$ **c** $= p + q$

d $= p + 2q$ **e** $= q - p$ **f** $= 2p + q$

g $= 3p + 3q$ **h** $= 3p + 2q$ **k** $= q - 3p$

l $= 2p - 2q$ **m** $= q - 2p$

18 $(5i - 5\sqrt{3}j)$ N

19 a $a + b$ **b** $2a + b$ **c** $2a - 3b$

d $\frac{11}{5}a - \frac{2}{5}b$ **e** $\frac{2}{5}a + \frac{11}{5}b$ **f** $2a - b$

20 a $(-112i + 384j)$ km/h

b $(140.8i - 374.4j)$ km/h

Exercise 4C   PAGE 90

1 $2i + 5j$, $-3i + 6j$, $0i - 5j$, $3i + 8j$.

2 a $-i - 2j$ **b** $i + 2j$

3 a $3i - 7j$ **b** $-i + 8j$ **c** $-2i - j$

4 a $3i - 4j$ **b** $-5i + 13j$ **c** $7i - 24j$
  d $15i - 20j$

5 a $\sqrt{58}$ units   **b** $\sqrt{5}$ units   **c** $\sqrt{61}$ units

6 a 5 units   **b** 5 units   **c** $\sqrt{17}$ units
  d $2\sqrt{17}$ units

7 a $\sqrt{37}$ units   **b** $\sqrt{34}$ units   **c** $3\sqrt{5}$ units

8 a $-i + 5j$ **b** $8i + 19j$ **c** $-3i - 23j$
  d 25 units   **e** $i + 2j$ **f** $16i + 45j$

9 $10i + 3j$

10 a $i + 10j$ **b** $3i + 4j$ **c** $2i - 6j$

11 a $3i + 5j$ **b** $4i - 3j$ **c** $i - 8j$
  d 13 units

12 a $(4i + 4j)$ m   **b** $(6i - j)$ m
  c $(22i - 41j)$ m 20 metres

13 a i $(7i + 6j)$ m   **ii** $(8i + 12j)$ m
  iii $(12i + 36j)$ m
  b 26 m   **c** 9

16 $10i + 9j$ **17** $2i + j$ **18** $4.6i + 6.8j$

Miscellaneous exercise four   PAGE 92

1 $\lambda = \dfrac{11}{17}, \mu = \dfrac{4}{17}$

2 150

3 Converse: If a positive whole number is a multiple of five then the number ends in a five.   False.

Contrapositive: If a positive whole number is not a multiple of five then the number does not end in a five.   True.

4 There are $95\,040$ ($= 12 \times 11 \times 10 \times 9 \times 8$, or $^{12}C_5 \times 5!$) possible different ordered lists.

5 $\triangle ABC \cong \triangle XWV$ (SAS), $\triangle GHI \cong \triangle BDC$ (SSS),
$\triangle MNO \cong \triangle TUS$ (RHS), $\triangle PQR \cong \triangle YZA$ (AA corres S).

6 $a = 0, b = 15$

7 a 144 **b** 3250

8 a 2.4 seconds   **b** 1.08 metres

Exercise 5A   PAGE 100

15 $x = 3$ $y = 13$

Exercise 5B   PAGE 105

11 $x = 12$ $y = 10$

12 $x = 30$ $y = 5$

Miscellaneous exercise five   PAGE 109

1 $1287, 40\,320$ **2** 45

3 $70, 30$ **4** $166°$

5 a $(-3400i - 9400j)$ N   **b** 3400 N

6 Compare your answer with those of others in your class.

7 a $2b$ **b** $\dfrac{4}{3}b$ **c** $a + b$ **d** $a + 2b$
  e $\dfrac{a + 3b}{2}$ **f** $\dfrac{3a - b}{6}, h = \dfrac{3}{7}, k = \dfrac{2}{7}$.

Exercise 6A   PAGE 117

1 a $2i + 3j$ **b** $4i + 5j$ **c** $i + 4j$
  d $-2i - 2j$ **e** $i - j$ **f** $-2i + 2j$
  g $2i - 2j$ **h** $3i + j$ **i** $-3i + 3j$
  j $3i - 3j$

2

3 $(-5i + 3j)$ km

4 B relative to A

5 $\sqrt{386}$ km

6 $\sqrt{10}$ km

7 $5i - j$

Exercise 6B   PAGE 122

1 $6i - 10j$ **2** $-3i + 3j$ **3** $-i - 9j$ **4** $3i + 3j$

5 $20\sqrt{3}$ km/h in direction $060°$.

6 10 km/h in direction $037°$.

7 15.8 km/h in direction $318°$.

8 28.8 km/h in direction $075°$.

9 11.8 km/h in direction $343°$.

10 17.3 km/h in direction $308°$.

11 a $(5i - 30j)$ km/h   **b** $(-5i + 30j)$ km/h

12 a 26.1 km/h in direction $253°$.
  b 26.1 km/h in direction $073°$.

13 174 km/h in direction $287°$

14 $i - 12j$ **15** $-2i + 6j$

17 10 km/h due South.   **18** 20 km/h due South.

19 160 km/h, $076°$ **20** $(13i + j)$ km/h

21 $(4\mathbf{i} - 2\mathbf{j})$ km/h   **22** 17.9 km/h from 279°.

23 $\sqrt{34}$ km/h from 211°.

24 B: 10 km/h due North.   C: 7 km/h due North.
D: 15 km/h due North.

25 Approximately 10 km/h from 208°.

26 F: at rest, G: 19.1 km/h, 030°, H: 31.1 km/h, 328°.

27 $6\mathbf{i} + 8\mathbf{j}$

28 14.8 km/h from N27°W.

29 6.2 km/h from S44°W.

Miscellaneous exercise six   PAGE 124

1 $\mathbf{p} = 3\mathbf{i} + 3\sqrt{3}\mathbf{j}$, $\mathbf{q} = -8\mathbf{i} + 8\mathbf{j}$,
    $\mathbf{r} = -5\mathbf{i} + 5\sqrt{3}\mathbf{j}$, $\mathbf{s} = 4\sqrt{3}\mathbf{i} - 4\mathbf{j}$.

2 $x = 5, y = \pm 1$

3 **a** 5040   **b** 720   **c** 120

4 Compare your response with that of others in
your class.

5 Compare your proof with that of others in your class.

6 **a** R is 10 units from Q.
    b R has position vector $5\mathbf{i} + 4\mathbf{j}$.
    c R is $\sqrt{41}$ units from the origin.

7 The contrapositive, if not Q then not P, must also
be true.

The other two statements, the converse and the
inverse, could be true or false.

8 13 300, 9 310   **9** 462, 194

10 **a** 56
    b Option I: 6, Option II: 20, Option III: 50.

Exercise 7A   PAGE 130

2 $\frac{1}{2}\mathbf{b}$, $\frac{1}{2}(\mathbf{c}-\mathbf{a})$, $\frac{1}{2}\mathbf{b}$, $\frac{1}{2}(\mathbf{c}-\mathbf{a})$.

3 $\frac{1}{2}(\mathbf{a}-\mathbf{c})$, $\mathbf{a}-\mathbf{c}$.

5 **b** The diagonals of a quadrilateral bisect each
other $\Leftrightarrow$ the quadrilateral is a parallelogram.

The diagonals of a quadrilateral bisect each other
if and only if the quadrilateral is a parallelogram.

6 **a** **i** $\mathbf{b}-\mathbf{a}$ **ii** $\frac{1}{2}(\mathbf{b}-2\mathbf{a})$

       iii $\frac{1}{2}(\mathbf{b}-\mathbf{a})$ **iv** $\frac{1}{2}(\mathbf{a}+\mathbf{b})$

       v $\frac{1}{3}(\mathbf{a}+\mathbf{b})$ **vi** $\frac{1}{3}(\mathbf{b}-2\mathbf{a})$

8 $h = \frac{2}{3}, k = \frac{2}{3}, \lambda = \frac{2}{3}$

10 **a** $\frac{1}{2}m\mathbf{a} - h\mathbf{a} + \frac{1}{2}m\mathbf{b}$ **b** $k\mathbf{b} - \frac{1}{2}m\mathbf{b} - \frac{1}{2}m\mathbf{a}$

Miscellaneous exercise seven   PAGE 133

1 **a** Now triangle is scalene $\Rightarrow$ triangle has three
different length sides,

and triangle has three different length sides
$\Rightarrow$ triangle is scalene.

Hence triangle is scalene $\Leftrightarrow$ triangle has three
different length sides.

Thus 'triangle has three different length sides'
and 'triangle is scalene' are equivalent statements
and so the 'if and only if' phrase can be used.
Hence given statement is correct.

    b Whilst it is true that if a positive whole number
ends with a 0 then it is a multiple of five, the
converse is false, because a multiple of 5 does
not have to end with a 0. Hence this is not an
'if and only if' situation. The given statement
is incorrect.

2 336

3 **a** 5005   **b** 720

4 73 256 400

5 Compare your proof with those of others in
your class.

6 **a** $5.8\mathbf{i} + 0.2\mathbf{j}$ **b** $9\mathbf{i} - 3\mathbf{j}$

7 Two seconds later the object is $6\sqrt{5}$ metres from
the origin.

8 $(\mathbf{a} + \mathbf{b})$ has magnitude $4\sqrt{5}$ units.

9 **a** 1 h 50 mins   **b** 2 h 13 mins

10 **a** $2\mathbf{b} - \mathbf{a}$ **b** $\frac{2}{3}\mathbf{b} - \frac{1}{3}\mathbf{a}$

     c $\frac{5}{3}\mathbf{b} + \frac{2}{3}\mathbf{a}$. $h = 1.5, k = 0.5$.

11 **a** 358 800   **b** 456 976   **c** 165 765 600
    d 308 915 776, 6 (including A and R to start with).

Exercise 8A   PAGE 140

1 $\frac{15\sqrt{3}}{2}$ **2** 0   **3** −6   **4** 10

5 $-5\sqrt{3}$ **6** 0   **7** 0   **8** 0

9 4   **10** 9   **11** $6\sqrt{2}$ **12** $6\sqrt{2}$

13 12   **14** −24   **15** 3   **16** 0

17 $-35\sqrt{3}$ **18** $-100\sqrt{2}$

19 a scalar   **b** scalar   **c** vector
   d vector   **e** vector   **f** scalar
   g scalar   **h** scalar   **i** vector
   j scalar

20 a 1   **b** 0   **c** 1

21 a $a^2 - b^2$ **b** $a^2 + 2\mathbf{a.b} + b^2$
   c $a^2 - 2\mathbf{a.b} + b^2$ **d** $4a^2 - b^2$
   e $a^2 + \mathbf{a.b} - 6b^2$ **f** a^2

23 b, d   **24** a, b, d   **25** $x_1 x_2 + y_1 y_2$

26 2.8   **27** 0.72   **28** c

29 a 62°   **b** 25   **c** 9
   d 20   **e** $2\sqrt{5}$

30 100

31 We can determine the scalar product of two vectors but not of a vector and a scalar.

Exercise 8B PAGE 145

1 a 3   **b** 3   **c** 8   **d** 4
2 a 7   **b** 14   **c** 14   **d** 18
3 a 8   **b** 48   **c** 22   **d** −12
4 a Not perpendicular   **b** Not perpendicular
   c Perpendicular   **d** Perpendicular
   e Not perpendicular   **f** Perpendicular
5 a 10   **b** −16   **c** 1   **d** −25
6 a 15   **b** 17   **c** 10   **d** −14
7 a −10   **b** 7   **c** $-3\mathbf{i}+\mathbf{j}$ −3
8 a 5   **b** 13   **c** −33, 121°
9 a $7\sqrt{2}$ **b** 17   **c** 49, 73°
12 a 24   **b** 16°
13 a 204   **b** 51°
14 a 4   **b** 60°
15 a 0   **b** 90°
16 a −75   **b** 180°
17 a 12   **b** 23°
18 $\lambda = 8, \mu = 10.5$
19 $w = 7, x = -5$
20 a $2\sqrt{5}$ **b** $4\mathbf{i}+2\mathbf{j}$
   c 2   **d** $1.2\mathbf{i}+1.6\mathbf{j}$
21 $\pm(20\mathbf{i}+15\mathbf{j})$ **22** $\pm\dfrac{1}{\sqrt{5}}(\mathbf{i}-2\mathbf{j})$
23 $5\mathbf{i}-2\mathbf{j}, -2\mathbf{i}-5\mathbf{j}$
24 a $4\mathbf{i}+2\mathbf{j}$ **b** $2\mathbf{i}-5\mathbf{j}$ **c** −2   **d** 95°
25 −8 and 2

Exercise 8C PAGE 148

1 a $\mathbf{b}-\mathbf{a}$
2 a 0   **b** $\mathbf{c}-\mathbf{a}, \mathbf{c}+\mathbf{a}$
4 a $\mathbf{c}-\mathbf{a}, \dfrac{1}{2}(\mathbf{a}+\mathbf{c})$ **5 a** $\mathbf{b}-\mathbf{c}, \mathbf{b}, \mathbf{b}+\mathbf{c}$

Miscellaneous exercise eight PAGE 150

1 If quadrilateral ABCD is a rhombus then it is a parallelogram but the converse is not true i.e. if quadrilateral ABCD is a parallelogram it does not have to be a rhombus. Hence the two way nature of the statement claimed by the use of the symbol '⇔' is not the case. The given statement is incorrect.

If PQRS is a rhombus then its diagonals will cut at right angles but the converse is not true i.e. if the diagonals of a quadrilateral cut at right angles the quadrilateral is not necessarily a rhombus, it could be a kite for example. Hence the two way nature of the statement claimed by the use of the symbol '⇔' is not the case. The given statement is incorrect.

If the diagonals of a parallelogram cut at right angles then the parallelogram is a rhombus. This was proved in question 3 of Exercise 8B. Also if a shape is a rhombus then its diagonals will cut at right angles. This was proved in example 7 of chapter 8. The given statement is correct.

2 a, b, d   **3** 7i   **4** 542 640
5 a $2\sqrt{29}$ units **b** $6\mathbf{i}+10\mathbf{j}$ **c** $\mathbf{i}+8\mathbf{j}$
6 a 125   **b** 60   **c** 24
   d 36   **e** 21
7 a 146°   **b** 4   **c** 9
   d 3   **e** $\sqrt{3}$
8 a 720 **b** 120 **c** 240 **d** 6
   e 36   **f** 6   **g** 144
9 a 339°   **b** 36 seconds
10 a 40 320   **b** 1440   **c** 384
11 1 : 3, 1 : 4
12 a 154 440   **b** 63 000
13 a $-6\mathbf{i}+\mathbf{j}$ **b** $5\mathbf{i}+5\mathbf{j}$ **c** −25 **d** 126°
14 $-\dfrac{14}{3}$ and 8   **15** 6, 13.5
16 $(8\mathbf{i}-3\mathbf{j})$ N, 54.2°   **17** $h = \dfrac{6}{7}, k = \dfrac{2}{7}$
18 a i 60 480   **ii** all of them
   b i 60 480   **ii** 55 440 of them
19 462
20 a $\dfrac{1}{2}\mathbf{c} - h\mathbf{c} - \dfrac{1}{2}\mathbf{a}$ **b** $\dfrac{1}{2}\mathbf{c} - k\mathbf{c} - \dfrac{1}{2}\mathbf{a}$

ANSWERS UNIT TWO

Exercise 9C PAGE 172

1 a $-\dfrac{24}{25}$ **b** $\dfrac{7}{25}$ **c** $-\dfrac{24}{7}$

2 a $\dfrac{120}{169}$ **b** $\dfrac{119}{169}$ **c** $\dfrac{120}{119}$

3 a $3\sin 2A$ **b** $2\sin 4A$ **c** $\dfrac{1}{2}\sin A$

4 a $2\cos 4A$ **b** $1\cos A$ **c** $1\cos 4A$

5 a $-\dfrac{336}{625}$ **b** $\dfrac{527}{625}$ **c** $-\dfrac{336}{527}$

6 $15°, 75°, 195°, 255°$

7 $-150°, -90°, -30°, 90°$

8 $0°, 75.5°, 180°, 284.5°, 360°$

9 $\dfrac{\pi}{8}, \dfrac{5\pi}{8}, \dfrac{9\pi}{8}, \dfrac{13\pi}{8}$ **10** $\dfrac{\pi}{3}, \dfrac{\pi}{2}, \dfrac{3\pi}{2}, \dfrac{5\pi}{3}$

11 $-\dfrac{5\pi}{6}, -\dfrac{\pi}{6}, \dfrac{\pi}{2}$ **12** $66.4°, 293.6°, 426.4°$

Exercise 9D PAGE 175

1 $5\cos(\theta + 53.1°)$ **2** $13\cos(\theta + 22.6°)$

3 $5\cos(\theta - 0.64)$ **4** $25\cos(\theta - 1.29)$

5 $13\sin(\theta + 67.4°)$ **6** $25\sin(\theta + 73.7°)$

7 $5\sin(\theta - 0.64)$ **8** $\sqrt{13}\sin(\theta - 0.98)$

10 a $\sqrt{2}\cos\left(\theta - \dfrac{\pi}{4}\right)$ **b** $\sqrt{2}, \dfrac{\pi}{4}$

11 $2.09, 6.05$ **12** $0.33, 1.88$

13 $1.36, 5.68$

Exercise 9E PAGE 178

1 $\dfrac{\pi}{3}, \dfrac{5\pi}{3}$

2 $\pm\dfrac{\pi}{3}, \pm\dfrac{2\pi}{3}$

3 $0°, 70.5°, 180°, 289.5°, 360°$

4 $\pm 60°$

5 $63.4°, 116.6°, 243.4°, 296.6°$

6 $\dfrac{5\pi}{12}, \dfrac{23\pi}{12}$

7 $0°, 120°, 240°, 360°$

8 $3.48, 5.94$

Exercise 9F PAGE 182

1 $\dfrac{1}{2}\cos 5x + \dfrac{1}{2}\cos x$ **2** $\dfrac{1}{2}\cos 2x - \dfrac{1}{2}\cos 4x$

3 $\dfrac{1}{2}\sin 8x + \dfrac{1}{2}\sin 6x$ **4** $\dfrac{1}{2}\sin 4x - \dfrac{1}{2}\sin 2x$

5 $2\cos 3x\cos 2x$ **6** $-2\sin 3x\sin 2x$

7 $2\sin 4x\cos 2x$ **8** $2\cos 4x\sin x$

9 $\dfrac{2+\sqrt{3}}{4}$ **10** $\dfrac{\sqrt{6}}{2}$

11 $12°, 24°, 84°, 96°, 156°, 168°$

12 $0, \dfrac{\pi}{6}, \dfrac{\pi}{5}, \dfrac{2\pi}{5}, \dfrac{3\pi}{5}, \dfrac{4\pi}{5}, \dfrac{5\pi}{6}, \pi$

13 $0°, 45°, 135°, 180°, 225°, 315°, 360°$

14 $-\pi, -\dfrac{5\pi}{6}, -\dfrac{\pi}{2}, -\dfrac{\pi}{6}, 0, \dfrac{\pi}{6}, \dfrac{\pi}{2}, \dfrac{5\pi}{6}, \pi$

There are often many ways of writing the answers to these questions but all correct versions will generate the same set of solutions for $n \in \mathbb{Z}$. For some of the questions 'common' alternatives are shown here.

1 $x = \begin{cases} 30° + n \times 360°, \\ 150° + n \times 360°, \end{cases}$ for $n \in \mathbb{Z}$.

2 $x = n \times 360°$, for $n \in \mathbb{Z}$.

3 $x = n \times 180° - 30°$, for $n \in \mathbb{Z}$.

Could be written as $x = n \times 180° + 150°$, for $n \in \mathbb{Z}$.

4 $x = n \times 180° + 30°$, for $n \in \mathbb{Z}$.

5 $x = \begin{cases} 35.2° + n \times 120°, \\ 4.8° + n \times 120°, \end{cases}$ for $n \in \mathbb{Z}$.

Could be written as $x = \begin{cases} 35.2° + n \times 120°, \\ 124.8° + n \times 120°, \end{cases}$ for $n \in \mathbb{Z}$.

6 $x = n \times 90° + 9.3°$, for $n \in \mathbb{Z}$.

7 $x = \begin{cases} \dfrac{7\pi}{12} + n\pi, \\ \dfrac{11\pi}{12} + n\pi, \end{cases}$ for $n \in \mathbb{Z}$.

Could be written as $x = \begin{cases} n\pi - \dfrac{\pi}{12}, \\ n\pi + \dfrac{7\pi}{12}, \end{cases}$ for $n \in \mathbb{Z}$.

8 $x = n\pi + \dfrac{\pi}{4}$, for $n \in \mathbb{Z}$.

9 $x = n\pi$, for $n \in \mathbb{Z}$.

10 $x = \begin{cases} \dfrac{2n\pi}{3} + \dfrac{\pi}{18}, \\ \dfrac{2n\pi}{3} + \dfrac{5\pi}{18}, \end{cases}$ for $n \in \mathbb{Z}$.

11 $x = \begin{cases} \dfrac{n\pi}{2} + 0.84, \\ \dfrac{n\pi}{2} + 1.16, \end{cases}$ for $n \in \mathbb{Z}$.

12 $x = n\pi \pm \dfrac{\pi}{6}$, for $n \in \mathbb{Z}$.

13 $x = \dfrac{n\pi}{3} + \dfrac{\pi}{4}$ for $n \in \mathbb{Z}$.

Could be written as $x = \begin{cases} \dfrac{2n\pi}{3} + \dfrac{\pi}{4}, \\ \dfrac{2n\pi}{3} - \dfrac{\pi}{12}, \end{cases}$ for $n \in \mathbb{Z}$.

14 $x = \begin{cases} \dfrac{8n}{3} + 0.44, \\ \dfrac{8n}{3} + 1.56, \end{cases}$ for $n \in \mathbb{Z}$.

1 a $y = 3 \sin x$ **b** $y = 4 \sin x$
   c $y = -3 \sin x$ **d** $y = -4 \sin x$

2 a $y = 3 \sin 2x$ **b** $y = 4 \sin \dfrac{2x}{3}$

   c $y = 4 \sin\left(\dfrac{2\pi}{5} x\right)$ **d** $y = -5 \sin\left(\dfrac{\pi}{3} x\right)$

3 a $y = 2 + 3 \sin x$ **b** $y = -2 - 4 \sin x$

4 a $y = 3 \sin\left(x - \dfrac{\pi}{2}\right)$ **b** $y = 4 \sin\left(x + \dfrac{\pi}{2}\right)$

5 a $y = 5 \sin\left(\dfrac{\pi}{4}(x - 2)\right)$ **b** $y = 4 \sin\left(\dfrac{\pi}{5}(x - 3)\right)$

6 a $y = 3 \sin\left(\dfrac{\pi}{4}(x - 1)\right) + 7$

   b $y = 2 \sin\left(\dfrac{\pi}{30}(x - 10)\right) + 7$

7 $h = 5 \sin\left(\dfrac{2\pi}{365} t\right) + 12$

8 a $d = -6 \cos\left(\dfrac{4\pi}{25} t\right) + 10$

   b $d = 6 \sin\left(\dfrac{4\pi}{25}\left(t - \dfrac{25}{8}\right)\right) + 10$

9 a $h = 3 \cos\left(\dfrac{\pi}{3}(t - 2)\right) + 6$

   b $h = 3 \sin\left(\dfrac{\pi}{3}\left(t - \dfrac{1}{2}\right)\right) + 6$

1 $\dfrac{\pi}{20}, \dfrac{3\pi}{20}, \dfrac{9\pi}{20}, \dfrac{11\pi}{20}, \dfrac{17\pi}{20}, \dfrac{19\pi}{20}$.

3 $30°, 120°, 210°, 300°$

---

4 $-\dfrac{\pi}{3}, 0, \dfrac{\pi}{3}$.

5 $x = \begin{cases} \dfrac{n\pi}{2} + \dfrac{\pi}{12}, \\ \dfrac{n\pi}{2} + \dfrac{\pi}{6}, \end{cases}$ for $n \in \mathbb{Z}$.

6 a $\sqrt{149}\,\sin(\theta - 0.96)$ **b** $-\sqrt{149}, 5.67$

7 a 'Eye-balling' the graph certainly suggests that a sinusoidal model could well be appropriate.

Taking the high of 27.2 and the low of 17.0 suggest an amplitude of 5.1.

Hence $a = 5.1$ and $d = 22.1$.

With a period of 12 units we have $\dfrac{2\pi}{b} = 12$.

Hence $b = \dfrac{\pi}{6}$.

Thus $T = 5.1 \sin\left(\dfrac{\pi}{6}(x \pm ?)\right) + 22.1$.

The typical 'start' of '$y = a \sin x + b$' seems to have been moved right 10 units (or left 2 units).

Thus $T = 5.1 \sin\left(\dfrac{\pi}{6}(x - 10)\right) + 22.1$.

(Or: $T = 5.1 \sin\left(\dfrac{\pi}{6}(x + 2)\right) + 22.1$).

Exercise 10A   PAGE 199

1 $A_{4\times 2}, B_{2\times 4}, C_{4\times 1}, D_{4\times 3}, E_{2\times 2}, F_{1\times 3}, G_{3\times 2}, H_{4\times 4}$

2 a 4   **b** -4 **c** 7

    d 7   **e** 3   **f** 0

3 a Cannot be determined   **b** $\begin{bmatrix} 3 & -1 \\ 1 & -9 \end{bmatrix}$

    c $\begin{bmatrix} 1 & -5 \\ 1 & -1 \end{bmatrix}$ **d** $\begin{bmatrix} 6 \\ 2 \\ -4 \end{bmatrix}$

    e $\begin{bmatrix} 9 & -3 \\ 6 & 12 \\ 0 & 9 \end{bmatrix}$ **f** Cannot be determined

    g $\begin{bmatrix} 2 & 4 \\ 0 & -8 \end{bmatrix}$ **h** $\begin{bmatrix} 0 & 7 \\ -1 & -3 \end{bmatrix}$

4 a $\begin{bmatrix} 5 & 3 & -1 \\ 1 & 3 & 3 \end{bmatrix}$ **b** $\begin{bmatrix} -1 & -1 & 1 \\ -1 & -5 & -3 \end{bmatrix}$

    c $\begin{bmatrix} 3 & 6 & 3 \\ 6 & 3 & 6 \end{bmatrix}$ **d** $\begin{bmatrix} 5 & 4 & -3 \\ 3 & 14 & 9 \end{bmatrix}$

5 a Cannot be determined   **b** $\begin{bmatrix} 6 & 12 \\ 3 & 9 \end{bmatrix}$

    c $\begin{bmatrix} 8 & 3 & 11 \end{bmatrix}$ **d** Cannot be determined

6 a Cannot be determined   **b** $\begin{bmatrix} 6 & 4 & 3 & 0 \\ 2 & 2 & 6 & 6 \\ 1 & 5 & 3 & 4 \end{bmatrix}$

    c $\begin{bmatrix} 6 & 2 & 8 \\ 4 & 2 & -6 \\ 0 & 2 & 4 \\ 2 & 0 & 0 \end{bmatrix}$ **d** $\begin{bmatrix} 0 & 14 & -3 & 6 \\ -2 & 4 & 6 & 12 \\ -1 & -5 & 3 & 20 \end{bmatrix}$

7 a No   **b** No   **c** Yes   **d** Yes

    e Yes   **f** No   **g** Yes   **h** No

8 Yes

9 Yes

10 $\begin{bmatrix} 1 & 2 & -3 \\ 1 & 0 & -2 \end{bmatrix}$

11 a

	P	A	B
Alan	40	20	4
Bob	37	15	14
Dave	47	19	9
Mark	39	21	3
Roger	39	19	16

    b

	P	A	B
Alan	10	5	1
Bob	9.25	3.75	3.5
Dave	11.75	4.75	2.25
Mark	9.75	5.25	0.75
Roger	9.75	4.75	4

12

	B	F	FL	G	GG
Centre I	6160	1925	2552	1947	4675
Centre II	3124	1397	1507	1122	2992
Centre III	5555	1617	3102	1408	2970
Centre IV	2409	1034	1672	924	1958

13 $\begin{bmatrix} 3 & 4 & 5 \\ 5 & 6 & 7 \\ 7 & 8 & 9 \end{bmatrix}$

14 $\begin{bmatrix} 1 & 1 & 1 & 1 \\ 2 & 4 & 8 & 16 \\ 3 & 9 & 27 & 81 \end{bmatrix}$

1 $\begin{bmatrix} 4 & 9 \end{bmatrix}$

2 Cannot be determined. Number of columns in 1st matrix ≠ number of rows in 2nd matrix.

3 $\begin{bmatrix} 2 & 10 \\ 1 & 4 \end{bmatrix}$ **4** $[7]$ **5** $\begin{bmatrix} 3 & 1 \\ 12 & 4 \end{bmatrix}$

6 $\begin{bmatrix} 13 & -4 \\ -14 & 7 \end{bmatrix}$ **7** $\begin{bmatrix} 2 & 3 \\ 1 & -1 \end{bmatrix}$ **8** $\begin{bmatrix} 1 & 4 \\ -1 & 3 \end{bmatrix}$

9 $\begin{bmatrix} 0 & 0 \\ 0 & 0 \end{bmatrix}$ **10** $\begin{bmatrix} 1 & 0 \\ 0 & 1 \end{bmatrix}$ **11** $\begin{bmatrix} 1 & 0 \\ 0 & 1 \end{bmatrix}$

12 $\begin{bmatrix} 1 & 0 \\ 0 & 1 \end{bmatrix}$ **13** $[8]$

14 $\begin{bmatrix} 3 & 2 & 3 \\ 4 & 3 & 1 \end{bmatrix}$ **15** $\begin{bmatrix} 1 & 0 & 5 \\ 10 & 2 & -2 \\ 6 & 1 & 4 \end{bmatrix}$

16 $\begin{bmatrix} 10 & 3 \\ 9 & 10 \end{bmatrix}$ **17** $\begin{bmatrix} 14 \\ 32 \end{bmatrix}$

18 $\begin{bmatrix} 2 & 4 & 1 \\ 5 & 7 & 18 \\ 12 & 8 & 22 \end{bmatrix}$

19 a $\begin{bmatrix} 0 & 2 & 1 \\ 0 & 1 & 5 \\ 2 & 0 & 1 \end{bmatrix}$ **b** $\begin{bmatrix} 2 & 2 & 3 \\ 4 & 0 & -1 \\ -2 & 1 & 0 \end{bmatrix}$

   c $\begin{bmatrix} 1 & -1 & -2 \\ 2 & 1 & -1 \\ 2 & 1 & 2 \end{bmatrix}$ **d** $\begin{bmatrix} 2 & -1 & 2 \\ 2 & 3 & 4 \\ -2 & -2 & 1 \end{bmatrix}$

20 No. Justify by showing example for which AB ≠ BA.

24 a Cannot be formed   **b** Cannot be formed
   c 3×3 **d** 2×2
   e Cannot be formed   **f** 1×2
   g 3×2 **h** 1×3

25 a Yes   **b** Yes   **c** Yes   **d** No
   e No   **f** No   **g** No   **h** Yes

26 Matrix A must be a square matrix.

27 AA, AC, BA, CB.

28 a $\begin{bmatrix} -1 & -2 \\ 4 & 0 \end{bmatrix}$ **b** $\begin{bmatrix} 2 & -2 \\ 7 & -3 \end{bmatrix}$

29 a 1st B, 2nd E, 3rd C, 4th D, 5th A.
   b 1st = B & C, 3rd E, 4th D, 5th A.

30 Initially: $\begin{matrix} \text{Client1} \\ \text{Client2} \\ \text{Client3} \end{matrix} \begin{bmatrix} \$15\,000 \\ \$15\,000 \\ \$15\,000 \end{bmatrix}$

Two years later: $\begin{matrix} \text{Client1} \\ \text{Client2} \\ \text{Client3} \end{matrix} \begin{bmatrix} \$17\,700 \\ \$19\,300 \\ \$18\,800 \end{bmatrix}$

31 $\begin{matrix} \text{Drink (mL)} & \text{Burgers} \\ \begin{bmatrix} 18\,125 & 55 \end{bmatrix} \end{matrix}$

32 a QP
   b $\begin{matrix} \text{Hotel A} & \text{Hotel B} & \text{Hotel C} \\ \begin{bmatrix} \$4610 & \$3680 & \$2665 \end{bmatrix} \end{matrix}$
   Displays total nightly tariff for each hotel when full.
   c Row 1 column 1 of PR would be
   Single rooms in A × Single room tariff +
   Single rooms in B × Double room tariff +
   Single rooms in C × Suite tariff
   Thus PR not giving useful information.

33 a $\begin{bmatrix} 3 & 1 & 2 \end{bmatrix}$
   b $\begin{matrix} \text{Poles} & \text{Decking} & \text{Framing} & \text{Sheeting} \\ \begin{bmatrix} 25 & 205 & 145 & 320 \end{bmatrix} \end{matrix}$

   Matrix shows number of metres of each size of timber required to complete order.

   c $\begin{bmatrix} \$4 \\ \$2 \\ \$3 \\ \$1.50 \end{bmatrix}$ Product will have dimensions 3×1. Matrix will display the total cost of timber for each type of cubby.

34 a $E = \begin{matrix} \text{A} & \text{B} & \text{C} \\ \begin{bmatrix} 800 & 50 & 1000 \end{bmatrix} \end{matrix}$
   b $\begin{matrix} \text{Model I} & \text{Model II} & \text{Model III} & \text{Model IV} \\ \begin{bmatrix} 4600 & 4900 & 6300 & 5600 \end{bmatrix} \end{matrix}$

   Matrix displays the total cost of commodities, in dollars, for each model type.

35 a RP
   b $\begin{bmatrix} 6700 & 7200 & 2300 \end{bmatrix}$
   c Matrix shows the number of minutes required for cutting (6700 minutes) assembling (7200 minutes) and packing (2300 minutes) to complete the order.

1 -2 **2** 10 **3** 7 **4** 7
5 5 **6** 0 **7** $-x^2$ **8** $x^2 - y^2$

9 $\begin{bmatrix} 1 & -1 \\ -1 & 2 \end{bmatrix}$ **10** $\begin{bmatrix} 3 & -2 \\ -4 & 3 \end{bmatrix}$

11 $\dfrac{1}{3}\begin{bmatrix} 1 & -1 \\ 1 & 2 \end{bmatrix}$

12 $\dfrac{1}{5}\begin{bmatrix} 2 & -3 \\ -1 & 4 \end{bmatrix}$

13 $\dfrac{1}{10}\begin{bmatrix} 3 & 1 \\ -1 & 3 \end{bmatrix}$

14 $\dfrac{1}{10}\begin{bmatrix} -3 & -1 \\ 1 & -3 \end{bmatrix}$

15 Singular

16 Singular

17 Singular

18 $\dfrac{1}{x}\begin{bmatrix} 1 & -y \\ 0 & x \end{bmatrix}, x \neq 0$

19 $\begin{bmatrix} -1 & 0 \\ 0 & -1 \end{bmatrix}$

20 $\begin{bmatrix} 1 & 0 \\ 0 & -1 \end{bmatrix}$

21 a True **b** True **c** Not necessarily
d True **e** True **f** True
g True **h** True **i** Not necessarily
j Not necessarily

22 $\begin{bmatrix} -1 \\ 4 \end{bmatrix}$

23 $\begin{bmatrix} 2 \\ -1 \end{bmatrix}$

24 $\begin{bmatrix} -1 \\ -2 \end{bmatrix}$

25 $\begin{bmatrix} 3 \\ -1 \end{bmatrix}$

26 7

27 5

28 a $\begin{bmatrix} -13 & 4 \\ 12 & -4 \end{bmatrix}$ **b** 10

c $\begin{bmatrix} 1 & 1 \\ 2 & 3 \end{bmatrix}$ **d** $\dfrac{1}{10}\begin{bmatrix} 5 & 5 \\ 14 & 16 \end{bmatrix}$

e $\begin{bmatrix} 4 \\ 3 \end{bmatrix}$ **f** $\begin{bmatrix} 6 & 1 \\ -4 & 1 \end{bmatrix}$

29 a $\begin{bmatrix} 2 & 1 \\ 3 & 2 \end{bmatrix}$ **b** $\begin{bmatrix} -1 & -1 \\ 3 & 4 \end{bmatrix}$

c $\dfrac{1}{6}\begin{bmatrix} 1 & -2 \\ 0 & 6 \end{bmatrix}$ **d** $\begin{bmatrix} 2 & -1 \\ -3 & 2 \end{bmatrix}$

e $\begin{bmatrix} 6 & -2 \\ -3 & 2 \end{bmatrix}$

30 $\begin{bmatrix} 2 & -1 \\ 17 & -9 \end{bmatrix}$

31 $\begin{bmatrix} 2 \\ -1 \end{bmatrix}$

32 a 8 **b** ± 4 **c** -4 or 5

33 $F = \begin{bmatrix} 1 & 3 \\ 2 & -3 \end{bmatrix}, G = \begin{bmatrix} 3 & -4 \\ 0 & 2 \end{bmatrix}$

34 $\begin{bmatrix} 1 & 0 & -1 \\ 2 & 1 & 3 \\ 1 & 4 & 1 \end{bmatrix}$

35 $\begin{bmatrix} 1 & 2 & -1 \\ 0 & 1 & 3 \\ 3 & 1 & 0 \end{bmatrix}$

36 a $\begin{bmatrix} \$24 & \$56 \\ \$16 & \$36 \end{bmatrix}$ **b** $\begin{bmatrix} 0.8 & 0 \\ 0 & 1 \end{bmatrix}$

37 $\begin{bmatrix} 2 & 1 \\ 3 & -1 \end{bmatrix}$

38 $\begin{bmatrix} 11 & 20 \\ 5 & 7 \end{bmatrix}$

39 $\begin{bmatrix} -1 & 0 \\ -15 & 2 \end{bmatrix}$

40 a $\begin{bmatrix} 6 & 5 \\ 8 & 7 \end{bmatrix}$

b $BA = \begin{bmatrix} 860 & 740 \end{bmatrix}$

i.e. $\begin{bmatrix} x & y \end{bmatrix}\begin{bmatrix} 6 & 5 \\ 8 & 7 \end{bmatrix} = \begin{bmatrix} 860 & 740 \end{bmatrix}$

c $\begin{bmatrix} x & y \end{bmatrix} = \begin{bmatrix} 860 & 740 \end{bmatrix}A^{-1}$
giving $x = 50$ and $y = 70$.

Exercise 10D PAGE 220

1 $\begin{bmatrix} 2 & 3 \\ 1 & -3 \end{bmatrix}\begin{bmatrix} x \\ y \end{bmatrix} = \begin{bmatrix} 5 \\ 0 \end{bmatrix}$

2 $\begin{bmatrix} -1 & 2 \\ 6 & -1 \end{bmatrix}\begin{bmatrix} x \\ y \end{bmatrix} = \begin{bmatrix} 6 \\ 4 \end{bmatrix}$

3 $\begin{bmatrix} 3 & 1 \\ 1 & -3 \end{bmatrix}\begin{bmatrix} x \\ y \end{bmatrix} = \begin{bmatrix} -2 \\ 1 \end{bmatrix}$

4 $\begin{bmatrix} 1 & 1 & 1 \\ 3 & -4 & 2 \\ 1 & -1 & -1 \end{bmatrix}\begin{bmatrix} x \\ y \\ z \end{bmatrix} = \begin{bmatrix} 2 \\ 6 \\ 4 \end{bmatrix}$

5 $\begin{bmatrix} 1 & 2 & 3 \\ 3 & -2 & 0 \\ 2 & 0 & -7 \end{bmatrix}\begin{bmatrix} x \\ y \\ z \end{bmatrix} = \begin{bmatrix} 5 \\ 4 \\ 0 \end{bmatrix}$

6 $\begin{bmatrix} 2 & -3 & 1 \\ 1 & 1 & -3 \\ 0 & -2 & 3 \end{bmatrix}\begin{bmatrix} x \\ y \\ z \end{bmatrix} = \begin{bmatrix} 1 \\ 0 \\ 4 \end{bmatrix}$

7 a $\dfrac{1}{2}\begin{bmatrix} 4 & 2 \\ 5 & 3 \end{bmatrix}$ **b** $x = -1, y = -3.5$

8 a $\begin{bmatrix} -2.5 & -2 & 0.5 \\ -2 & -2 & 1 \\ 1 & 1 & 0 \end{bmatrix}$ **b** $x = -1, y = 5, z = 2$

9 a $x = 3, y = -7$ **b** $x = 5.5, y = -8.5$

10 a $\begin{bmatrix} 7 & 0 & 0 \\ 0 & 7 & 0 \\ 0 & 0 & 7 \end{bmatrix}$ **b** $A^{-1} = \dfrac{1}{7}B$

    c $x = 3, y = -1, z = 1$

11 a $\begin{bmatrix} 1 & 1 & 1 & 1 & 1 \\ 1 & -1 & 1 & 2 & -1 \\ 2 & -1 & 3 & -1 & 2 \\ 3 & 2 & -1 & -1 & -2 \\ 0 & 2 & 0 & 3 & -1 \end{bmatrix}\begin{bmatrix} v \\ w \\ x \\ y \\ z \end{bmatrix} = \begin{bmatrix} 1 \\ 13 \\ 2 \\ 4 \\ 8 \end{bmatrix}$

    b $v = 1, w = -1, x = 3, y = 2, z = -4.$

Miscellaneous exercise ten   **PAGE 222**

1 a $B = \begin{bmatrix} -2 & 0 \\ 4 & 3 \end{bmatrix}$ **b** $C = \begin{bmatrix} -1 & 0 \\ 4 & 4 \end{bmatrix}$

2 a $E = \begin{bmatrix} 0 & 0 \\ 0 & 0 \end{bmatrix}$ **b** $F = \begin{bmatrix} 5 & -1 \\ 2 & 0 \end{bmatrix}$

    c $G = \begin{bmatrix} 6 & -1 \\ 2 & 1 \end{bmatrix}$ **d** $H = \begin{bmatrix} 5 & -1 \\ 2 & 0 \end{bmatrix}$

    e $K = \begin{bmatrix} 5 & -1 \\ 2 & 0 \end{bmatrix}$

3 $\dfrac{\pi}{12}, \dfrac{5\pi}{12}$

4 $\theta = 0, p, \pi, (\pi + p), 2\pi$

7 a $2y^2 + y - 1$ **b** $-\dfrac{11\pi}{6}, -\dfrac{7\pi}{6}, -\dfrac{\pi}{2}, \dfrac{\pi}{6}, \dfrac{5\pi}{6}, \dfrac{3\pi}{2}$

8 a $\sqrt{29}\cos(\theta - 68.2°)$ **b** $-\sqrt{29}, 248.2°$

9 a Cannot be determined. A and B are not the same size and so $A + B$ cannot be formed.

    b $\begin{bmatrix} 0 & -1 & 1 \\ 5 & -2 & 5 \end{bmatrix}$

    c Cannot be determined. The number of columns in A ≠ the number of rows in B.

    d $\begin{bmatrix} 5 & -2 & 8 \\ 2 & -1 & 3 \end{bmatrix}$

    e Cannot be determined. The number of columns in A ≠ the number of rows in C.

    f $\begin{bmatrix} 5 \\ 2 \end{bmatrix}$

10 a XY

    b $\begin{bmatrix} 420 \\ 410 \\ 430 \end{bmatrix}$

    c $\begin{bmatrix} \text{Commodity costs (\$) to produce one model A} \\ \text{Commodity costs (\$) to produce one model B} \\ \text{Commodity costs (\$) to produce one model C} \end{bmatrix}$

11 Of the four products in the list only XZ can be formed. It shows the total points obtained by each team.

$$XZ = \begin{matrix} & \text{Points} \\ \begin{matrix} A \\ B \\ C \\ D \\ E \end{matrix} & \begin{bmatrix} 13 \\ 10 \\ 9 \\ 10 \\ 15 \end{bmatrix} \end{matrix}$$

12 a $y = 4\sin 8x$ **b** $y = -3\sin\left(\dfrac{2\pi}{5}x\right)$

13 a $y = 2\sin\left(\dfrac{\pi}{2}(x-1)\right)$ **b** $y = 20\sin\left(\dfrac{\pi}{15}(x+5)\right)$

14 a $y = 5\sin\left(\dfrac{\pi}{5}(x-2)\right) + 10$

    b $y = 10\sin\left(\dfrac{\pi}{50}(x-20)\right) + 40$

15 $x = -1, y = -2, p = -5, q = 7, r = -7, s = 2.$

16 $\begin{bmatrix} -1 & 1 \\ 3 & -5 \end{bmatrix}$

Exercise 11A   **PAGE 229**

1 Rotate 180° about origin.

2 Rotate 90° anticlockwise about origin.

3 Reflect in the x-axis.

4 Reflect in the y-axis.

5 Reflect in the line $y = x$.

6 Reflect in the line $y = -x$.

7 Dilation parallel to x-axis, scale factor 2.

8 Dilation parallel to y-axis, scale factor 3.

9 Dilation parallel to x-axis, scale factor 2 and dilation parallel to y-axis, scale factor 3.

10 Dilation parallel to x-axis, scale factor 3 and dilation parallel to y-axis, scale factor 3.

11 Shear parallel to *x*-axis, scale factor 2.

12 Shear parallel to *y*-axis, scale factor 3.

13 Results of **a** and **b** should lead you to conclude
$$\frac{\text{Area O'A'B'C'}}{\text{Area OABC}} = |\text{determinant of matrix}|.$$

Exercise 11B   PAGE 234

1 a $A = \begin{bmatrix} 0 & 1 \\ -1 & 0 \end{bmatrix}$ $B = \begin{bmatrix} -1 & 0 \\ 0 & -1 \end{bmatrix}$ $C = \begin{bmatrix} 0 & -1 \\ 1 & 0 \end{bmatrix}$

2 a $\begin{bmatrix} 1 & 0 \\ 0 & -1 \end{bmatrix}$ **b** $\begin{bmatrix} -1 & 0 \\ 0 & 1 \end{bmatrix}$ **c** $\begin{bmatrix} -1 & 0 \\ 0 & -1 \end{bmatrix}$

3 $\begin{bmatrix} 0 & -1 \\ -1 & 0 \end{bmatrix}$ **4** $\begin{bmatrix} 3 & 0 \\ 0 & 1 \end{bmatrix}$

5 a $\begin{bmatrix} 0 & 1 & 3 & 2 \\ 1 & -2 & -5 & -5 \end{bmatrix}$

  b A'(0, 1), B'(1, −2), C'(3, −5), D'(2, −5)

6 A(1, 3), B(1, 1), C(4, −3)

7 A(1, 3), B(−1, 2), C(0, 2)

8 $\begin{bmatrix} 4 & 1 \\ 1 & 0 \end{bmatrix}$

9 a $\begin{bmatrix} 1 & 0 \\ -2 & -1 \end{bmatrix}$ **b** $\begin{bmatrix} 1 & 0 \\ -2 & -1 \end{bmatrix}$

10 $\begin{bmatrix} 3 & 1 \\ -1 & 0 \end{bmatrix}$ **11** $\begin{bmatrix} 0 & 1 \\ -1 & 3 \end{bmatrix}$

12 *a* = 2, *b* = 5, *c* = 1, *d* = 3

13 a $\begin{bmatrix} 1 & 4 \\ 2 & 9 \end{bmatrix}$ **b** $\begin{bmatrix} 2 & 9 \\ -1 & -4 \end{bmatrix}$

  c $\begin{bmatrix} 1 & 0 \\ -2 & 1 \end{bmatrix}$ **d** $\begin{bmatrix} 0 & -1 \\ 1 & 2 \end{bmatrix}$

15 a 36 square units

  b O'(0, 0), A'(12, 3), B'(8, 5), C'(−4, 2)

  c Diagram not shown here.

16 a Diagram not shown here.

  b 8 square units

  c 40 square units

  d Diagram, not shown here, should have A'(−2, 2), B'(−4, −6), C'(2, −2), D'(4, 6).

18 *y* = 3*x* − 1

20 a (10, 5)   **b** *y* = 0.5*x*

21 $m_2 = \dfrac{m_1 + 2}{3}$. $-3 + \sqrt{10}$ and $-3 - \sqrt{10}$

Exercise 11C   PAGE 238

1 a $\begin{bmatrix} \frac{\sqrt{3}}{2} & -\frac{1}{2} \\ \frac{1}{2} & \frac{\sqrt{3}}{2} \end{bmatrix}$ **b** $\begin{bmatrix} \frac{1}{\sqrt{2}} & -\frac{1}{\sqrt{2}} \\ \frac{1}{\sqrt{2}} & \frac{1}{\sqrt{2}} \end{bmatrix}$

  c $\begin{bmatrix} \frac{1}{2} & -\frac{\sqrt{3}}{2} \\ \frac{\sqrt{3}}{2} & \frac{1}{2} \end{bmatrix}$ **d** $\begin{bmatrix} 0 & -1 \\ 1 & 0 \end{bmatrix}$

2 a $\begin{bmatrix} \frac{1}{2} & \frac{\sqrt{3}}{2} \\ \frac{\sqrt{3}}{2} & -\frac{1}{2} \end{bmatrix}$ **b** $\begin{bmatrix} -\frac{1}{2} & \frac{\sqrt{3}}{2} \\ \frac{\sqrt{3}}{2} & \frac{1}{2} \end{bmatrix}$

  c A repeat reflection will return us to the original position. Hence the square of each matrix is the identity because the repeat reflection leaves the final position identical to initial position.

3 $\begin{bmatrix} \cos\theta & \sin\theta \\ -\sin\theta & \cos\theta \end{bmatrix}$

6 $\alpha = 2(\phi - \theta)$

7 a $\begin{bmatrix} x' \\ y' \end{bmatrix} = \begin{bmatrix} -1 & 0 \\ 0 & -1 \end{bmatrix} \begin{bmatrix} x \\ y \end{bmatrix} + \begin{bmatrix} 6 \\ 4 \end{bmatrix}$

  b $\dfrac{1}{\sqrt{13}} \begin{bmatrix} 3 & 2 \\ -2 & 3 \end{bmatrix}$

  c $O''(2\sqrt{13}, 0), A''\left(\dfrac{23\sqrt{13}}{13}, \dfrac{2\sqrt{13}}{13}\right),$

  $B''\left(\dfrac{21\sqrt{13}}{13}, -\dfrac{\sqrt{13}}{13}\right), C''\left(\dfrac{24\sqrt{13}}{13}, -\dfrac{3\sqrt{13}}{13}\right).$

Miscellaneous exercise eleven   PAGE 239

2 $0, \pi, \dfrac{7\pi}{6}, \dfrac{11\pi}{6}, 2\pi$

3 *a* = 4, *b* = 0, *c* = −3, *d* = 0

4 a $\begin{bmatrix} 0 & -1 \\ 1 & 0 \end{bmatrix}$, $|\det| = 1$. ✓

b $\begin{bmatrix} -1 & 0 \\ 0 & -1 \end{bmatrix}$, $|\det| = 1$. ✓

c $\begin{bmatrix} 1 & 0 \\ 0 & -1 \end{bmatrix}$, $|\det| = 1$. ✓

d $\begin{bmatrix} 0 & 1 \\ 1 & 0 \end{bmatrix}$, $|\det| = 1$. ✓

e $\begin{bmatrix} 1 & 4 \\ 0 & 1 \end{bmatrix}$, $|\det| = 1$. ✓

f $\begin{bmatrix} 1 & 0 \\ 3 & 1 \end{bmatrix}$, $|\det| = 1$. ✓

5 BAC, $\begin{bmatrix} 5 & 5 & 2 & 4 \end{bmatrix}$

6 a Cannot be determined. A and B are not of the same size.

b $\begin{bmatrix} 3 & 1 \\ 0 & 5 \end{bmatrix}$

c Cannot be determined. Number of columns in A ≠ Number of rows in C.

d $\begin{bmatrix} 1 & 3 & 0 \\ 1 & -3 & -2 \end{bmatrix}$ **e** $\begin{bmatrix} 1 & 2 \\ -2 & 6 \end{bmatrix}$ **f** $\begin{bmatrix} 5 & 5 \\ 5 & 10 \end{bmatrix}$

g Cannot be determined. BA can be formed but cannot be added to C as of different size.

7 To be singular we require determinant to be zero. For given matrix, determinant $= 2x^2 + 4$ which is ≥ 4 for all real x. Thus determinant cannot be zero for real x. Thus not a singular matrix.

8 $k = 3$, $p = 12$, $q = -9$

9 a $[0]$ **b** $\begin{bmatrix} 2 & -4 & 4 \\ 0 & 0 & 0 \\ -1 & 2 & -2 \end{bmatrix}$

10 $x = 5$, $y = -2$

11 y must equal zero, no restrictions necessary on x and z.

12 $a = 4$, $b = 3.5$, $c = -2$, $d = -0.5$

13 a $\begin{bmatrix} 3 & -2 \\ -1 & 1 \end{bmatrix}$ **b** $\begin{bmatrix} 3 & -1 \\ 1 & 1 \end{bmatrix}$

14 $\begin{bmatrix} 1 & 1 \\ 0 & 3 \end{bmatrix}$

15 $x = \dfrac{n\pi}{2} + 2.08$ for $n \in \mathbb{Z}$.

Exercise 12A **PAGE 245**

1 to 10 Answers not given here. Compare your answers with those of others in your class.

11 a $\dfrac{5}{9}$ **b** $\dfrac{25}{33}$ **c** $\dfrac{7}{11}$

d $\dfrac{743}{333}$ **e** $\dfrac{2083}{9000}$

12 If we assume that $\sqrt{2}$ can be written in the form $\dfrac{a}{b}$ for integer a and b, $b \neq 0$, and a and b having no common factors then $\quad \sqrt{2} = \dfrac{a}{b}$

It therefore follows that $\quad 2 = \dfrac{a^2}{b^2}$

and so $\qquad\qquad 2b^2 = a^2$

Thus a^2, and hence a, is even.

We could therefore write a as $2k$, k an integer.

Hence $\qquad\qquad 2b^2 = (2k)^2$

$\qquad\qquad\qquad 2b^2 = 4k^2$

$\qquad\qquad\qquad b^2 = 2k^2$

and so b^2, and hence b, is even.

But if a and b are both even they have a common factor, 2. Hence we have a contradiction.

Our original premise, or underlying assumption, about $\sqrt{2}$ must be false.

Hence $\sqrt{2}$ is irrational.

Exercise 12B **PAGE 246**

5 Yes always a multiple of ten. Compare your justification with others in your class.

No, not always a multiple of twenty. Justify using a counter example.

7 John's conjecture is not correct. $6^3 - 6 = 210$ which is not divisible by 12.

A possible alternative conjecture:

For any integer x, $x \geq 2$, $x^3 - x$ is always divisible by 6. (Proof not given here.)

Miscellaneous exercise twelve **PAGE 254**

1 a Cannot be determined **b** $\begin{bmatrix} -1 \\ 7 \end{bmatrix}$

c $\begin{bmatrix} -3 & 3 & -3 \\ -3 & 3 & -3 \end{bmatrix}$ **d** Cannot be determined

e $\begin{bmatrix} 0 & 1 & 2 \\ 6 & 5 & 4 \end{bmatrix}$

2 $B = \begin{bmatrix} 5 \\ 1 \end{bmatrix}$, $C = \begin{bmatrix} -1 \\ 5 \end{bmatrix}$, $D = \begin{bmatrix} 5 & 4 \end{bmatrix}$, $E = \begin{bmatrix} 1 & -3 \end{bmatrix}$.

3 $\begin{bmatrix} -1 & 3 \\ 2 & 5 \end{bmatrix}$

4 a XY, ZX   **b** ZX

  c

No. of Aus stamps required	No. of RoW stamps required
18 150	14 850

5 $p = 0$, $q = 12$, $x = -3$

8 a $\sqrt{34} \cos(\theta + 0.54)$ **b** $-\sqrt{34}$, 2.60

9 BAC, $\begin{bmatrix} 10 & 0 & 10 & 10 \\ 8 & 0 & 8 & 8 \end{bmatrix}$

10 Either $x = -3$, $y = 6$, $p = 0$ and $q = 24$
    or $x = 2$, $y = -4$, $p = 0$ and $q = 24$.

11 No conflict. Final proof showing A = B is quite correct **provided A^{-1} exists**. In example 1, matrix A is not a square matrix so A^{-1} does not exist. In example 2, det A = 0, so A^{-1} does not exist.

12 AC, AD, BD, CB, DC, DD

13 A′(2, 0), B′(11, 3), C′(13, 3)

  A shear parallel to x-axis, scale factor 3.

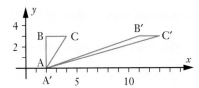

15 $x = \begin{cases} 2n\pi + 0.64, \\ 2n\pi + 2.21, \end{cases}$ for $n \in \mathbb{Z}$.

Exercise 13A   PAGE 263

1 $5i$ **2** $12i$ **3** $3i$

4 $7i$ **5** $20i$ **6** $\sqrt{5}i$

7 $2\sqrt{2}i$ **8** $3\sqrt{5}i$

9 a 3   **b** 5

10 a -2 **b** 7

11 a 3   **b** -1

12 $-1 + 2i$, $-1 - 2i$

13 $-1 + \sqrt{2}i$, $-1 - \sqrt{2}i$

14 $-2 + \sqrt{2}i$, $-2 - \sqrt{2}i$

15 $-1 + 3i$, $-1 - 3i$

16 $2 + \sqrt{2}i$, $2 - \sqrt{2}i$

17 $\dfrac{1}{4} + \dfrac{\sqrt{7}}{4}i$, $\dfrac{1}{4} - \dfrac{\sqrt{7}}{4}i$

18 $-\dfrac{1}{4} + \dfrac{\sqrt{7}}{4}i$, $-\dfrac{1}{4} - \dfrac{\sqrt{7}}{4}i$

19 $-\dfrac{3}{2} + \dfrac{1}{2}i$, $-\dfrac{3}{2} - \dfrac{1}{2}i$

20 $\dfrac{1}{2} + \dfrac{7}{2}i$, $\dfrac{1}{2} - \dfrac{7}{2}i$

21 $\dfrac{1}{5} + \dfrac{8}{5}i$, $\dfrac{1}{5} - \dfrac{8}{5}i$

22 $\dfrac{1}{2} + \dfrac{\sqrt{3}}{2}i$, $\dfrac{1}{2} - \dfrac{\sqrt{3}}{2}i$

23 $\dfrac{3}{10} + \dfrac{\sqrt{11}}{10}i$, $\dfrac{3}{10} - \dfrac{\sqrt{11}}{10}i$

Exercise 13B   PAGE 266

1 $7 + 2i$ **2** $3 - 10i$ **3** $-3 + 4i$

4 $7 - 2i$ **5** $-3 + 2i$ **6** $7 - 2i$

7 $13 + 4i$ **8** $12 + 7i$ **9** $13 + 2i$

10 $7 + 8i$ **11** $3 - 8i$ **12** $10 - 15i$

13 7 **14** 5 **15** $-4 + 19i$

16 $-3 + 11i$ **17** $3 - i$ **18** $-13 + 13i$

19 $\dfrac{1}{2} - \dfrac{1}{2}i$ **20** $\dfrac{1}{5} + \dfrac{7}{5}i$ **21** $\dfrac{2}{5} - \dfrac{6}{5}i$

22 $\dfrac{8}{17} + \dfrac{2}{17}i$ **23** $0 + 1i$ **24** $\dfrac{17}{13} - \dfrac{7}{13}i$

25 a $9 + i$ **b** $1 - 5i$ **c** $7 - 12i$

    d $26 + 7i$ **e** $7 + 24i$ **f** $\dfrac{14}{25} - \dfrac{23}{25}i$

26 a 4 **b** $-2 - 10i$ **c** $6 - 10i$

    d $28 - 10i$ **e** $-16 + 30i$ **f** $-\dfrac{11}{13} + \dfrac{10}{13}i$

27 a $24 + 7i$ **b** 48 **c** 625

    d $\dfrac{527}{625} - \dfrac{336}{625}i$

28 a $4 - 9i$ **b** $18i$ **c** $20 - 9i$

    d $-4 + 45i$ **e** 97 **f** $-\dfrac{65}{97} + \dfrac{72}{97}i$

29 $c = 3$, $d = 2$ **30** $a = -5$, $b = -12$

31 $c = -10$, $d = 4$ **32** $a = 15$, $p = 78$

33 a Yes, statement is correct for all complex z and w.

    b No, eg $z = 3 + 2i$ and $w = 5 - 2i$: $\text{Im}(z) = -\text{Im}(w)$ but $w \neq \bar{z}$.

34 **a** $(x-2-3i)(x-2+3i)$

   b $(x-1-3i)(x-1+3i)$

   c $(x-3-2\sqrt{2})(x-3+2\sqrt{2})$

   d $(x+5+i)(x+5-i)$

   e $(x+7-2i)(x+7+2i)$

   f $(x+2+\sqrt{7})(x+2-\sqrt{7})$

35 **b** $b=-6, c=13$ **c** $d=-10, e=34$

36 **a** -1 **b** i **c** i

37 $0.25, 4$ and $-1, -1$

38 **a** 1

40 **a** $(2,3)$ **b** $(-5,6)$ **c** $(0,7)$

   d $(3,0)$ **e** $(1,9)$ **f** $(6,0)$

   g $(3,3)$ **h** $(0,14)$ **i** $(-10,6)$

   j $(10,0)$ **k** $(0.3,0.6)$ **l** $(-\frac{55}{73}, -\frac{48}{73})$

41 $-\dfrac{5}{53} - \dfrac{9}{53}i$

Exercise 13C   **PAGE 270**

1 $Z_1 = 7 + 2i$ $Z_2 = 2 + 4i$ $Z_3 = 0 + 6i$ $Z_4 = -5 + 3i$
   $Z_5 = -7 - 5i$ $Z_6 = 0 - 4i$ $Z_7 = 3 - 6i$ $Z_8 = 6 - 3i$

2 $Z_1 = (6,0)$ $Z_2 = (7,5)$ $Z_3 = (-3,6)$ $Z_4 = (-5,0)$
   $Z_5 = (-6,-3)$ $Z_6 = (-3,-6)$ $Z_7 = (0,-6)$ $Z_8 = (7,-7)$

3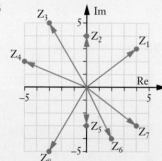

4 $Z_1 = 1 + 2i$
   $Z_2 = -3 - 2i$
   $Z_3 = -3 + 2i$
   $Z_4 = 3 - 2i$

5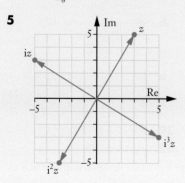

6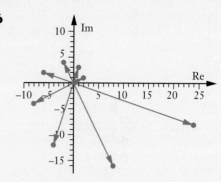

Miscellaneous exercise thirteen   **PAGE 271**

1 **a** 29 **b** 10 **c** 40

   d $-7 + 24i$ **e** $\dfrac{3}{10} - \dfrac{11}{10}i$ **f** $\dfrac{3}{13} + \dfrac{11}{13}i$

2 **a** $-1 + 2i$ **b** $9 + 19i$ **c** $2 + 3i$

   d $9 - 19i$ **e** $-5 - 12i$

   f $-280 + 342i$ **g** $2 - 5i$

3 **a** $p = 2, q = 1, r = -2$ **b** $0.5, 1 + \sqrt{2}i, 1 - \sqrt{2}i$

4 **a** 10 **b** 22

5 **a** $-5\sqrt{2}i$ **b** -50 **c** $-49 + 10\sqrt{2}i$

6 $a = 3$ and $b = -2$, $a = -3$ and $b = 2$

7 **a** $(2,-1), (-2,4), (-4,0)$

   b Hint for part **b**:

   To find the area of each triangle draw a rectangle around each one, find the area of the rectangle and subtract appropriate areas. Then confirm Area $\triangle A'B'C' = |\det T|$ Area $\triangle ABC$.

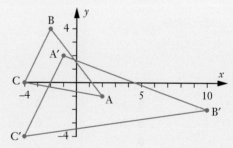

8 $a = 3, b = -1, c = 2, d = 1$

9 $2 - 2i, 2 - 12i$

10 **a** Areas multiplied by zero. Thus determinant equals zero.

   b $(0,0)$ has image $(0,0)$. Thus $(0,0)$ lies on the line. Thus line passes through origin.

11 $a = 3, b = -5$

12 **a** $-119 - 120i$ **b** 12

13 $x^2 - 4x + 13 = 0$ **14** $1 + 5i$

15 $\dfrac{7}{13} + \dfrac{17}{13}i$ **16** $-5 + 3i, 16 - 30i$

18 $0°, 60°, 120°, 180°, 240°, 300°, 360°$

20 a $\begin{bmatrix} -12 & 20 \\ -3 & 5 \end{bmatrix}$ **b** $[-7]$

21 a B **b** B, D **c** A, B, F
  d A, C **e** A, B, C, D **f** A, C
  g E **h** A, B, F

22 a BC^2B^{-1} **b** BC^3B^{-1} **c** BC^nB^{-1}

27 $x = n\pi + \dfrac{\pi}{12}$ for $n \in \mathbb{Z}$, $n\pi + \dfrac{5\pi}{12}$ for $n \in \mathbb{Z}$,

$n\pi + \dfrac{\pi}{2}$ for $n \in \mathbb{Z}$.

28 a $\dfrac{1}{2}\begin{bmatrix} \sqrt{3} & -1 \\ 1 & \sqrt{3} \end{bmatrix}$ **b** $\dfrac{1}{2}\begin{bmatrix} 1 & -\sqrt{3} \\ \sqrt{3} & 1 \end{bmatrix}$

  c $\dfrac{1}{2}\begin{bmatrix} 1 & \sqrt{3} \\ \sqrt{3} & -1 \end{bmatrix}$

  d $\dfrac{1}{2}\begin{bmatrix} 1 & -\sqrt{3} \\ \sqrt{3} & 1 \end{bmatrix} \times \dfrac{1}{2}\begin{bmatrix} 1 & \sqrt{3} \\ \sqrt{3} & -1 \end{bmatrix} = \dfrac{1}{2}\begin{bmatrix} -1 & \sqrt{3} \\ \sqrt{3} & 1 \end{bmatrix} \neq \dfrac{1}{2}\begin{bmatrix} \sqrt{3} & -1 \\ 1 & \sqrt{3} \end{bmatrix}$

  Rotating square 1 anticlockwise 30° about the origin makes A go to C″ and C go to A″.

  So while the rotated image occupies the same space as square 3, it is not the same image because A does not go to A″ and C does not go to C″.

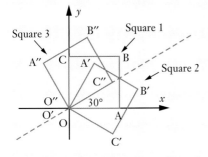

INDEX

A

$a \cos \theta + b \sin \theta$ 173–5
absolute value viii
acceleration 53
adding matrices 197
addition principle 22–3
addition of vectors 56–60, 62–3
additive and multiplicative reasoning 32–7
algebraic expressions ix
amplitude 156, 157, 159
angle sum and angle difference identities 161, 168
   proof 162–3
angles in circles 98–109
   angle at the centre of a circle is twice the angle at the circumference 99–100, 101
   angles in the same segment are equal 101
   angles in a semicircle are right angles 100, 101
   opposite angles of a cyclic quadrilateral are supplementary 101
Argand diagrams 269–70
arrangements *see* permutations
axioms 95

B

bearings ix, 51–2

C

circle properties 98–109
column matrix 77, 196
combinations 37–45
   and Pascal's triangle 46
completing the square method 259
complex number arithmetic 264
complex numbers 261–3
   Argand diagrams 269–70
   conjugate of 265
   equal 265
congruent triangles xii–xiii
conjecture 3, 6, 244–5, 253
conjugate of a complex number 265
contrapositive statements 7, 8, 10
converse statements 7, 8, 9–10
cosec θ 176–9
cosecant 176
cosine function 157, 158
cosine rule ix
cot θ 176–9
cotangent 176
counter examples 6, 244
cycle (periodic functions) 156, 157
cyclic quadrilaterals 5, 7, 101

D

deductive proof 95, 243
definitions (true statements) 95
determinant
   2×2 matrix 213
   3×3 matrix 221
   transformation matrix 230, 231
diagonal matrix 196
dilation, $y = \sin x$ 159
discriminant 261
displacement 53–4, 56, 115
   see also relative displacement
distance 53
double angle identities 170–2

E

elements (matrix) 196
elements (set) xiii
equal complex numbers 265
equal matrices 198
equal vectors 61
equivalent statements 7
even functions 157
exact values ix, 160, 172, 182
expanding and simplifying, algebraic expressions ix

F

factorial notation 16–7
factorising, algebraic expressions ix
force 53, 54, 55–8

G

geometric proofs x–xiii, 95–109
   angles in circles 98–104
   tangents and secants 104–9
   using scalar product 147–8
   using vectors 129–32

H

horizontal component (vector) 76

identity matrix 211
  additive 210
  multiplicative 211
'If and only if' (iff) 7
imaginary numbers 260
implied statements 7
inclusion - exclusion principle 23–4, 26
inductive proof 243, 248–50
invariant point 228
inverse (square matrix) 212–217
  3×3 matrix 221
  to solve systems of equations 218–21
inverse (transformation matrix) 230
inverse statements 8, 10
invertible matrices 213
irrational numbers 243, 261

leading diagonal (matrix) 196
linear factors of quadratic
  polynomials 266–7

magnitude (vectors in component
  form) 77–82
matrices 195–7, 198–201
  adding and subtracting 197
  column 196
  diagonal 196
  equal 198
  invertible 213
  multiplicative identity 211
  multiplicative inverse of a square matrix
    212–17, 218–21
  multiplying 201–9
  multiplying by a scalar 197
  row 196
  singular 213
  size or dimensions 196
  square 196, 213
  transformation 227–39
  zero 209–10
modulus (vectors in component form)
  77–82
multiplication
  of a matrix by a matrix 201–9
  of a matrix by a scalar 197
  of a vector by a scalar 62
multiplication principle 15–6
multiplicative identity matrices 211

multiplicative inverse of a square matrix
  212–7, 218–21
multiplicative reasoning 27–31
  and additive reasoning 32-7
mutually exclusive events 32

negation of a statement 8
negative of a vector 61

odd functions 156, 159
opposite angles of a cyclic quadrilateral are
  supplementary 101

parallel vectors 62
Pascal's triangle 46
period 156, 157, 159
period motion, modelling 188–91
periodic functions 156–9
permutations 15, 39
  addition principle 22–3
  additive and multiplicative
    reasoning 32–7
  factorial notation 16–7
  inclusion–exclusion principle
    23–4, 25–6
  multiplication principle 15–6
  multiplicative reasoning 27–8
  of objects from a group of objects, all
    different 17–20
  of objects, not all different 21, 24
  of objects with some restriction imposed
    27–37
  see also combinations
phase (periodic functions) 157, 159
pigeon-hole principle 8, 9
position vectors 88–91, 115–17
  relative displacement 115–17
product to sum and sum to product
  180–2
proof by contradiction 8, 243, 246
proof by deduction 95, 243
proof by exhaustion 243, 246–7
proof by induction 243, 248–52
proofs 95, 243
  trigonometric identities 161–2, 166–9
  see also geometric proofs
Pythagorean identity 160, 166
  identities 176

QED 97
quadratic equations 259–61
  complex solutions 260–1, 262–3
  discriminant 261
quadratic formula 259, 260
quadratic polynomials, linear factors 266–7

radian measure 156
rational numbers 243
real numbers viii, 243
reflection
  transformation matrix 237, 238–9
  $y = \sin x$ 159
relative displacement 115–17
relative velocity 113–14, 118–24
resultant 56–9
rotation, transformation matrix 237, 238–9
row matrix 196

scalar multiplication
  of a matrix 197
  of a vector 62
scalar product 137–8
  algebraic properties 138–42
  from the components 142–6
  proofs using 147–9
scalar projection 137
scalar quantities 53
sec θ 176–9
secants 105, 108–9, 176
selections *see* combinations
shear 228
similar triangles ix, x–xii
sine function 155, 156
sine rule ix
singular matrices 213
speed 53
square matrices 196
  determinant 213
  inverse 212–3, 218–21
statements, true or false 3–11
subtracting matrices 197
subtraction of one vector from another 63
sum to product and product to
  sum 180–2
summation notation 252

T

tangent function 156, 159
tangents to a circle 104–8
  angle between tangent and chord equals
    angle in alternate segment 106, 107
  angle between a tangent and a radius is a
    right angle 104–6, 107
  two tangents drawn from a point to a
    circle are of equal length 106, 107
  see also secants
theorems 95, 253
'there exists' 6
transformation matrices 227–39
  combining transformations 232–6
  determinant 230, 231
  determining the matrix for a particular
    transformation 230, 234
  general reflection in a line that passes
    through the origin 237
  general rotation about the
    origin 237
  inverse 230
transformations
  $y = \cos x$ 159
  $y = \sin x$ 159
  $y = \tan x$ 159
triangles
  area ix
  congruent xii–xiii
  similar ix, x–xiii
trigonometric equations
  general solutions 183–6
  modelling periodic motion 188–91
  obtaining the rule from the graph 187
  solving 163, 171, 172, 175, 176, 177, 178,
    181, 182
trigonometric graphs 155–9
  modelling periodic motion 188–91
  obtaining the rule from the graph 187
trigonometric identities 161, 168
  $a \cos \theta + b \sin \theta$ 173–5
  angle sum and angle difference
    identities 161–3
  double angle identities 170–2
  involving $\sec \theta$, $\csc \theta$ and $\cot \theta$ 176–9
  product to sum and sum to
    product 180–2
  proofs 161–2, 166–9, 170
  Pythagorean identities 161, 166, 176
  trigonometric ratios ix
true by definition (statements) 7
true or false statements 3–11

U

unit circle definitions of trigonometric
  functions ix, 155–9
unit vectors 75–6, 82
universal set xiii

V

vector product 137
vector projection 137
vector quantities 52–55
  mathematical representation
    60–9
vectors
  addition 56–60, 62–3
  equal 61
  for geometric proofs 129–32
  multiplication by a scalar 62
  negative of a vector 61
  parallel 62
  subtraction of one vector from
    another 63
  zero 63
vectors in component form 73–88
  horizontal component 76
  magnitude or modulus 77–82
  position vectors 89–91, 115–17
  vertical component 76
velocity 53, 58–60
  see also relative velocity
Venn diagrams xiii
vertical component (vector) 76

Y

$y = \cos x$ 157
  transformations 159
$y = \sin x$ 155, 156
  transformations 159
$y = \sin x + d$ 159
$y = \sin bx$ 159
$y = \sin [b(x + c)]$ 159
$y = a \sin x$ 159
$y = \tan x$ 158
  transformations 159

Z

zero factorial 18
zero matrices 209–10
zero vector 63